The Tiger's Way

(Source: Corel Gallery Clipart, Totem Graphics, Animal 04C137)

The Tiger's Way

A U.S. Private's Best Chance for Survival

Illustrated

H. John Poole

Foreword by
Maj.Gen. Ray L. Smith USMC (Ret.)

Posterity
Press

Copyright © 2003 by H. John Poole

Published by Posterity Press
P.O. Box 5360, Emerald Isle, NC 28594
(www.posteritypress.org)

Protected under the Berne Convention. All rights reserved. No part of this book may be reproduced or utilized in any form or by any means, electronic or mechanical, including photocopying or recording by any information storage-and-retrieval system without full permission in writing from the publisher. All inquiries should be addressed to Posterity Press, P.O. Box 5360, Emerald Isle, NC 28594.

Cataloging-in-Publication Data

Poole, H. John, 1943-
The Tiger's Way.
Includes bibliography and index.
1. Infantry drill and tactics.
2. Military art and science.
3. Military history.

I. Title. ISBN 0-9638695-6-6 2003 355'.42
Library of Congress Control Number: 2003093619

This publication contains copyrighted material from other sources. Permission for its use is recorded in the "source notes" at the back of the book. This material may not be further reproduced without the consent of the original owner. Every effort has been made to reach all copyright holders. Any who may have been missed can identify themselves now to negotiate the appropriate permission agreements.

Cover art © 2003 by Edward Molina
Edited by Dr. Mary Beth Poole
Proofread by William E. Harris

Second printing, United States of America, January 2004

To that filthy, sweaty, dirt-encrusted, foot-sore, camouflage-painted, ripped-trousered, tired, sleepy, beautiful little son of a b— who has keep the wolf away from the door for over two hundred years.

(Source: Courtesy of H.G. Duncan, from *Green Side Out,* © 1980)

Contents

List of Illustrations ix
List of Tables xv
Foreword xvii
Acknowledgments xix
Introduction xxi

Part One: *A Growing Threat at 75 Yards*

Chapter 1: *American Units Must Further Disperse* 3
Chapter 2: *Orphaned Squads Are at Greater Risk* 19
Chapter 3: *U.S. Riflemen Will Need More Skill* 37

Part Two: *The New "Basics"*

Chapter 4: *Microterrain Appreciation* 47
Chapter 5: *Harnessing the Senses* 57
Chapter 6: *Night Familiarity* 69
Chapter 7: *Nondetectable Movement* 75
Chapter 8: *Guarded Communication* 93
Chapter 9: *Discreet Force at Close Range* 101
Chapter 10: *Combat Deception* 111
Chapter 11: *One-on-One Tactical Decision Making* 131

Part Three: *What the "Eastern" Soldier Does*

Chapter 12: *When Told to Hold* 143
Chapter 13: *At the Listening Post* 177
Chapter 14: *With Contact Patrolling* 185
Chapter 15: *On Point* 193
Chapter 16: *About Tracking an Intruder* 203
Chapter 17: *While Stalking a Quarry* 221
Chapter 18: *To Reconnoiter an Enemy Position* 228

Chapter 19: *In the Rural Assault* 237
Chapter 20: *For Attacking Cities* 267
Chapter 21: *During an Urban Defense* 279

Part Four: *The Winning Edge*

Chapter 22: *The Rising Value of the "Little Picture"* 299
Chapter 23: *How the Tiger Is Born* 309
Chapter 24: *Field Proficiency Has No Substitute* 321

Appendix A: *Casualty Comparisons* 347
Appendix B: *Enemy Entry-Level Training* 349
Appendix C: *Advised U.S. Battledrills* 355

Notes 365
Glossary 429
Bibliography 433
About the Author 447
Name Index 449

Illustrations

Figures

Chapter 1: *American Units Must Further Disperse*

1.1: Kill Zone for New Russian "Shoulder-Fired" FAE 6
1.2: Soviet Airborne Private, 1982 12
1.3: Communist Chinese Paratrooper, 1982 14

Chapter 2: *Orphaned Squads Are at Greater Risk*

2.1: Russian Commando 23
2.2: Key Engagements Won through Troop Initiative 24
2.3: Japanese Infantry Private, 1941 34

Chapter 3: *U.S. Riflemen Will Need More Skill*

3.1: The Price of Pride without Thinking 42

Chapter 4: *Microterrain Appreciation*

4.1: Grasslands and Swamp 48
4.2: Gullies and Rocks 49
4.3: Bushes and Manmade Passageways 50
4.4: Mounds and Trees 51
4:5: There is More Cover and Concealment Down Here 52
4.6: Rural Russians Are Closer to Nature 54
4.7: Damp Areas Provide Best Gap in Enemy Lines 55

Chapter 5: *Harnessing the Senses*

5.1: Putting One's Ear to the Ground 60
5.2: Odor Can Carry a Long Way 63

5.3: Potential Threats to Sensory Balance 67

Chapter 6: *Night Familiarity*

6.1: The Telltale Silhouette 73

Chapter 7: *Nondetectable Movement*

7.1: The Eastern Art of Shadow Walking 88
7.2: Russian Crawling Techniques 91

Chapter 8: *Guarded Communication*

8.1: There's No Talking on a "World-Class" Patrol 97

Chapter 9: *Discreet Force at Close Range*

9.1: Bayonets Work 104

Chapter 10: *Combat Deception*

10.1: The Visible Defender—"Straw-Man" Decoy 129

Chapter 11: *One-on-One Tactical Decision Making*

11.1: What Is Often the Best Decision 135

Chapter 12: *When Told to Hold*

12.1: WWI Germans Depart from Linear Defense 144
12.2: WWII Germans Rely on Squad Strongpoint Matrix 148
12.3: WWII Soviets Add Fake Positions to Same Model 149
12.4: Modern-Day Soviets Use Same Squad Strongpoint 149
12.5: Current North Korean "Defensive Stronghold" 158
12.6: North Vietnamese "Double-Line" Defense 158
12.7: Japanese Strongpoint Had Built-In Firesacks 160
12.8: Viet Cong "Wagon Wheel" Outpost 160
12.9: WWII Japanese Link Squad Forts to Reverse Slope 162
12.10: Chinese Defend Ridgelines Same Way in Korea 163
12.11: Russian Diagram of North Korean Difference 164
12.12: Many Trenchlines Were Covered 164

12.13: One Escape Tunnel for Every Few Positions 165
12.14: Manned Sections Became Strongpoints 165
12.15: Gun Bunkers Linked by Covered Trench 166
12.16: Gun Pits Linked by Lateral Tunnel 166
12.17: Evidence of Ingenious Escape Route 166
12.18: Possible Vietnamese "Military-Fortress" Segment 172
12.19: The Eastern Way to Kill a Tank 173

Chapter 13: *At the Listening Post*

13.1: Russian Observation Posts 180
13.2: An Outpost Can Fight without Revealing Itself 182

Chapter 14: *With Contact Patrolling*

14.1: The Eastern Soldier Is Elusive 189

Chapter 15: *On Point*

15.1: Point Security Is a Two-Man Job 198

Chapter 16: *About Tracking an Intruder*

16.1: America's Best Trackers Are of Eastern Origin 204
16.2: Optimal Viewing Conditions 210
16.3: A Recent Print in Soft Soil 212
16.4: The Effects of Aging 213
16.5: The Average-Pace Method for Counting Heads 214
16.6: Russian Tracker 216
16.7: Water Will Not Obscure All Sign 218

Chapter 17: *While Stalking a Quarry*

17.1: The Western Approach to Stalking 225
17.2: Eastern Refinements 227

Chapter 18: *To Reconnoiter an Enemy Position*

18.1: The Well-Hidden "Prepared Enemy Position" 231
18.2: Eastern Sappers Can Crawl through Barbed Wire 234

Chapter 19: *In the Rural Assault*

19.1: Russian Scouts 239
19.2: How Russians Cross Protective Wire 242
19.3: Barbed Wire Can Be Silently Cut 243

Chapter 20: *For Attacking Cities*

20.1: Every Town Has Underground Drainage 271

Chapter 21: *During an Urban Defense*

21.1: The Russian Appreciation for Urban Microterrain 282
21.2: The "Soft" Eastern Building 283

Chapter 22: *The Rising Value of the "Little Picture"*

22.1: More Can Be Accomplished below Ground 304
22.2: Smaller Teams Generate More Surprise 306

Chapter 23: *How the Tiger Is Born*

23.1: The Battledrill Instructors Are Junior NCOs 319

Chapter 24: *Field Proficiency Has No Substitute*

24.1: Iraqi Special Forces Trooper, 1982 329
24.2: Muslim Irregular, 1982 330
24.3: The Cost of Inferior Skill at Close Range 344

Maps

Chapter 7: *Nondetectable Movement*

7.1: Truong Son Network of Strategic Routes, 1965-68 80
7.2: Truong Son Network of Strategic Routes, 1969-73 81
7.3: Truong Son Network of Strategic Routes, 1973-75 82
7.4: Coastal Branch of Ho Chi Minh Trail 84

Chapter 10: *Combat Deception*

10:1: Spots in the "Leopard Skin" Formation 120

Chapter 12: *When Told to Hold*

12.1: Chinese Punitive Expedition of 1979 169
12.2: Extent of the Chinese Penetration 170

Chapter 19: *In the Rural Assault*

19.1: Tightening Circle of Trenchlines at Dien Bien Phu 259
19.2: Subsurface Fortification Encircles Khe Sanh 263
19.3: Evidence of Feeder Tunnels at Khe Sanh 264

Chapter 20: *For Attacking Cities*

20.1: The North Vietnamese Tet Attack on Hue City 273
20.2: The North Vietnamese Tet Attack on Saigon 274

Tables

Part One

Chapter 2: *Orphaned Squads Are at Greater Risk*

2.1: The "Maneuver Warfare" Difference 21
2.2: Partial List of *Ninjutsu* Techniques 29-32
2.3: VC "*Ninja*-Like" Training Schedule 35

Part Three

Chapter 16: *About Tracking an Intruder*

16.1: The Various Types of Trail Detail 208-209

Part Four

Chapter 24: *Field Proficiency Has No Substitute*

24.1: A Few of the Additional Capabilities of *Ninpo* 337

Appendices

Appendix A: *Enemy Loss Comparisons*

A.1: Participant's Highest Est. of Own Losses in Korea 347
A.2: Participant's Lowest Est. of Foe Losses in Korea 347
A.3: Participant's Highest Est. of Own Losses for . . . Tet 348
A.4: Participant's Lowest Est. of Foe Losses for . . . Tet 348

Foreword

When I was a Lieutenant going through The Basic School in early 1967, the instructors there talked about the Viet Cong and North Vietnamese Army as if they were ghosts and able to appear and disappear at will—completely invisible at night and only making themselves known when they had you and your platoon in their killing zone.

When I arrived in Vietnam in the early fall of 1967 and took over my first platoon, I found that the "old-timers" in the platoon talked about the enemy in the same way. I soon learned why—both the Viet Cong irregular forces and North Vietnamese Army regulars were outstanding light infantrymen. Always camouflaged, even on the move, they dug in whenever they weren't moving, and their prepared positions were invariably almost invisible until you were right on top of them—more often than not, much closer than you wanted to be. Their fire discipline was superb; rarely did a North Vietnamese soldier give away his position by firing too soon or too much. In the attack, they were masters of infiltration and able to penetrate even alert and well-prepared positions—including U.S. Marine positions— clandestinely, often announcing their presence with a volley of concussion grenades. When you drove them from one of their prepared positions or found it abandoned because they had chosen not to fight you there, a study of their positions always impressed you with the discipline they displayed. Every fighting position was dug precisely—deep and small, only big enough for the one or two fighters it held. The spoils from their digging were always hidden—their positions covered and camouflaged perfectly. Even in the urban battle in Hue City, where they were clearly no longer in their "natural" element, their positions were very difficult to find, and their individual movement was superb.

In my later tour as an advisor, the North Vietnamese Army crossed the DMZ and attacked Quang Tri Province with three infantry divisions, two armored regiments and four or five artillery regiments. This Corps-sized force remained invisible until they launched their attacks across the entire front.

Of course, the North Vietnamese Army had been disciplined, and they had to remain invisible. If they did not, our superior firepower ended their chances. But I soon came to realize that they were not ghosts and not superhuman. They were not invisible, and they could not see at night—they were simply superbly trained and disciplined light infantrymen. As much as I hated communism and the political system that these opponents sprang from, I had to grudgingly admit that they were formidable adversaries and admire them for their professional skills.

Over the years, as our own forces modernized and we have been able to apply better and better technology to our intelligence collection capability and as our weapons systems became more and more accurate and lethal, I have often thought of what an awesome force it would be if we could marry that high technology in the air and in command and control with the superb light infantry that I saw in our NVA enemy. What a force we could be if an enemy could not see us on the ground while he could not challenge us in the air.

Clearly we face a different challenge than did our enemy in Southeast Asia. Our requirement is to project force wherever the Nation needs it, at any time and any place. This requires a rapid buildup of logistics ashore that does not allow for an "invisible" ground combat element. But, in the final analysis, where the "rubber meets the road" in those last few hundred yards, warfare still comes down to infantrymen against infantrymen. With our overwhelming combined-arms capabilities, our infantrymen will win that fight—of that I have no doubt.

The question is, can we win that fight with fewer casualties? Poole's answer, in this heavily researched study of the Asian infantryman, is yes! I often asked myself those many years ago, how do they do it? Poole has researched the answer and provides it in this book. He also gives some good suggestions on how we might learn to do the same thing.

American fighting men may not like everything they read in this book. Some of it may make you mad—it did me. But all of it will make you better prepared for the future fight. I recommend it to all infantrymen and infantry leaders.

MAJ.GEN. RAY L. SMITH USMC (RET.)

Acknowledgments

Upon reverting to E-7 Gunnery Sergeant after 20 lackluster years as a commissioned officer, the author joined Camp Lejeune's Advanced Infantry Training Company in the fall of 1986. While struggling to function in the much more difficult world of the enlisted "Camp Geiger Tiger," he often heard the phase "eye of the tiger." It was what each infantry private would need to survive in combat. Of course, to develop the "eye of the tiger," that private must first practice "the tiger's way." As the weapons become more lethal, battles will necessarily be fought by tiny, semi-independent groups of highly skilled enlisted personnel. To prepare the lower ranks for this more challenging role, the U.S. military must meet or exceed the short-range combat capabilities of other nations. To do so, it must study the small-unit tactics and training of armies that can win despite little firepower or control over subordinate units.

Since that uncertain winter of 1986, a lot of water has passed under the bridge. Much has been learned, and much more remains to be learned. Yet, this humble Marine will always be indebted to those thousands of NCOs who belatedly taught him what the "system" had failed to mention. If what he can remember of their collective wisdom could now save one life, his fondest dream would be realized. *Semper fidelis.*

Introduction

With the aid of overwhelming preemptive bombardment, U.S. ground forces have just occupied a Third-World desert nation. Insofar as American precision-guided munitions destroyed Saddam Hussein's strategic assets, they were productive. Still, for the U.S. ground combat commander, a heavy reliance on firepower has its disadvantages. For one thing, it gives his small-unit leaders license to operate the same way. After encountering their first serious pocket of resistance, they may start to shoot at anything that moves. As their bullets begin to stray, they may become less certain of their "warrior skills" and moral mandate. Additionally, they may not get to evaluate their ability to withstand a temporary firepower deficit in more normal terrain. Finally, whatever those courageous front-line fighters learn the hard way will have little chance of entering into their organization's tactical memory.

While it is the commander's job to destroy strategic targets from a distance if possible, it is his troops' responsibility to exercise minimal force at close range. While small contingents of enemy soldiers are often difficult to distinguish from civilians, they do not normally constitute strategic targets. No amount of talk about "weapons of mass destruction," "human shields," "guerrilla activity," or "terrorism," will dispel the long-term effects of preventable civilian casualties. To exercise minimal force, U.S. enlisted personnel will need more skill than was evident in the TV coverage of their short-lived Iraqi adventure. Against a well-motivated foe under more normal circumstances, those overcontrolled, overloaded, and undercautious mechanized infantrymen would have paid a heavier price for waddling across streets and filling doorways. Unfortunately, the problem is not new and has only been aggravated by America's most recent fling with military technology.

> Since 1941 . . . our forces were not as well trained as those of the enemy, especially in the early stages of the fighting.

> After the buildup of forces, when we went on the offensive, we did not defeat the enemy tactically. We overpowered and overwhelmed our enemies with equipment and fire power.[1]
>
> — Lt.Gen. Arthur S. Collins U.S. Army (Ret.)

This book has been written for the U.S. enlistee and those sworn to protect him (or her). It's about ground combat at 75 yards or less. Admittedly, this is only one facet of an immense subject. Still, it is the facet that will most greatly influence that enlistee's chances of survival. By official estimate, America dominates every aspect of military endeavor. Yet, one wonders what the world's other major armies might have learned from their relatively greater combined experience. The Germans, Russians, Chinese, Japanese, North Koreans, and North Vietnamese have often had to function at a deficit in technology and firepower. Perhaps that's why their small units have—during those periods of deficit—conducted similar maneuvers at short range. Yet, these are not the maneuvers that U.S. squads or platoons have done under similar circumstances. Why the disparity in tactics? Could those armies (henceforth called "Eastern") have collectively refined frontline procedures to compensate for a firepower deficit? Is there adrift in the "Third World" a portfolio of tactical techniques with which to contest a superpower? Or even more disturbingly, might those Eastern armies have evolved tactically while their Western counterparts did not? One hates to think what the latter possibility might have already cost in casualties. Either way, every U.S. citizen needs to get interested. Their sons and daughters deserve to know of every threat to their survival. That the Soviets/Russians have since sacrificed much of their short-range-combat proficiency to technology in no way detracts from this assessment. A nation can briefly embrace tactical reform without being able to permanently institutionalize it.

That the best of the Eastern armies now serve Communist regimes is purely coincidental. The difference in how they train and fight is cultural—predating the "ism's" by thousands of years. The author does not intend to endorse the politics of "Eastern-Bloc" nations, only to study the subcultures of their armies.

It has been said that each army develops its own tactical techniques based on unique assets and requirements. Those with superior firepower can sometimes win with very little tactical expertise. Those without it must overcome the same battlefield conditions. As those conditions grow more lethal, their tactical methodologies should converge. However, they will only converge for the nations that treat infantry tactics as an ongoing science as well as an impromptu art. As sophisticated tactics can often compensate for inferior firepower, the scientific method of learning might have more appeal to undergunned armies. At their lowest echelons, they might be more likely to conduct research through "collective opinion" polls and "trial and error" experiments. This may help to explain why there appears to have been German, Russian, and Asian influences on most of the Eastern trends in technique. For example in the 1950's, the guerrilla tactics of the Viet Minh were largely of Soviet and Chinese origin.[2] Within these similarities may lie the seeds of tactical innovation for the West—of how better to dodge the all-seeing munitions of the future. Whether or not the U.S. military ever joins the trend, it minimally has the obligation to apprise its lower ranks of it. Just by being aware of those tactical similarities, America's young men and women could as much as double their chances of survival against future adversaries.

> When you are ignorant of the enemy but know yourself, your chances of winning and losing are equal.[3]
>
> — Sun Tzu

As the Eastern difference began before the time of Christ, Sun Tzu is a good place to start. He lived in northern China during the fourth century B.C. Because of the empire building and mass migrations of the period, his influence over the Asian, Russian, and German regions began almost immediately.

> [S]ongs, legends, stories, and teachings have always survived primarily by word of mouth. Commercial intercourse by boat also existed among China, Egypt, and Africa; students may have passed on his [Sun Tzu's] teachings, probably in mutilated form, to various incursive tribes such as the Hiong Nu, ". . . who attacked the empire of China in the second and first centuries B.C.," before turning west (Gordon, *The Age of Attila,* 1960). . . .
>
> The Huns may have stemmed from the . . . Hiong Nu.[4]

By the first half of the fourth century A.D., the most famous of the pre-Germanic "Huns"—Attila—was running unchecked through

Western Europe. His people were Mongol nomads who, at the height of their plundering, controlled everything from the Caspian steppes to the Rhine River.[5] Included therein was much of present-day Austria and Germany.

> The Huns, a Mongolian people from Asia, had crossed into Eastern Europe above the Black Sea. . . .
>
> Their race was Mongolian, as revealed by their squat figures, the wheat-color of their skins, their slit eyes, and the short black hair that covered the heads. . . . Their [most recent] homeland . . . was the vast, unproductive expanse that lay between the Ural and Altai Mountains [of southern Russia]. . . .
>
> Exactly how extensive this [Hun] empire was [before Attila took over] no one will ever know, but it surely included what is present-day Hungary, Roumania *[sic],* and Southern Russia. No trace remains of . . . their headquarters on the Theiss River in Hungary from which they ruled scores of German and Slavic tribes they had subjugated. . . .
>
> Early in 451, Attila moved westward with a half-million men, most of them Germanic and Slavic mercenaries. . . .
>
> His first objective was Orleans [in the middle of what is now France].[6]

Upon Attila's death in 454 A.D., many of the Huns settled in the lower Danube Valley of what is now Bulgaria and Romania. Then, several centuries later, more Mongol nomads entered Eastern Europe.

> The Hungarians or Magyars were the only Asiatic nomadic invaders of the early Middle Ages who retained their identity and established a country of their own. When they appeared in the ninth century, they were as savage as any of their Asiatic predecessors including the Avars whom they absorbed. For about fifty years they carried their plundering raids across Central Europe.[7]

By 1211, Genghis Khan had united the Mongol tribes from Siberia, Manchuria, and Mongolia and penetrated the Great Wall of China.[8] Soon thereafter, he moved his seat of power to Beijing (his successors were to rule China as the Yuan Dynasty until 1368).[9]

While one grandson—Kublai Khan—completed the conquest of China, another—Batu Khan—led the "Golden Horde" (also known as Tatars or Tartars) into Russia in 1223. Fully aware of Sun Tzu's wisdom, Batu relied heavily on military intelligence, spies, and advance scouting parties. His armies lived off the land. At full speed, his horsemen were highly accurate with their arrows. They would often converge on the same objective from several different directions.

> These Mongols were from east-central Asia where the modern province of Outer Mongolia is located. . . . The Mongols were not unlike such earlier Asiatic tribes as the Huns, Avars, and Magyars.[10]

In every region Batu conquered, the male population learned to fight Tatar style. "They [the Tatars] reinforced their numbers by using the forces of conquered countries to help them in their further campaigns."[11] By 1240, Batu had captured most of Russia,[12] Poland, Hungary, and Bohemia (western Czechoslovakia).[13]

> The Mongol Empire was the largest land empire in history. At its height it encompassed China, central Asia, Persia, Mesopotamia [modern-day Iraq], Poland, Hungary, Bulgaria, Serbia, and all of Russia with the exception of Novgorod [a city in the extreme northwest].[14]

Then, by intermarrying with the Russian aristocracy, Batu's successors controlled the region for 250 years.[15] In 1380, Moscow's Russian prince renounced Mongol tribute. Soon, another of Genghis Khan's descendants—Tamerlane[16]—sent his generals to sack Moscow and reimpose Tatar rule.[17] By the 15th Century, most of those descended from the Mongol/Tatars had converted to Islam.[18] Their former homelands would become integral parts of the Tzarist Empire and the Union of Soviet Socialist Republics. To imagine Russian tactics not shaped by the teachings of Sun Tzu would be to disregard a full 250 years of Mongol influence.

> The Tatar [Mongol] invasion was to have lasting political, social and economic effects on Russia's subsequent development. . . . [T]he Tatars . . . influenced the Russian . . . military (especially cavalry) development.[19]

The Russian army's assimilation of Asian tactics can also be traced to more recent events. The former Soviet Union extended all the way to the Pacific Ocean, and its 20th-Century victories can largely be attributed to its Oriental military heritage. It was, after all, the Soviets who had beaten the Japanese in Manchuria in 1939 that turned the tide of WWII, not those who had been embarrassed by the Finns. It was Zhukov's Far Eastern Command that defeated the Nazis at Moscow, Stalingrad, and Berlin.

Since being partially absorbed by the Mongol Huns of Attila, Germany has been continually exposed to Eastern influences. It has fought the Soviet Far Eastern Command, and it has provided tactical advisors to Japan off and on since 1872.[20] As late as 1939, Japanese officers were receiving instruction on Prussian Major Jakob Meckel's "soft-spot" (gap) tactics.[21] Several generations of German advisors must have witnessed the ways in which an Asian can sneak through enemy lines (an apparent surface). Whether any took that short-range infiltration technique back to Germany is very much in question. More probably, the highly decentralized German army was able to sufficiently refine Boer tactics to arrive at something similar. Germans have a gregarious nature that enables them to opt for collective effort (an Eastern tradition) over individuality (a Western trend).

Meanwhile in America, most of the riflemen who have survived an Eastern conflict will readily admit to having been inadequately trained for their ordeal. While few of their commanders may agree with that assessment, those commanders did little of the short-range fighting and thus lack the same frame of reference. One segment of the military population experiences battle from high overhead, through binoculars, or via some rear-area "war" room; and the other gets a much closer look. While the infantry officer plans his next move on a 1:50,000 map, the "grunt" private is often busy diving behind mounds, rocks, or trees. This wonderful, little American "snuffy" inhabits that unforgiving and underreported place where the "rubber meets the road." His chances of survival will depend more on what he is allowed to do, than on what he is told to do.

Many recently commissioned Americans may honestly believe their Quantico or Benning pep talk—that victory will depend more on their personal decisions and leadership than on any other factor. If this were true, their privates would need only to know how to operate their equipment and follow orders. What if the preparatory fire fails? To survive a one-on-one encounter, each American private must be able to think and act on his own. He has far more God-given growth potential than his superiors would have him believe. His mental processes and survival instincts are alive and well. Yet, he gets little out of the task force exercises, flag rank demonstrations, and bureaucratic requirements that monopolize his time. His requirements are simple. He wants to know how to survive the same nightmarish conditions that his father and grandfather had to face: (1) all-around, close-range fire from invisible, below-ground opponents; (2) precision, long-range fire from who knows where; and (3) having to run upright into enemy machinegun fire.

What U.S. veterans have been complaining about is the tactical sophistication of their immediate adversary—that invisible guy behind the bullet. They can't understand why their parent organization didn't better prepare them for him.

The reasons are many. Human beings have difficulty identifying and correcting their own deficiencies. They subconsciously strive to look good at the things they aren't. Where they suspect a personal shortcoming, they often verbalize its solution. Then they conclude that preaching a good game is tantamount to playing a good game. Societies follow the same pattern. Each professes to be good at something it isn't and then tries to bridge the gap with semantics. Western society is no exception. While mainstream Americans warn their children about drug abuse, they personally pop pills for everything from obesity to the "blues." Sadly, this all-too-predictable disparity between intentions and results can have dire consequences with military issues.

A unit's final casualty count in war will depend more on the tactical techniques of its corporals and privates than on the operational philosophies of its commissioned officers and senior staff noncommissioned officers (SNCOs). With the first shot, modern battle degenerates into a myriad of tiny, widely separated, life-or-death struggles over which no commander has much control. The side winning the majority of these tiny struggles will suffer the fewest casualties. For each encounter, the victory will generally go to the participant with the most field skill, decision-making experience, and initiative. Those who have constantly to master new equipment will lack field skills. Those whose commanders are enamored with their own wisdom will lack tactical decision-making experience. Those conditioned always to follow orders will lack initiative. In other words, the privates who have—for whatever reason—been overprotected will be at a distinct disadvantage. Through inflexible training and oversupervision, America's military leaders may have jeopardized thousands of recent enlistees unnecessarily.

To limit the bloodshed in the most obscene of man's activities, the U.S. military has traditionally opted for overwhelming force. To minimize casualties, it has concentrated on long-range warfare. As a consequence, its short-range proficiency has suffered; and that's not all. In an attempt to compensate for this chink in its armor (to win wars without "world-class" infantry), it has had to use too much force. This has put its normal *modus operandi* into direct conflict with moral theology. The United States has unwittingly created ground forces with too little small-unit skill to exercise minimal force.

> [T]he hypothesis of legitimate defense, which never concerns an innocent but always and only an unjust aggressor, must respect the principle that moralists call the *principium inculpatae tutelae* (the principle of nonculpable defense). In order to be legitimate, the "defense" must be carried out in a way that causes the least damage and, if possible, saves the life of the aggressor.[22]
>
> — Pope John Paul II

For all the "right reasons," the once-proficient Soviets made the same mistake in Afghanistan.

> During the war in Afghanistan, like the United States in Vietnam, the Soviets chose to expend massive firepower in order to save Soviet lives and to compensate for their lack of infantry. It was an expensive, indiscriminate, and . . . ineffective practice.[22]
>
> — Editor's note, U.S. Dept. of Defense (DoD) published Soviet military academy study

The Tiger's Way sheds new light on this emerging portfolio of tactical techniques—the individual and small-unit maneuvers with which to compensate for a deficit in technology and firepower. It also shows how each of America's past adversaries has contributed to this body of knowledge. Sadly, its frontline soldiers appear to have been, for the most part, better prepared to survive than their U.S. counterparts.

> Japanese army training of all kinds proceeds along more exact and minute lines than is usual in our own [American] service.[24]
>
> — Translator's preface
> Japanese night-fighting manual, 1913

Should this training deficit go unaltered, American teenagers will continue to have trouble winning their life-or-death, one-on-one encounters. As there are pronounced trends in how Eastern privates fight, one wonders why this is just now being explored. One suspects the elitism that sometimes plagues Western bureaucracy. Eastern armies have more closely listened to their enlisted men. They have also more readily delegated authority. One would hate to think that top-echelon U.S. officials have been ignoring this training shortfall so as not to incur the obligation to delegate their power. To create riflemen of commensurate ability, they would have to decentralize control over infantry training and operations.

As the weapons grow more lethal, wars will necessarily be waged by well-dispersed ground forces. The set-piece battles of old will be replaced by thousands of tiny encounters. Indisputably, the maneuver element and defensive-fort leader of the future will be a noncommissioned officer (NCO). As in any Asian army, that NCO's secret weapon will be his individual rifleman. Yet, to meet this challenge, the American rifleman will need vastly more skill than he presently receives at entry-level training or his first line unit. He will need all of the cunning and wood's smarts of a Third-World guerrilla. Many of his Eastern counterparts already have it.

> Japanese strength . . . lay in small units and the epitome of Japanese doctrine was embodied in small-unit tactics. Night attacks . . . and the willingness to engage in hand-to-hand combat were the hallmarks of the Japanese infantryman. Indeed such tactics were very successful against Soviet infantry [in 1939].[25]
>
> — U.S. Army, *Leavenworth Papers No. 2*

Whether the U.S. military has failed to evolve tactically or just to recognize how other armies confound firepower, American privates will be at unnecessary risk. To stand the best chance of survival anywhere east of Berlin, they must first be familiar with "the tiger's way."

H. John Poole

Part One

A Growing Threat at 75 Yards

It will be a war between an elephant and a tiger.
If the tiger stands still, the elephant will crush him,
so the tiger never stands still. — Ho Chi Minh, 1946

(Source: *The Battle of Dien Bien Phu,* © 1988 by New Star Video, videocassette #4010)

1 American Units Must Further Disperse

- *What's it like to be a U.S. infantry private in combat?*
- *Why must 21st-Century soldiers spread out?*

(Source: *FM 7-8* (1984), p. 2-1)

Hurry Sunrise

Darkness descends over the rain forest like an undertaker's cloak. Not even the stars shine overhead. As the mist rises from the jungle floor, the croaking of frogs, buzzing of insects, and calling of animals blend into a dull roar. It's so dark that the U.S. perimeter guard sees only shapes in his assigned sector, so loud that he can't hear himself talk. While this young American has buddies less than 20 yards away on either side, he feels isolated and uncomfortable. There are so many shadows, plants, and ground irregularities between those buddies and him, that those 20 yards might as well be 2000. His distant squad leader has the only night vision goggles (NVGs), and his even-more-distant platoon leader has the only thermal-imaging device. Under these conditions, neither piece of technology would help him anyway. The former needs ambient light, and latter can't see through bushes.

Raised in the city and with only six months in service, the U.S. perimeter guard feels out of place in the woods. He has been repeatedly told that he is the best in the world, but he has heard stories. His uncle and grandfather have talked about German, Japanese, North Korean, Chinese, and North Vietnamese soldiers who could crawl up on a wide-awake sentry. The young American is tired. For weeks on end, he has been patrolling all day and staying awake most of every night. Though only 130 pounds soaking wet, he has been lugging 100 pounds of mostly ammunition through every bog and tree fall in the area. At first, he tries to analyze every sound and shadow. Then, his mind wanders back to Trish and home. For hours, he sits erect in his hole, moving in and out of "the here and now." Then it happens. He doesn't notice that the twelve bushes to his front have turned to thirteen. There's a "whoosh," a "thunk," indescribable pain, a suspicion of betrayal, a gasping for breath, and then nothing at all. Private Robert B. "Squirt" Ryan, Jr.—the pride of Cedar Rapids—is gone. He will not have that family of which he and Trish had dreamed. He will not spend his Saturdays fishing with his best friend Bill. He will not become that fireman who would save all those other people.

It is over now for Private Ryan, but questions linger. Did he really have to die? Was his organization somehow remiss? Could his leaders have saved him? How did his opponent get so close? What can the U.S. private do to help himself? These are complex questions that will take many chapters to answer.

What Happened?

It's clear that Private Ryan's survivability had somehow been compromised. His ability to defend himself had been less than his opponent's ability to kill him. Perhaps that was his fault. After all, he was very young, inexperienced, and trusting. Yet, he was still a child of God and, as such, had tremendous growth potential. Perhaps, for whatever reason, his organization had not helped him to fully achieve his warrior potential. It could have told him about the legendary stalking ability of his Eastern counterpart. His defensive training could have covered the counting of bushes. Instead, he was sent out into someone else's backyard knowing little more than how to use his own equipment and follow orders. Then, he was forced to carry three-fourths his body weight on daily patrol and to stay awake much of every night on ambush, listening post, or guard duty. This is the world of the U.S. infantry private. To save him, one must never lose sight of its parameters. These parameters are not adequately depicted by satellite image or war room map. If, for whatever reason, Private Ryan is overprotected, he will not have the field skills and self-sufficiency to survive in the real world. To have the best chance, he and his low-ranking buddies must be allowed to collectively influence how they will be trained and employed.

Where Might the Fault Lie?

It has become fashionable for "upper-echelon" Americans to strive for a broad range of knowledge. With so many responsibilities, they can't afford to dwell on any particular subject. To gain "perspective," they settle for an overview. Unfortunately, the "big-picture" manager tends to discount the contributions of his lower-echelon subordinates. Having never personally performed their menial functions, he views those functions as elementary or without substance. He often comes to believe himself more capable of each task than the task providers are collectively. At this point, his decisions become less feasible.

What if America's gifted elite were to start grooming middle managers in their image? Before long, the technicians would be standing around talking about their trades instead of practicing them. In industry, such a scenario would lead to doctored books, lower profit margins, and possible bankruptcy. In the military, it equates to glowing battle chronicles, additional casualties, and occasionally losing a war. Short-range infantry combat—that on which most casualties are based—is disjointed and quick moving. It can best be performed by low-ranking, but highly skilled, tactical "technicians" who have been allowed to make many of their own decisions.

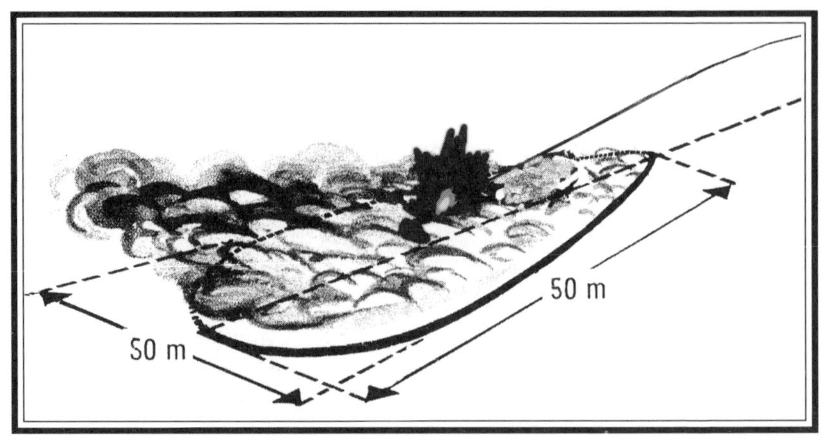

Figure 1.1: Kill Zone for New Russian "Shoulder-Fired" FAE
(Source: *Soldateni falt 2* (1972), p. 13)

Infantry Combat Has Become Much More Dangerous

More deadly than conventional munitions are the fuel-air-explosive (FAE) or "thermobaric" variety. They kill by overpressure and incineration. The Russians have used the "Tanin" FAE-tipped, rocket-propelled-grenade (RPG) round in Chechnya since 1996. They have also sold it on the international market.[1] So even guerrilla units may now have grenade-throwing machineguns, laser-guided munitions, and shoulder-fired FAEs. Also for sale by the Russians are larger gun- and rocket-launched thermobaric rounds.[2] Just one could take out a closely arrayed rifle company. To counter this trend in weapons lethality, U.S. ground forces must more widely disperse.

> [T]he most important [Russian] equipment development was the fuel air explosives (both the jet propelled [Shmel] "Bumblebee" 93mm [shoulder-fired] flamethrower capable of shooting [see Figure 1.1] a thermobaric round; and the TOS-1, a flamethrower mounted on a T-72 chassis and capable of shooting a thermobaric round over 3.5 km).[3]
>
> — "Russian Lessons Learned from Grozny Battles"
> *Marine Corps Gazette,* April 2000

Unfortunately, most U.S. line units lack the composite skills to spread out without getting destroyed piecemeal. In other words, their buddy teams, fire teams, and squads can't handle being temporarily outnumbered and surrounded. This creates a dilemma for any army that has concentrated more on standoff weaponry than on short-range infantry technique.

Other Nations Have Had to Develop Better Infantry Skills

To claim that the U.S. military is the best in the world is to infer that other nations have less competent riflemen. This is simply not the case. With fewer material resources, many have had to rely more heavily on their infantry. Even the firepower- and centralized-control-loving Russians do two things for which Americans have never been noted—study their foe and refine small-unit technique:

> The Red Army consciously studied both its own methods and those of its opponents, and from the 1930's fighting against the Japanese through the Winter War with Finland and throughout the Great War against Germany, was constantly collecting data, analyzing and modifying its tactics and techniques at all levels.[4]
>
> — Translator's introduction to
> *Soviet Combat Regulations of November 1942*

That Western nations more greatly value their soldiers' lives is also a myth. How much training each infantryman receives is the best indicator. Throughout WWII, the Russian tactics manual devoted a whole section to the individual rifleman.[5]

The Western Classifications of Infantry

Western analysts like to delineate between "heavy" and "light" infantry. Implicit in their choice of words is that the former better facilitates power projection. The light is supposedly specialized—not appropriate for general deployment or open terrain. Still, one could argue that the light has greater utility. According to a U.S. Defense Department study, "It is the most suitable for all types of close terrain (forests, mountains, tundra, and cities)."[6]

> Light infantry is a surprise and terrain dependent force. These protect it from tanks and artillery, compartmentalize its opponents, and mask its movements. . . .
>
> In central Europe, the operational purpose of light infantry is to control close terrain and to set up the battle for the tank forces. . . . It operates within the stone villages and forests so as to neutralize the effects of large-caliber weapons (direct and indirect). Its tactics are the ambush and the counterstroke against enemy forces attempting to pass through or clear the close-terrain space.
>
> The characteristic of light-infantry tactics everywhere is infiltration in the attack and ambush and counterstroke in the defense.[7]
>
> — U.S. Defense Agency study

Technology-conscious Americans like to think of mechanized infantry as heavy, and foot infantry as light. However, the real difference is more subtle. While not all U.S. rifle companies continually ride around, none—in the eyes of the world—may qualify as light infantry.

> Positional (foot) infantry of the type still predominant in the U.S. force structure is everywhere [in Europe] being phased out. The [German] Bundeswehr converted their last positional infantry (actually, strongpoint antitank infantry) to Panzer grenadiers in 1980.
>
> In an era when the tank has finally (at least partially) been offset by enhanced antitank weaponry, this shift from a fighting to a hit-and-run infantry seems paradoxical. The reasons are artillery and the increased standoff range and lethality of direct-fire weapons. To survive, infantry must be mobile or inconspicuous. . . . Dismounted armed infantry cannot survive in forward slope fighting positions as the Americans plan to do.
>
> Light infantry survives differently. It protects itself from both artillery and armor by fighting mainly in close terrain; in open terrain, it is a night and special mission force. . . .
>
> . . . On the modern battlefield, infantry can only move and survive by employing stealth and stalking techniques.[8]
>
> — U.S. Defense Agency study

As better transportation becomes available, armies lose sight of what only highly skilled "ground-pounders" can accomplish. To qualify as "mobile" foot infantry, their grunts cannot ride around all day and then be unable to patrol. The word "mobile" is being applied in its Maoist sense. According to the officially sanctioned chronicle, "Afghanistan was a light-infantryman's war—and the Soviets did not have light infantry."[9] The inescapable conclusion is that vehicular-borne riflemen tend to lose their infantry identity.

The Eastern Version of Light Infantry

In Asia (possibly for any true maneuver warfare proponent), the distinction blurs between light infantry, special forces, and scouts. "[S]ometime during the early 1990's . . . the [North Korean] VIII Special Corps was renamed The Light Infantry Training Guidance Bureau."[10] Only part of its mission is to conduct unconventional and special-warfare operations.[11]

> The [North Korean light-infantry] battalions focus their operations in the forward zone, from a point some 15-30 kilometers in the ROK/U.S.'s rear, back toward the forward edge of the battle area (FEBA), whereas the light infantry brigades focus their operations . . . (deeper into the ROK/U.S.'s rear area).[12]

This Bureau also provides training on conventional and special-warfare operations to the rest of the army.[13] As such, it may publish guidelines to, and perform instruction for, each rifle company.

> [All North Korean] soldiers are trained in both conventional and unconventional warfare.[14]

It would make perfect sense to let special-forces personnel help line infantry units with their small-unit technique. "Unconventional warfare" is, after all, a small-unit evolution. North Korea's primary emphasis is on small-unit capabilities.

> To the [Korean Peoples Army (KPA)], a squad of men who

> can march 50 kilometers with a 40 kilogram pack in 24 hours over mountainous terrain is worth more than a battalion of road-bound mechanized infantry.[15]
> — U.S. Army, "OPFOR: North Korea," *TC-30-37*

The North Korean Light Infantry Bureau has a large training establishment and 22 special purpose brigades.[16] Of the 14 tasked with light infantry duties, 11 have already been placed under the operational control of an army corps.[17] This may indicate that "light infantry brigade" personnel are attached to each company. There they could provide a scouting and short-range-infiltration capability that the company would not otherwise have. They could also providing advanced instruction to the individual line infantryman. Either way, the Light Infantry Bureau commandos appear to be getting enough real-life experience to make them extremely dangerous.

> Another defector, Lt. Yu Tae-yun, stated [around 1987] that nearly all members of the [North Korean] light infantry battalions had [previously] infiltrated up to 40 kilometers south of the DMZ [demilitarized zone] in preparation for missions against ROK [Republic of Korea]/U.S. installations and lines of communications within the forward area.[18]
> — *Inside North Korea*
> ROK Institute of Internal and External Affairs

There Are Things That Only Small Units Can Do

An Asian distracts his opponent with ordinary forces and beats him with extraordinary forces.[19] Often in modern war, big units make too lucrative a target to fully commit. When they are committed, they need small units to prepare an unobstructed path through enemy defenses. Big units are also difficult to run. Squad-sized units don't require five paragraph orders and charismatic leadership. They are like football teams.

Tiny contingents of highly skilled infantrymen can also more easily accomplish the following: (1) generate surprise, (2) fight in close terrain, (3) infiltrate, (4) avoid bombardment, and (5) prevail at short range.

Generate Surprise

It is virtually impossible for anything large or powered by internal combustion to secretly approach an objective. Surprise can generally be accomplished by no more than 14 skilled ground-pounders. By U.S. determination, the initial penetration for German armor during WWII was generally made by a tiny contingent of infantrymen.[20]

Russian soldiers are by reputation tenacious on defense and inflexible on offense. Reputations can be misleading. By 1942, the U.S. War Department knew that Russian infiltration, encirclement, and hasty defense were often facilitated by small, mixed teams with significant firepower.[21] How the Russians stymied an entire German army at Stalingrad with loosely controlled reinforced squads is still a thing of legend.

That was long ago, and the modern Russian army has new problems, including a counterproductive distrust of NCOs. Still, the contemporary Russian fighting style is not without foundation or merit. In open terrain, the Russian commander's operational goals will still dictate his tactics. If possible, he directs an avalanche of tanks and supporting arms against a narrow sector of enemy lines and exploits any breakthrough. Where Stormoviks once provided his tanks with overhead fire support, he now has HIND helicopters. In the modern display at Moscow's Armed Forces Museum, a wall-length mural depicts paratroopers (Figure 1.2) operating well ahead of the armor. In addition, the armor has a protective contingent of BMP-borne infantrymen. "Motorized rifle troops . . . 'have replaced the traditional foot soldier' (Lomov, *Nauchno-Teknicheskiy Progress,* 1973, 106)."[22] For most of the participants of such an attack, there may seem little opportunity for initiative. "Individual initiative could ruin the whole plan."[23] However, a small-unit leader can be good at following orders and still able to function without them.

> The . . . uncompromising accomplishment of orders is the foremost task of the leader in battle. In the case of a sudden change in the situation when it is not possible to wait for (new) orders, the leader must handle the situation independently. The absence of orders can never be used by the leader as an excuse for inactivity.[24]
>
> — *Soviet Combat Regulations of November 1942*

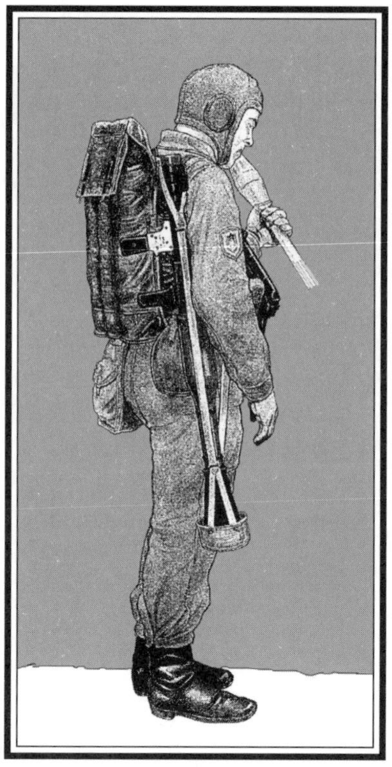

Figure 1.2: Soviet Airborne Private

(Source: Courtesy of Cassell PLC from *Uniforms of the Elite Forces,* ©1982 by Blandford Press Ltd., Plate 15, No. 43)

For a hasty attack in open terrain, each Russian squad may simply leap off its respective BMP and assault on line ahead of it (as depicted in separate paintings in the Army Forces Museum in Moscow).[25] That way each squad would enjoy its own source of overhead machinegun fire. Near a prepared enemy position or in close terrain, the Soviet attack method consisted of narrow penetrations. By U.S. determination, the Soviets used an infantry column formation in the assault on fortified positions during WWII.[26] Archival film footage in the same museum shows a flamethrower operator at the head of the column during a hasty attack through the Stalingrad rubble. Of note, Soviet antitank guns were creating a diversion nearby. For a heavily fortified objective, the Russian spear tip would have to be more sophisticated. Very probably, the Russians have discovered how to sharpen this tip. By Soviet doctrine, every operation was rehearsed down to the tactics of the individual soldier.[27] When that individual soldier displayed initiative, he was reinforced with more soldiers.

> A successful advance, even if only by a single rifleman, must be immediately supported by the other soldiers and the squad by all possible means. In battle align yourself only on the most advanced soldier.[28]
>
> — *Soviet Combat Regulations of November 1942*

Of course, in close terrain, Russian squads or platoons get even more play. If Soviet commanders value the collective opinions of their men as much as they claim, their small-unit assault techniques may be quite advanced. "In a combat division, while a sergeant must not be overfamiliar with his senior soldiers, he must at least respect them and take their opinions into account."[29]

Having been heavily influenced by the Russians, the tactics of North Korea show similarities.

> In the rougher mixed terrain of Korea, North Korean light infantry is the spearhead, behind which their main forces follow and consolidate. In pure open terrain, it is a raid and special operation infantry.[30]
>
> — study contracted by the
> U.S. Defense Advanced Research Project Agency

> The standard [North Korean] light infantry units have the capability . . . to perform the following missions:
>
> - Enveloping . . . attacks in support of [other units] . . .
> - The . . . destruction of [enemy strategic assets] . . .
> - [I]nterdiction . . . of . . . reinforcement or resupply . . .
> - Seizure of . . . topographic [and man-made] features . . .
> - Augmenting . . . [other unit] reconnaissance assets
> - Unconventional warfare operations
> - Rearguard and delaying operation
> - Counter [enemy] special operations [31]

It must have been "light-infantry" attachments that spearheaded the Chinese assaults at the Chosin Reservoir in the fall of 1950.[32] Perhaps the North Vietnamese used their "scouts" for the same purpose. That would explain why the rear guard in Hue City so easily vanished. One wonders to what extent those skills have been retained (Figure 1.3).

Handle Close Terrain

There is evidence that the WWII Russians were more willing to decentralize control in woods, swamps, and mountains. There they depended on self-sufficient task forces as small as a reinforced platoon. *TM 30-340* confirms that a Russian submachinegun platoon could infiltrate enemy lines, ambush, and counterattack.[33] As a Soviet client state, North Korea also made full use of small units in close terrain.

> During the early days of the [Korean] war, the North Korean Peoples Army [NKPA] never varied its tactics. It never had any need to do so. Its general maneuver was to press

Figure 1.3: Communist Chinese Paratrooper

(Source: Courtesy of Cassell PLC, from *Uniforms of the Elite Forces,* ©1982 by Blandford Press Ltd., Plate 28, No. 82)

> the ROK or American forces closely, engage them by means of a frontal holding attack, while at the same time turning the enemy flank and infiltrating troops to the enemy rear.[34]

While the Russians and North Koreans now talk of light infantry, the Chinese and North Vietnamese have distinguished only between scouts and other soldiers. The Chinese have become expert at dispersing along a front and concealing themselves before an attack.[35] This is the modern equivalent to the ancient cloud formation.[36]

> Disperse or concentrate one's own forces swiftly on a wide and flexible battlefield . . . . One's own forces must be assembled in secrecy and must attack at the time and place which the enemy least expects.[37]
>
> — Mao Tse-tung quote from U.S. Army, *PAM 30-51*

Infiltrate an Enemy Position

According to *TM 30-340,* the concepts of infiltration and encirclement had started to crystallize in Soviet infantry tactics by 1941.[38] As both are central to Chinese military philosophy, one might suspect that the change came out of Soviet Asia. Or it may have been something Zhukov had learned in his hard-fought victory over the Japanese in Manchuria in 1939.

After penetrating enemy lines from the front, the WWII Russians liked to exploit adjacent sectors from the rear. Their "buttonhook" maneuver was quite effective.

> In a breakthrough through the (enemy) front, the unit breaks into the enemy defense position and afterwards assaults the newly created flanks or the rear of the enemy.[39]
>
> — "Fire and Maneuver of Infantry"
> *Soviet Combat Regulations of November 1942*

Confront Superior Firepower

Even the tactically deprived Russians will withdraw to slow a more powerful opponent or set a trap.[40] To attack a stronger force at its weakest point, an infantry unit has only to let it pass overhead.

WWII Russian soldiers would automatically spread out, entrench, and camouflage every time they stopped.[41] So doing gave them an automatic ambush opportunity. Just by advancing into their midst, their opponents would find themselves surrounded. Once, on the grassy steppes of Russia, Soviet foot infantry destroyed a whole regiment of German tanks protected by an armored infantry brigade. They did so by establishing a myriad of camouflaged dugouts, letting the Germans overrun them, and then firing handheld antitank weapons from the back. Each time a tank would turn to fire at one position, another position would open up, causing the quarry to lose track of its initial target.[42]

Succeed at Close Combat

Even the WWII Soviets preferred close combat to the standoff variety. That their manuals directed uninterrupted observation of the enemy and assaults by bayonet should be adequate evidence.[43]

> The infantry carries out the most arduous of tasks: close combat man-to-man with the enemy.[44]
>
> — "Principles of Infantry Battle"
> *Soviet Combat Regulations of November 1942*

Unfortunately in Afghanistan, the more technologically advanced Russian army lost much of its taste for close combat and fared accordingly.

The East Has Learned How to Harness Collective Wisdom

The German tactical success late in WWI was the product, not of a dominant personality, but of a corporate effort.[45] To have any chance of winning, the Germans had to respond to changing circumstances and think collectively. Ludendorff directed the corporate effort toward the problem of tactical change.[46] Winning armies experience less urgency in this regard. Change doesn't come easily to a "top-down" bureaucracy.

> The process of developing principles to obtain . . . [defeat the enemy] was a collective or corporate effort. . . . Individual talents and personalities were essential, but the doctrine emerged in an atmosphere where ideas were discovered and shared, not invented and arbitrarily imposed.[47]
> — U.S. Army, *Leavenworth Papers No. 4*

The German Army's real secret, of course, was that it didn't publish doctrine at all, but rather guidelines.[48]

While "truth is the first casualty of war," "pride is the original sin" any time. Eastern infantrymen may have been required to keep their own capabilities in better perspective. Every WWII Russian soldier was required to study the tactical techniques of his enemy counterpart.[49] He was also expected to be able to kill tanks with his bare hands and to operate as a scout/sniper.[50] More importantly, every Russian commander was required to keep track of the relative training status of his soldiers,[51] and to submit "truthful" after-action reports to higher headquarters.[52] In combat as in sports, the edge will go to the team that studies its opposition's capabilities and admits to its own shortcomings.

Even today, Russian commanders are expected to consider collective wisdom in reaching their decisions.[53] Soviet military writings described the official position in these words: "The collective always possesses more experience, wisdom, and insight than any commander. It can do what one man is powerless to do."[54] Unfortunately, much of this emphasis on "the collective opinion" is used as an excuse to enforce doctrine.[55] As with most American units, Russians units are heavily officer controlled and their NCOs—those with the most close-combat experience—relegated to subservient status.[56]

Interviews with captured North Vietnamese troops make evident the same joint nature of mission planning.

> Question: Were the fighters given a chance to discuss and criticize a plan of operation before the operation?
>
> Answer: Yes, they were given the chance to discuss and criticize. The idea was to get unity of command and action during the operation. Before any operation, a few among us would be sent out to make a study and survey of the

battlefield, and then a plan of operation would be drawn up and presented to all the men in the unit. Each would then be given a chance to contribute ideas and suggestions. Each squad, each man, would be told what action to take if the enemy was to take such-and-such a [defensive] position . . . but it was also the fighters' duty to contribute to the plan by advancing suggestions or criticizing what had been put forward. Thus, the final decision concerning an operation or attack was very often the result of a collective discussion in which each member had contributed his opinion or suggestion.[57]

— U.S. Defense Agency study

2 Orphaned Squads Are at Greater Risk

- *Are Eastern small-unit tactics different from Western?*
- *What can a soldier who studies ninjutsu do?*

GERMAN PRIVATE ON RUSSIAN FRONT, 1942

(Source: Courtesy of Cassell PLC, from *World Army Uniforms since 1939*, © 1975, 1980, 1981, 1983 by Blandford Press Ltd., Part I, Plate 103; *Soldateni falt 2* (1972), p. 9)

The Threat Is Still One of Being Overrun

As enemy weapons systems become more destructive, U.S. ground units must spread out to avoid being severely hurt. However, with this dispersal comes additional risk. The lone infantry squad cannot pack enough firepower to keep from being encircled/annihilated by a larger unit. Unless that squad comes to depend on surprise, escape, and evasion, it will not survive.

For the orphaned U.S. squad, even a highly responsive supporting-arms community can offer only so much protection. If too many other squads get into trouble at the same time, it can offer none.

Supporting Arms Have Their Limitations

Opposition strongpoints can escape harm from standoff weaponry for a number of reasons—foolproof construction, dummy positions, faulty intelligence, or lack of precision. There are ways—e.g., solid-rock excavations, blast doors, pressure-release shafts, and rear entrances—to build below-ground fortifications that are still virtually impregnable. Target accuracy is no less a problem in the 21st Century. On 25 March 2003, the Associated Press (AP) reported that the Iraqis were receiving Global Positioning System (GPS) jamming equipment from the Russians.[1] Unfortunately, almost all U.S. targeting is done by GPS coordinates. According to *Newsweek* on 7 April 2003, all it may take to foil a GPS guidance system is a "$50 radio jammer kit, easily purchased."[2]

Just to counter a temporary deficit in firepower on offense, an infantry unit has only to subdivide into more parts than its opponent has guns, planes, and expensive ordnance to bombard simultaneously. To disperse successfully, that unit will need advanced small-unit skills. Where might those skills be found? One must study the armies that have routinely had to operate at a firepower disadvantage: (1) the Germans of WWI; (2) the Japanese, Germans, and Russians of WWII; (3) the North Koreans and Chinese of the Korean War; and (4) the Viet Cong (VC) and North Vietnamese Army (NVA) of later conflicts.

Eastern Armies Have More Experience with Surprise

In the mid-1980's, the United States Marine Corps named its new doctrine "maneuver warfare." That its tenets are ancient may have been intentionally downplayed to give them a better chance of survival. The new doctrine differs from its predecessor in many ways. That its central precepts are largely opposite to those of attrition or "positional" warfare is no coincidence. As maneuver warfare is based almost entirely on surprise, it will often mimic its more famous "firepower-dependent" cousin as a type of deception. It will then do the opposite of what an attrition-oriented opponent might expect. Accordingly, it may spare the lives of victor, vanquished, and bystander alike. As important to a lasting peace is its capacity to leave intact the socio-economic infrastructure of the country being fought over. (See Table 2.1.)

Maneuver Warrior	Attrition Warrior
Tries to Bypass or Demoralize Foe	Attempts to Kill Opponent
Depends on Surprise	Relies on Firepower
Focuses Outward on Enemy	Concentrates on Self
Harnesses Reconnaissance Pull	Relies on Command Push
Moves through Undefended Gaps	Hits Enemy Concentrations
Exploits Breakthroughs	Stays on Line with Others
Captures Centers of Gravity	Seizes Hilltops
Employs Combined Arms	Uses Biggest Weapon
Prefers Nonilluminated Offense	Assaults under Illumination
Seeks High Tempo	Is Slow and Methodical
Defends Only As a Trap	Holds His Position Often
Sometimes Pulls Back	Never Retreats
Enjoys Delegated Authority	Is Centrally Controlled

Table 2.1: The "Maneuver Warfare" Difference

What was not heavily advertised during the mid-1980's is that Eastern armies have been practicing maneuver warfare for over 85 years. That there is evidence to the contrary is easy to explain. Those armies are proficient at both styles of warfare and will use the style that best matches the situation. They are also devotees of deception and will routinely use an attrition tactic as a preliminary feint. That Eastern small units fight somewhat differently in no way contradicts this assessment. Many learn and operate "from the bottom up." Under decentralized control, platoons from the same company are free to develop slightly different techniques.

By following the same line of reasoning, one can see that Eastern riflemen may be trained to support both styles of warfare. According to Mao Tse-tung, Chinese units can alternate between positional, mobile, and guerrilla warfare.[3] He goes on to say that their mobile techniques have grown out of guerrilla techniques.[4] That would indicate that the average rifleman within a Chinese line company may have the initiative and "woods-smarts" of a seasoned guerrilla.

A Different Focus

That U.S. opponents have preferred surprise to firepower needs no proof. That's why they have invariably attacked at night. That's why they have sent short-range infiltrators through the back door while their main forces demonstrated at the front. Their aim was not to kill U.S. soldiers *per se,* but rather to destroy strategic U.S. assets. All they needed to keep the Western-style defender honest was an occasional human-wave assault. In Afghanistan, the Russians used their special forces more for combat than intelligence gathering (Figure 2.1). As a result, they seldom surprised anyone. They did however acknowledge that "the *mujahideen* always strove to achieve surprise."[5]

Asians routinely practice the "false face" and art of delay. They will let an opponent see only what they want him to see. After Afghanistan, the Soviets admitted to often failing "to find their opposition unless the *mujahideen* wanted them to."[6] The reason is simple. The Asian bases his sucker move on Western tactics and decisive jab on *yin/yang* antithesis. In other words, his secret maneuver is routinely the opposite of what Western troops expect and would do under similar circumstances. "Asymmetric warfare" is the product of ingenious design, not organizational laxity. It may do nothing more than opt for frontline common sense over headquarters' philosophy.

Soft-Spot Tactics on Offense

Orientals have been using soft-spot tactics since Sun Tzu promoted taking the path of least resistance in his 5th Century B.C. classic—*The Art of War.*

> The ground, the state of the enemy, and the weak points of his distributions must be known.[7]
>
> — Japanese night-fighting manual, 1913

In fact, every American opponent since WWI has attacked through gaps in U.S. lines. Even the Soviets—with their love of brute force—try not to attack at an enemy's strongest point.

The Soviets practice soft-spot tactics mostly at the operational level. By U.S. estimate, most of their WWII attacks involved pen-

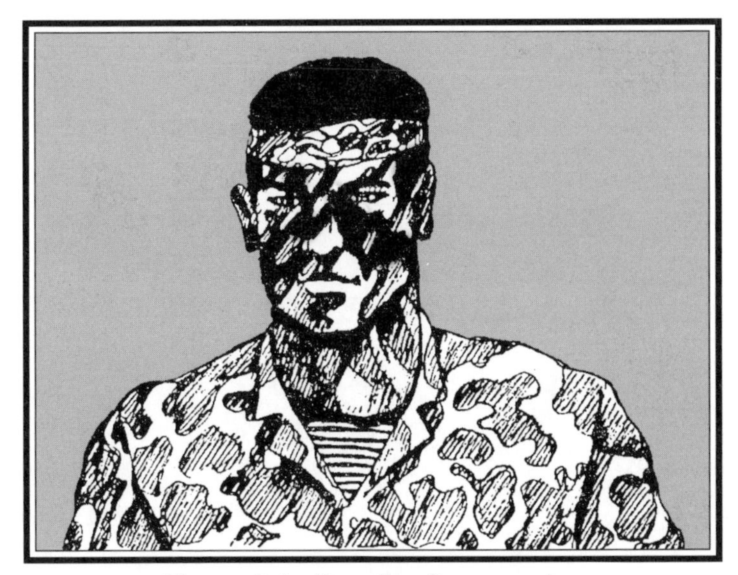

Figure 2.1: Russian Commando

(Source: *Podgotovka Razvedcheka: Systema Speznaza Gru,* © 1998 by A.E.Taras and F.D. Zaruz, p. 330)

etration and a subsequent mission.[8] The same source claims the typical Soviet assault group consisted of a reinforced platoon.[9] In other words, they were using platoons at the point of their spears. That spearpoint point had to be sharp. By U.S. admission, WWII Soviet infantry could follow as closely as 100 yards behind the barrage.[10]

Decentralized Control on Offense

To handle unforeseen circumstances, assault troops require leeway and (Figure 2.2). Contrary to popular opinion, Asian commanders routinely decentralize control over their maneuver elements:

> Each detachment must be given an independent objective, and absolute uniformity will not be blindly adhered to. Each detachment must act firmly and independently.[11]
>
> — Japanese night-fighting manual, 1913

Figure 2.2: Key Engagements Won through Troop Initiative
(Source: *FM 100-5* (1994), p. 52)

The classic example of decentralized control on offense belongs to the German stormtroopers of WWI. Each squad conducted its own "reconnaissance pull" and independent assault. Every line infantry squad that participated in the Spring Offensives of 1918 knew the stormtrooper technique.[12] Presumably many spearheaded their parent unit's attack. American forces have encountered stormtrooper-style assaults in every war since then.

Soviet commanders may have only decentralized control over their assault troops during the actual penetration of enemy lines.[13] Still, some flexibility is better than none.

Strongpoint Matrix on Defense

One of the major tactical innovations to come out of WWI was the German strongpoint defense. Since then, many Eastern opponents have pretended to use linear defenses, but none actually has. That is a distinction that only Western armies can claim.

Clearly by WWII, even the Russians had replaced the linear defense with the strongpoint variety:

> The position of a squad set up in the depth of the main battle position must have a good field of fire over the entire front, into the intervals between the squads set in front and on their flanks, and the alternate position must have a field of fire to the rear.[14]
>
> —*Soviet Combat Regulations of November 1942*

Decentralized Control on Defense

By the end of WWI, the Germans had put NCOs in charge of the tiny forts that constituted their strongpoint defense.[15] According to the U.S. War Department, the Soviets could mount either a "centralized" or a "decentralized" defense by 1942.[16] *TM 30-430* goes on to say that the Russian centralized defense used strongpoints, dummy positions, and false fortifications to canalize an enemy attack into a preplanned firesack into which the overall commander would direct the counterattack.[17] The decentralized defense allowed each unit down to the smallest (presumably the squad) to create its own semi-independent strongpoint. Like the Germans before them, the Soviet squads still needed permission to pull back very far to anything but a preplanned fallback position. According to *TM 30-340,* the Russian squads did however have the authority to counterattack on their own initiative.[18]

Asian Armies Need Only One Man to Practice Maneuver War

The ancient Japanese arts of *ninjutsu* (stealth) and *taijutsu* (unarmed fighting) do little more than apply the core precepts of maneuver warfare to close combat.[19] Wars are won by destroying a foe's strategic assets. To do so, the Asian private needs only to accomplish one of the three *ninja* missions—sabotage, espionage, and assassination.[20] The word *"ninja"* means "one who sneaks in."[21]

> Suggestion took the place of force, deception replaced confrontation, and the adversary was guided into unknowingly doing the *ninja's* bidding instead of being crushed in a humiliating [and politically embarrassing] defeat.[22]

It is generally believed that *ninjutsu* was carried to Japan and Southeast Asia around 900 A.D. by soldiers of the collapsing Chinese T'ang Dynasty.[23] Not only were its first practitioners totally familiar with Sun Tzu's *The Art of War,* but they also took much of their inspiration from Chapter 13—"On the Use of Spies."[24] Maneuver warfare depends upon gaps. Another translation of Sun Tzu's "spies" is "gap men." One author goes so far as to say that *ninjutsu* is based on a single passage from *The Art of War*.[25]

> Centuries before the advent of the *shinobi ninja* in Japan, China had its own *ninja*-like spies, the *moshuh nanren*. The *moshuh nanren* plied their trade of spying and assassination for centuries, beginning in China's Warring States Period (453-221 B.C.E.) . . . It was during this tumultuous period . . . that Sun Tzu wrote *Ping Fa* ("Art of War"), which outlined . . . the use of *k'ai ho* (literally "gap men,") . . .
>
> Over the years, techniques of [the] displaced *moshuh nanren* . . . found their way to Japan, Korea, Indonesia, and Malaysia [along the southern trade route]. There they influenced the development of such night fighters as the *ninja* of Japan, the *hwarang* ("flower nights") of Korea, and the "nightsider" criminal gangs of Malaysia.[26]

Contemporary *"ninjutsu"* is a synthesis of many Japanese traditions and separate schools *(ryus).* During the reign of the fourth Tokugowa Shogun, Yasuyoshi Fujibayashi compiled a ten-volume set of encyclopedias on *ninja* lore entitled the *Bansenshukai.* The third *(Yo-nin)* showed how to apply the more creative dimensions of intellect to a search for information. The fifth, sixth, and seventh volumes (collectively referred to as the *In-nin)* contained the methods of stealth, disguise, night fighting, infiltration, and deception. The eighth volume *(Tenji)* covered the workings of nature.[27] The

Koyo, Ninko, and *Takeda ryus* dealt with intelligence gathering through wandering monks and merchants.[28] The *Fuma ryu ninpo* specialized in guerrilla warfare.[29] The *Yoshitsune ryu* taught a blend of espionage methods.[30] Unfortunately, all died out long ago. The only authentic schools to survive now fall under grandmaster Masaaki Hatsumi's *Bushinkan* organization.[31] Most interesting is the 900-year-old *Togakure ryu,* the school of the legendary shadow warriors.[32] It specializes in climbing, stalking, evasion, and other infiltration skills.[33] While its techniques are seldom still taught to

(Source: *Corel Gallery Clipart,* Totem Graphics, Man Historical, 28T009)

the civilian community, they almost certainly survive (along with those from discontinued schools) in the organizational memories of Asian armies. As they enable those armies to defeat an opponent with "extraordinary forces," they are carefully protected. Only after Dr. Masaaki Hatsumi had the foresight to make some public, have others surfaced. His instructor (and former grandmaster of the *Togakure ryu)* was Toshitsugu Takamatsu, a Japanese special operator (in the Chinese theater) during WWII.[34] While not all Oriental night fighters are technically *ninjas,* they have all had the same Chinese background and Japanese exposure. For the sake of simplicity, similar techniques—from whichever Asian army—will henceforth be referred to as *ninja.*

The Essence of *Ninjutsu*

Ninjutsu is the art of stealth.[35] The term can best be understood by looking at its derivatives. *Nin* means both concealment or sneaking in, and endurance.[36] *Jutsu* means technique.[37] Whereas most of the martial arts are about confronting an enemy, *ninjutsu* is about avoiding him.

> The *ninja's* guiding philosophy was to choose the dark, quiet, and subtle method over the bold, active, and forceful. In this way, the natural order of events was disturbed as little as possible.[38]

The *ninja* works hard to comply with the laws of nature.[39] He escapes detection by merging into the background and by being acutely responsive to changing circumstances.[40] In many ways, he emulates a tiger—moving silently and trying to see in the dark.

Only by concentrating on the actual, can one hope to use deception in war. The *ninja* learns and defends his weak point first. For, he can never be sure that his opponent's weak point is not a decoy.[41] Through studying his own mind, he acquires a better understanding of others and a sharper awareness of the environment. Contained in Table 2.2 is only a sampling of his many techniques. As the *ninja* will normally only fight when cornered, his close-combat methods have been intentionally omitted. Suffice it to say that his sentry removal techniques are distressingly lethal.

Techniques for sensory enhancement

- Seeing at night *(ankokutoshijutsu)* [42]
  - Fully adapting one's eyes to the darkness [43]
  - Keeping one eye closed near (and not looking at) any light [44]
  - Scanning an object without looking directly at it [45]
  - Watching for unexplained shadows *(ankokutoshi no jutsu)* [46]
  - Enhanced peripheral vision through defocusing exercises [47]
- Hearing
  - Listening for stealthy footsteps at 30 yards [48]
  - Cupping hands behind ears while tilting head with open mouth [49]
- Smelling
  - Sniffing while closing mouth and concentrating on upper nose [50]
- Tasting [51]
  - Sampling the air with one's tongue [52]
- Touching
  - Sensing a person nearby without actually touching him [53]

Techniques for moving through enemy country

- Predicting the weather *(ten-mon)* [54]
- Knowing and using terrain features *(chi-mon)* [55]
  - Staying off the horizon [56]
  - Blending with one's background [57]
  - Fully utilizing all available cover
    - Lost-track pivot for darting behind cover [58]
    - Entering pivot for turning corners [59]
- Land navigation [60]
  - Celestial dead reckoning (determining direction from the stars)
  - Terrain association (following linear terrain features like streets)
- Crossing open areas
  - In full moonlight
    - Predicting the shadows of natural and man-made objects [61]
    - Moving quickly from shadow to shadow [62]
      - Semiprone rushing stance [63]
  - In subdued light
    - Moving slowly and steadily from spot to spot [64]
    - Motion so slow as to be indistinguishable to the naked eye [65]
    - Rolling-travel method *(taihenjutsu)* [66]
- Turning step for continuously scanning for enemy [67]
- Walking like the wind over long distances *(taijutsusosoku shugyo)* [68]
  - Scouting in actual battle *(jissensekko gijutsu)* [69]
- Crossing a road or trail at low spot, curve, or otherwise covered place [70]
- Changing direction in tall grass to avoid its unnatural motion [71]
- Moving about in the mist *(muton yuho)* [72]
- Traversing an area in pitch darkness [73]
  - Stealthy step or running on tiptoe [74]
- Remaining perfectly still under sudden, artificial illumination [75]
- Avoiding one's reflection in standing water while near the enemy [76]

Table 2.2: Partial List of *Ninjutsu* Techniques

Techniques for waylaying an enemy scout

- Moving around the opposite side of a tree as the scout passes it [77]
- Hiding behind a tree at the bend in a trail [78]
- Pretending to be a bush near the trail [79]
- Hiding within plain view (see techniques for hiding) [80]
  - Remaining perfectly still [81]
  - Not looking directly at the victim [82]
- Dropping on the enemy from above [83]
- Assaulting the enemy from below [84]

Techniques for stealthily crossing different surfaces *(ninja-aruki)* [85]

- Walking across wooden floor on unrolled sash *(obi)* [86]
- Sweeping step for planks and straw mat *(nuki ashi)* [87]
- Sweeping step for clearing away obstacles in grass [88]
- Small step for shallow water and dry leaves *(ko ashi)* [89]
- Rolling step for rocks and loose dirt [90]
- Side step for moving through narrow spaces or a threshold [91]
- Cross step for a building's shadow or tight passageway *(yoko aruki)* [92]
- Using cat's claws *(neko te)* to crawl along rafters [93]
- Hands-and-knees crawl [94]
- Forearms-and-toes crawl [95]
- Serpent crawl [96]
- Dragon crawl [97]

Techniques for bypassing an enemy sentry

- Discovering the sentry's weaknesses *(nyudaki no jitsu)* [98]
- Deceiving the sentry's senses [99]
  - Avoiding eye contact with the sentry [100]
  - Knowing the limits of the sentry's peripheral vision [101]
  - Directing sentry's attention elsewhere *(yojigakure no jitsu)* [102]
  - Preconditioning sentry to the sounds and sights of infiltration [103]
  - Covering one's noise with wind's rustling of leaves [104]
  - Masking footfalls with natural sounds or enemy conversations [105]
  - Staying downwind of sentry so as not to project any odor [106]
  - Blinding sentry with a flash or keeping bright light to one's rear [107]
- Manipulating the sentry's thought processes [108]
  - Interchanging truth with falsehood *(hojutsu)* [109]
  - Altering the perception of truth and falsehood *(kyojitsu tenkan ho)* [110]
  - Playing on the sentry's needs and fears [111]
- Predicting sentry's next move by fixing one's gaze at base of his neck [112]
- Actual passing strategies
  - Not casting a shadow into the sentry's field of vision [113]
  - Cross-stepping behind sentry so as to watch him [114]
  - Staying below sentry's shoulders to escape his peripheral vision [115]
  - Melting around a corner [116]
  - Tapping sentry on shoulder and doing "lost-track" pivot [117]
  - Pinching sentry on ankle and rolling behind his resting place [118]

Table 2.2: Partial List of *Ninjutsu* Techniques (Continued)

Techniques for hiding ***(inton-jutsu)*** [119]
- Camouflage skills
  - Donning white clothing for snow or smoke [120]
  - Wearing black clothing at night [121]
  - Using various disguises *(hensojutsu)* [122]
    - Impersonation *(gisojutsu)* [123]
  - Covering reflective skin with cloth or ashes [124]
- Other presence-obscuring skills *(tonkei no jutsu)* [125]
  - Continually monitoring one's own shadow [126]
  - Constantly alert to the possibility of reflection (as in a mirror) [127]
  - Hiding in the shadows *(jinton no jutsu)* [128]
  - Trying not to reflect any rays of light [129]
  - Shallow breathing [130]
  - Hiding in a place too small to normally hold a human being [131]
  - Hiding within plain sight *(joei-on no jitsu)* [132]
    - Distorting one's silhouette (as with a cape) [133]
    - Balling up like a stone [134]
    - Posturing motionless in front of obvious cover [135]

Techniques for entering an enemy fortress ***(chiku jo gunryaku heiho)*** [136]
- Using rain, snow, or fog to restrict the defender's vision and hearing [137]
- Choosing an overcast night [138]
- Leaping over low bushes (or obstacles) [139]
- Crawling over a low wall or fence [140]
- Ramparts crossing
  - Through sewers, air shafts, other openings deemed inaccessible [141]
  - Pole or limbless-tree climbing without any equipment [142]
  - Climbing tree with spiked hand and foot bands *(shuka & ashika)* [143]
  - Leaping from tree to tree [144]
  - Rope throwing *(nawanage)* [145]
  - Pulling occupied treetop over wall with grappling hook *(kaginawa)* [146]
  - Ascending a building's trellis or drain pipe [147]
  - Jumping from rooftop to rooftop [148]
  - Moving along a horizontal pole or "leaping ladder" *(tobi bashigo)* [149]
  - Crawling or using a pulley *(kasha)* to cross an outstretched rope [150]
- Ramparts scaling
  - Vertical-surface running *(shoten no jutsu)* [151]
  - Running up an inclined plank [152]
  - Rock climbing [153]
  - Stone or brick wall climbing [154]
    - While facing either towards or away from the wall [155]
  - Rope climbing [156]
  - "Loop," "spider," and "hanging-rope" ladders *(bashigo)* [157]
- Entering an inaccessible structure *(shinobi-iri)* [158]
  - Effortlessly moving through a window opening [159]
  - Climbing stairs along the wall [160]
  - Opening a door so as not to let in any light [161]

Table 2.2: Partial List of *Ninjutsu* Techniques (Continued)

Techniques for collecting intelligence
- Observing the foe from his perimeter *(monomi-no-jitsu)* [162]
- Memorizing defensive details [163]
- Scanning a room through the crack of a gradually opening door [164]

Techniques for leaving without a trace [165]
- Natural-element escape strategies *(gotonpo)* [166]
  - "Fire escape arts" *(katonjutsu)* [167]
    - Diversionary fire and other fire/explosive methods *(kajutsu)* [168]
    - Firecrackers to simulate rifle fire [169]
    - Masking one's movement with smoke [170]
  - "Metal escape arts" *(kintonjutsu)* [171]
    - Climbing the underside of a metal stair [172]
    - Disorienting pursuers with a metal blinding mirror [173]
  - "Wood escape arts" *(mokutonjutsu)* [174]
    - Conforming to the crook of a tree [175]
    - Using low vegetation for concealment *(moku ton jutsu)* [176]
    - Climbing to a high place *(tanuki gakure no-jitsu)* [177]
  - "Earth escape arts" *(dotonjutsu)* [178]
    - Hiding under soft soil or the uprooted roots of a tree [179]
    - Stone concealment *(sekiton jutsu)* [180]
    - Hiding in gap between two objects *(uzura gakure no-jitsu)* [181]
  - "Water escape arts" *(suitonjutsu)* [182]
    - Breathing underwater through a hollow reed [183]
    - Stealth swimming and flotation pots *(suijutsu)* [184]
- Other escape strategies
  - Diversions
    - Blinding powders *(metsubushi)* [185]
    - Releasing a drugged or pet animal when first noticed [186]
    - Leaving behind rice or booby traps to warn of pursuit [187]
    - Imitating the sound of a cat or other small mammal [188]
    - Ventriloquism [189]
    - Scurrying or bounding away like a wild animal [190]
  - Movements
    - Ducking through or hiding behind a door [191]
    - Ascending the overhang of a roof [192]
    - Rolling, leaping, or tumbling away *(taihenjutsu)* [193]
      - Handspring methods [194]
      - Rolling methods [195]
        - Reverse-shoulder roll *(chigari)* [196]
      - Leaping methods *(hichojutsu)* [197]
        - Vaulting the foe [198]
        - Dodging to one side of and pivoting behind foe [199]
      - Body drop methods [200]
        - Forward breakfall *(zempo ukemi)* [201]
        - Sideways-flowing rear-body drop *(yokonagare)* [202]
        - Upright-flowing rear-body drop *(tachinagare)* [203]

Table 2.2: Partial List of *Ninjutsu* Techniques (Continued)

How the U.S. Soldier Is Affected by All of This

Ninjas are not the remnant of a shadowy counterculture, but rather the product of a highly systematic and scientific method of secret espionage, danger prevention, and advanced combat.[204] Historically, the U.S. military has generally ignored what a single *ninja* can accomplish. After all, what little he destroys can easily be replaced. However, the cumulative damage of tens of thousands of *ninjas* could be enough to discourage a powerful opponent from pursuing a protracted war.

> Often a few black-garbed *ninja* could do by subtle means what it would have taken hundreds of armored soldiers to accomplish.[205]

For this reason, most of the "low-tech" Asian armies still give their commandos *ninja*-like training. Unfortunately, many also use their commandos as training cadre for line infantry outfits. That means all Asian privates may be regularly trained in *ninja* techniques.

Which Eastern Infantrymen to Avoid

Japan has been instructing its riflemen on sensory-enhancement, shadow-walking, and other decidedly "*ninja*" techniques since 1913.[206] (Notice the footwear in Figure 2.3.) The "split-toe" is also worn by modern-day *ninjas*.[207]

The Korean *hwarang* were used as scouts, guerrillas, shock troops, and night infiltrators.[208] Their legacy is the current North Korean "Light Infantry Training Guidance Bureau." As suggested by its name, this bureau undoubtedly detaches—as training cadre to each line infantry company—several of its 100,000 commandos.[209]

By 1970, every local-force Viet Cong may have been practicing monthly how to cross dry leaves and swampy ground.[210] (See Table 2.3.) The similarities between the ground-crossing techniques in Tables 2.2 and 2.3 are far from coincidental.

In 2001, a People's Republic of China (PRC) magazine showed an ordinary conscript sailing five feet above and parallel to the ground.[211] One contemporary author actually identifies the Chi-

Figure 2.3: Japanese Infantry Private, 1941

(Source: Courtesy of Cassell PLC, from *World Army Uniforms since 1939*, © 1975, 1980, 1981, 1983 by Blandford Press Ltd., Part I, Plate 155)

nese names of many of the *ninjutsu* techniques.[212] It would appear that the methods of the *moshuh nanren* have survived in more that just the criminal *t'ongs.*

What does all of this mean to the average American private? It means that—while walking through the woods some evening or just back to his rear-area bunk—he may find himself on the short end of a lethal one-on-one encounter. Fighting a *ninja* is not like fighting a martial artist. That *ninja* can suddenly appear out of nowhere and then, for all practical purposes, read his opponent's mind.[213]

Through mental discipline, the *ninja* has rid himself of selfish intentions, preconceived notions, and bad habits.[214] He can there-

Table 11.2 ***C2 D445*** **training program, 13–27 August 1970**

DATE IN AUG	*MORNING*	*NO. HOURS*	*AFTERNOON*	*NO. HOURS*	*EVENING*	*NO. HOURS*
13	Crossing dry leaves	4	Individual mine training	3	Flare training	3
14	Crossing trenches	3	Crossing enemy wire	3	Crossing dry leaves, mine warfare and flares	3
15	Crossing swamps	3	Obstacle crossing	3	Obstacle crossing	4
16	Range practice—infantry weapons	2	Obstacle crossing	3	Swamp crossing under flares	3
17	Trench crossing	3	Revision of weaker subjects	2	Enemy obstacle crossing	4
18	Crossing swamps	2	Range practice	3	Revision of weaker subjects	2
19	Section attacks on bunkers and bridges	4	Crossing dry leaves and wire	3	Night firing	3
20	Attacking points from tunnels	3	Shooting infantry weapons at aircraft	3	Attack on bunkers and bridges	4
21	Revision of weaker subjects	2	Recce and guerilla warfare	3	Attack from tunnels	3
22	Section attacks on bunkers	3	Range practice	3	Crossing enemy obstacles	4
23	Revision of weaker subjects	2	Section tactics	3	Attack on bunkers and bridges	3
24	Recce and guerilla warfare training	3	Weapon handling	3	Revision of weaker subjects	2
25	Practice in company raids	4	Recce and guerilla warfare	3	Attack on bunkers and bridges	3
26	Practice in company raids	4	Sapper recce	3	Revision of weaker subjects	2
27	Practice in company raids	4	Sapper and recce methods lecture	3	Course critique	2

Table 2.3: VC "*Ninja*-Like" Training Schedule

(Source: Courtesy of Allen & Unwin Publishing, from *Conscripts and Regulars*, by Michael O'Brien, © 1995 7th Battalion, Royal Australian Regiment Assoc. Inc.)

fore easily adapt to changing circumstances. He can blend with his environment, disguise his true intent, and recognize the hidden motives of others.[215] He can capitalize on his foe's emotional weaknesses (laziness, anger, fear, sympathy, vanity) and fundamental needs (security, sex, wealth, pride, and pleasure).[216] Almost unbelievably, the *ninja* can tell what his adversary is about to do from that adversary's body signals, voice intonations, facial features, and personality quirks alone.

To make matters worse, most Eastern soldiers have been fully apprised—from the nonrestricted U.S. manuals—of the standard close combat moves of a U.S. soldier. He knows that Americans like to make the first move. To win, all he has to do is to dodge it.[217] Unfortunately, that is the *ninja's* stock in trade.

> When you are sensitive enough to detect this intention of harmful action, you can fight back by simply not being where the attack will take place.[218]

A *ninja*-trained soldier can easily remove any opposition with a star-shaped throwing blade *(shuriken)*,[219] blinding powder *(metsubushi)*,[220] nervous-system pinch *(koshijutsu)*,[221] or any number of other armed and unarmed techniques. In mortal combat, he will almost always win.

The American special operator who thinks he can sneak up on a *ninja* master should think again. That *ninja* master can hear stealthy footfalls 30 yards behind him.[222] He can smell wind-born human scent at 200.[223]

U.S. Riflemen Will Need More Skill

- *Is the Eastern soldier better prepared to survive?*
- *Can the Eastern private exercise more initiative?*

SOVIET AIRBORNE SNIPER, 1982

(Source: Courtesy of Cassell PLC, from *Uniforms of the Elite Forces*, ©1982 by Blandford Press Ltd., Plate 15, No. 45; *FM 21-76* (1957), p. 100)

How Well Can the U.S. Private Really Take Care of Himself?

As most Eastern armies are relatively "low-tech," they must more carefully train their soldiers and squads. To counter Western planes and tanks, they have to rely on the skill and initiative of their frontline fighters. As was painfully evident in Vietnam, they have correctly deduced two things: (1) that a lone human being can outsmart a machine, and (2) that tens of thousands of lone human beings can collectively make a difference. As a result, the Eastern infantryman has gotten more field skills and decision-making practice. One on one, he has a decided edge over his Western counterpart. As battles are comprised of engagements, and engagements of encounters, there is nothing to be gained from individual deficiencies.

The Eastern Edge in Tactical Decision Making

Today's U.S. private comes from the city. He won't fare as well as his predecessor in a rural conflict against a woods-wise opponent. Those deficiencies noted at the outset of WWII have only grown more pronounced.

> My message . . . to the troops training for this type of warfare is to go back to the tactics of the French and Indian days.[1]
>
> — Maj.Gen. Vandegrift after Guadalcanal
> *FMFRP 12-110*

U.S. privates have been traditionally encouraged to let their leaders do much of the thinking. In boot camp, their societal individuality has been largely replaced with a blind obedience to orders. Later, to keep from being punished in an error-free environment, most will do only what is minimally expected of them. They wouldn't dream of telling a superior something he didn't already know. If they are too closely supervised, many will lose any semblance of initiative. This problem may have grown worse since WWII. For the American private between the World Wars, a corporal was God and an officer even higher. Now, majors are doing things that any good corporal could do.

> If I had to train my regiment over again, I would stress small group training and the training of the individual. . . . Our basic training is all right. . . . In your training put your time and emphasis on the squad and platoon rather than on the company, battalion and regiment.
>
> In your scouting and patrolling, . . . have the men work against each other. Same thing [goes] for squads and platoons in their problems. . . .
>
> . . . With proper training, our Americans are better [than the Japanese], as our people can think better as individu-

> als. Encourage your individuals and bring them out.[2]
> — Col. Merritt A. Edson after Guadalcanal
> *FMFRP 12-110*

Recently, a Dutch Naval Infantry officer witnessed a German private being punished for not showing enough initiative.[3] Almost all of the Eastern armies—with the possible exception of the Russians—rely heavily on squad combat. That means their NCOs are getting plenty of tactical-decision-making experience.

The Eastern Edge in Perceptivity

As U.S. ground forces attack predominantly during the day and other armies do not, past adversaries have honestly thought Americans to be night blind. Routinely operating at night sharpens one's senses.

The "undersupported" soldier must depend more on his God-given abilities. He soon learns that having too few bullets helps him to generate surprise. He discovers that having only rice to eat causes him to give off less of a telltale odor. He knows that—from upwind—no affluent Westerner can sneak up on him.

The Eastern Edge in Movement

While the Western sapper's role is to breach enemy obstacles, the Eastern sapper's mission is to sneak through those obstacles and then blow up some target of strategic significance. In the East, a sapper is defined as an "infiltration and demolition commando."[4] Misunderstood within America to this day is the extent to which VC and NVA sappers influenced the Vietnam War. Their numbers alone should tell the tale.

> The special training requirements and role that sappers were to play in the Tet Offensive were recognized with the establishment of the Sapper Branch [of the People's Army of Vietnam (PAVN)] . . . on 19 March 1967, with direct control over *Brigade 305, Regiment 426,* (naval) *Group 126,* and nine battalions. The Branch primarily trained sappers for

> attachment to other PAVN commands and units. In its first year it trained and dispatched over 2,500 sapper officers and troops to the South, and it doubled this output within two years.
>
> Southern Regional Headquarters maintained its own sapper department, known as J16, and for a time it trained as many sappers as the [northern] branch. In 1973, B2 controlled eight sapper "groups" equivalent to regiments, including one in Cambodia. On the eve of the 1975 Spring Offensive, 12 sapper regiments or their equivalents, 36 battalions, 121 companies, 15 groups of *biet dong,* and hundreds of squads, teams, and cells were scattered across the South. The entire *dac cong-biet dong* "brigade" 316 was operating inside Saigon or its outskirts by war's end [italics added].[5]
>
> — *Encyclopedia of the Vietnam War*

Like the North Korean Light Infantry Training Guidance Bureau, the North Vietnamese Sapper Branch may have been sending a few of its 40,000 members to each NVA and VC rifle company to act as scouts, short-range infiltrators, and training cadre. That would mean the average Communist conscript was receiving advanced instruction in "unconventional warfare." Table 2.3 reflects the monthly training schedule for a regional-force "infantry company" or local-militia "security detachment." It is not the training schedule for a main-force sapper unit.[6] There is additional evidence that the parent VC battalion had NVA soldiers attached.[7]

The Eastern Rifleman Must Learn to Dodge Bombardment

The members of a "firepower-deficient" military force must become expert at building dummy positions and donning camouflage. For it is only through deception and concealment, that they will be able to withstand the massive preparatory fire that precedes almost every Western assault.

Undergunned forces must also be proficient at spreading out and digging in. To whatever extent Western firepower is hurting them, they will have to disperse further or excavate deeper. That this confounds the well endowed makes it no less true.

The Eastern Rifleman Must Learn to Fight Tanks

The member of an "armor-poor" force must learn how to kill tanks with his bare hands. Only for the overcontrolled enlistee does such a task seem impossible. Lacking sufficient armor at Nomonhan [Manchuria], the Japanese were forced to infiltrate Russian lines and knock out opposition tanks with hand-held explosives.[8] *Soviet Combat Regulations of November 1942* decreed that "the Russian soldier must know how he (personally) can fight tanks."[9]

The other Eastern nations have required tank killing of their individual riflemen as well. Either they all undervalue human life, or Americans have been misled on how easy it is to kill tanks. The "bare-handed" techniques will be discussed in Chapter 12.

Casualty Totals Can Have More Than One Interpretation

Many of the opposition soldiers to have become casualties in WWII, Korea, and Vietnam did so while approaching the front lines or serving as rear-area support personnel. For it is there that they were most vulnerable to disease, rudimentary medicine, and U.S. air power. Their frontline comrades—largely through deception, excavation, and tactical withdrawal—did not suffer nearly as heavily as purported by U.S. battle chronicles.

To make the point, the combat losses for the Korean War and Vietnam Tet Offensive have been closely examined. Appendix A contains each side's highest estimate of casualties suffered and lowest estimate of casualties inflicted. By comparing their most conservative totals, one eliminates exaggeration. Each side's actual casualties would then lie somewhere between its own count and its opponent's estimate.

As documented in Appendix A, the Communist casualty total for Korea doesn't look nearly as lopsided when compared to the United Nations (U.N.) total (that included ROK losses). Two officially endorsed U.S. historians have estimated Communist casualties at 1,500,000+. In 1993, the Chinese as much as admitted to having lost 1,010,700 (from all causes). At the war's end, the Associated Press reported that U.N. (including ROK) forces had suffered 1,474,269 casualties. While the Chinese Army initially estimated U.N. losses at 718,477, the North Koreans counted twice that many U.N. soldiers lost. It would appear that the casualties for the two

Figure 3.1: The Price of Pride without Thinking

(Source: *FM 100-5* (1994), p. 56)

sides may have been roughly equivalent—around 1,500,000 apiece. Yet one side had vastly more supplies, armor, artillery (at least initially), and air cover. At the Chosin Reservoir, the Chinese did not even have sufficient cold-weather clothing. How did the Communist armies manage to compensate for those deficits? The most plausible explanation is that they enjoyed more tactical skill at the front lines. For U.S. traditionalists now to claim that only the ROK units were deficient is to ignore who trained them.

The same conclusion can be more easily drawn from the first 60 days of the 1968 Tet Offensive (Appendix A). This time, the Communists were not only badly outgunned, but also badly outnumbered. An officially endorsed U.S. historian has estimated that the VC/NVA lost a full half of the 80,000 troops committed. Hanoi admits to losing 18,560+ out of the 67,000 soldiers participating. The records from the Army of the Republic of Vietnam (ARVN) show the Allied total for this period to have been 45,000+ (of which 24,013 were American). The North Vietnamese Military History Institute claims the Allies (including the ARVN) lost 147,000 of 1,179,000 troops available. It purports that 43,000 of the Allied injured or killed were Americans. So this time, it would appear that the U.S. forces alone suffered as heavily as their attackers. With anything that moved beyond danger-close range of American positions being plastered by U.S. supporting arms, the Communists just outside the wire must have enjoyed a tactical advantage.

The Eastern Soldier Enjoys an Edge in Field Skill

So, what has been called "suicide tactics by uncaring nations" may be just another political exaggeration. The frontline Asian soldier has been holding his own with much less firepower. Whereas the U.S. private learns how to "shoot," "move," and "communicate" (with the emphasis on orchestrated shooting and one-way communication), the Eastern soldier learns much more. His "basic" training includes all or part of the following: (1) microterrain appreciation, (2) harnessing the senses, (3) night familiarity, (4) nondetectable movement, (5) guarded communication, (6) discreet force at close range, (7) combat deception, and (8) one-on-one tactical decision making. Each will have its own chapter.

"Low-tech" Eastern armies are well aware of their privates' wider capability. They downplay it in the media but readily admit it to their own personnel. The following are excerpts from pamphlets distributed to the Chinese People's Volunteer Army during the Korean War. While those pamphlets grossly underestimated the American fighting man's courage, they accurately assessed his organizational burden (Figure 3.1). In effect, Chinese intelligence had discovered that U.S. soldiers were so good at using such fine equipment that they had become machine dependent.[10] These excerpts could be discounted as pure propaganda, if other foes hadn't noted similar deficiencies.

> The coordinated action of mortars and tanks is an important factor. . . . Their firing instruments are highly powerful. . . . Their artillery is very active. . . . Aircraft strafing and bombing of our transportation have become a great hazard to us. . . . Their transportation system is magnificent. Their rate of infantry fire is greater, and the long range of that fire is even greater.[11]
>
> — Untitled Chinese Communist Forces pamphlet

> Cut off from the rear, they abandon all their heavy weapons. . . . Their infantrymen are weak, afraid to die, and have no courage to attack or defend. They depend always on their planes, tanks, artillery. . . . They specialize in day fighting. They are not familiar with night fighting or hand-to-hand combat. If defeated, they have no orderly formation. Without the use of their mortars, they become completely lost. They become dazed and completely demoralized. They are afraid when the rear is cut off. When transportation comes to a standstill, the infantry loses the will to fight.[12]
>
> — *Primary Conclusions from the Battle of Unsan*
> Chinese Communist Forces pamphlet

In effect, "low-tech" Eastern armies have had to capitalize on the field skills of the individual soldier. Part Two will fully reveal what he can do. (See Appendix B for a detailed analysis of his entry-level training.)

Part Two

The New "Basics"

The best form of "welfare" for the troops is first-class training, for this saves unnecessary casualties. — Rommel

(Source: *The Rommel Papers,* © 1953 by B.H. Liddell Hart, p. 226)

4 Microterrain Appreciation

- *What is microterrain?*
- *How could it expand one's tactical opportunity?*

JAPANESE INFANTRY PRIVATE, 1942

(Source: Courtesy of Cassell PLC, from *World Army Uniforms since 1939*, © 1975, 1980, 1981, 1983 by Blandford Press Ltd., Part I, Plate 156; *FM 21-76* (1957), p. 70)

No Cover Required for a Western-Style "Push"

Throughout WWI, WWII, Korea, Vietnam, and the Gulf Wars, the American method of attack has remained the same: (1) haphazardly hit an objective with artillery and air strikes, and then (2) hastily assault it with troops and tanks. This was thought to be the best way to maximize force while minimizing fratricide. Whether or not the U.S. shells hit anything, they theoretically kept the defenders' heads down. Whether or not the perfectly aligned U.S. troops were fully exposed, they theoretically maintained fire superiority. Sadly, little things like overall surprise, squad techniques, microterrain utilization, and enemy tactics seldom entered into the equation. For the young Americans dodging lead, this approach had its downside.

Being overly simplistic on offense carries with it a price. Every time a U.S. company was given until dark to capture an objective, a few fully exposed privates had to rush an enemy machinegun. The company's lane was often too narrow to provide a covered avenue of approach. There wasn't enough time to crawl the whole way. It was broad daylight. In other words, there was no way the assigned unit could get within assaulting distance without being observed. To make matters worse, the last 75 yards were often virtually level, obstructed by barbed wire and claymores, and swept by grazing machinegun fire.

Figure 4.1: Grasslands and Swamp

(Source: *FM 90-5* (1982), p. 1-5)

Figure 4.2: Gullies and Rocks

(Source: *FM 90-30* (1977), p. 2-7)

With the "big-picture" approach to war comes an overemphasis on pronounced terrain features. Unfortunately, when the bullets and shrapnel start flying, it is not macroterrain, but rather microterrain that matters (Figures 4.1, 4.2, 4.3, and 4.4). Through cover and concealment, the latter offers individuals, fire teams, and squads their best means of survival (Figure 4.5). For the prone combatant, the smallest ground swell can obscure his movement or deflect a bullet. As a cruel irony of war, almost every attack objective has a covered avenue of approach that is only a few feet (or inches) deep. In the country, it is often some combination of road shoulder and watershed ditch. In the city, it is a mix of subsurface drain and sidewalk curb. Unfortunately, to U.S. aerial or binocular reconnaissance, neither avenue exists. On the other side of the rural objective are small bushes and waist-high grass.

Figure 4.3: Bushes and Manmade Passageways

(Source: *FM 90-30* (1977), p. 2-11)

On the other side of the urban objective are piles of litter. By paying so little attention to microterrain, low growth, and refuse around its objectives, U.S. units have unnecessarily limited their tactical options.

"Top-down" management can create just as disastrous a result on defense. Every time a company commander equally distributes his defensive sector to his three platoons (a common practice within the U.S. military), he risks penetration by sappers. An even third of his lines may contain 25 places where a sapper could sneak in, whereas the other two thirds contain 10. Each fighting hole must have two occupants, so that one can sleep. To be assured of securing the porous sector, the company commander would have had to assign more than a platoon to it.

The Eastern Emphasis on Microterrain

Units deficient in firepower must depend more on surprise. Surprise can be most easily generated by tiny elements that approach their objectives through microterrain. Where there are too few folds in the ground, Easterners will create some. In May of 1940, the Germans needed only a few combat engineers to defeat Fort Eban-Emael—the linchpin of the Belgian defense line.

> . . . The German heavy artillery fired, not in a vain attempt to destroy the fort, but to create craters in the flat terrain. . . .
>
> When darkness fell, the German engineers crossed, in rubber boats, an artificial lake that separated them from Eban-Emael. Using the shellholes made by their own guns for cover, they crept forward. At dawn, flamethrowers sent streams of burning oil onto the embrasures from which the machineguns responsible for the close defense of the fort were expected to fire. Reeling from the heat and blinded by the smoke, the machinegunners failed to see the small team that had rushed forward with a huge shaped charge. A few seconds later, the charge went off. . . . Other explosions followed. . . . By the end of the morning, the fort was defenseless and surrendered.[1]

Figure 4.4: Mounds and Trees

(Source: *Soldateni falt 2* (1972), p. 101)

Figure 4.5: There's More Cover and Concealment Down Here
(Source: *FM 21-75* (1967), p. 26)

At fully utilizing microterrain, the Russians were even more adept.

> In World War II, as in preceding wars, the Russian soldier demonstrated that he was closer to nature than his West European counterpart. This was hardly surprising since most of the Russian soldiers were born and raised far from big cities. . . . The Russian was able to move without a sound and orient himself in the darkness. . . . [He] performed particularly well as a night observer. Stern discipline and self-constraint enabled him to lie motionless for hours and observe the German troops at close range without being detected. . . . Infiltration by small detachments, as well as by larger units up to an entire division, was probably the most effective Russian method of night combat. . . . Time and again their troops slipped through a lightly held sector during the night and were securely behind the German front by the next morning.[2]
>
> — U.S. Army, *DA Pamphlet 20-236*

Russian special forces still have a healthy respect for microterrain (Figures 4.6 and 4.7). Well illustrated in their 1997 manual is how one crosses uneven ground under fire—he runs bent over through depressions and crawls (or rolls) over rises.[3]

The masters at uneven ground utilization were, of course, the East Asians. After the Japanese had honeycombed Iwo Jima and parts of Okinawa, the Chinese held their own clinic on the Korean Peninsula.

> Chinese infantrymen [in Korea] also knew how to use every fold in the terrain for cover and concealment. They seldom got disoriented in the darkness and showed an unerring appreciation of ground in the choice of their routes, placement of their machineguns and mortars, and in their selection of fighting positions.[4]
>
> — U.S. Army, *Leavenworth Research Survey No. 6*

How Eastern Attackers Make Use of Ground Detail

During WWI, the German stormtroopers were able to approach Allied lines through the shallow communications trenches that often crisscross a heavily contested battlefield.

It is the Asians who have made an art form out of using the natural terrain. They can blend in so perfectly with their surroundings as to cross, unnoticed, ground that is virtually flat. In full moonlight during Operation Maui Peak in late 1968, NVA soldiers successfully "stalked" the Marine defenders in their lanes.[5] They did so by attaching leaves, staying behind bushes, and moving slower than perceptible to the naked eye. The position that had been placed behind the military crest to protect it from incoming recoilless-rifle fire had no chance against the incoming grenades. Those same North Vietnamese soldiers would have had little difficulty with more mundane routes: (1) marshes, (2) runoff ditches, (3) depressions, (4) rotting tree trunks, or (5) uncleared vegetation. When there was no surface abnormality, they could have used a shadow.

How Eastern Defenders Use the Folds in the Earth

To avoid being silhouetted, WWII German defenders piled dirt behind their entrenchments.[6] At Dien Bien Phu, the Viet Minh transplanted grass to obscure their approach trenches.

Figure 4.6: Rural Russians Are Closer to Nature

(Source: *Podgotovka Razvedcheka: Systema Speznaza Gru,* © 1998 by A.E.Taras and F.D. Zaruz, pp. 144, 145, 152)

Of course, Asians are also more masterful at concealing their bunkers. On Iwo Jima, machinegun apertures would literally disappear when not in use.

> One of the wounded Marines we received aboard when the hospital ships were overtaxed, declared the Japanese had camouflaged trapdoors all over the face of the cliff and that they would open a door, pour out a murderous volley, and close the door again, leaving the attackers staring at an apparently blank area of mountainside.[7]
>
> — "Seventeen Days at Iwo Jima"
> *Marine Corps Gazette,* February 2000

In Vietnam, enemy bunker complexes were so well hidden that U.S. troops would routinely assault them from less than 50 yards after thinking themselves ambushed.[8]

How Eastern Defenders Vanish

In the South Pacific, whole Japanese units may have climbed trees to escape entrapment. They did, after all, have easy access to *ninja* pole-climbing skills,[9] *shuka* hand-fitting claws,[10] and *ashika* foot-mounted spikes.[11]

Figure 4.7: Damp Areas Provide Best Gap in Enemy Lines

(Source: *FM 21-76* (1957), p. 38)

When cornered, the Viet Minh would often practice one of the *ninja's suitonjutsu* "water escape arts."[12]

> They had probably resorted to one the favorite guerrilla tricks: to submerge in the swamp and stay underwater breathing through hollow sections of cane until it was dark enough for them to withdraw to the woods.[13]

The VC simply vanished below ground. As U.S. troops couldn't easily spot a camouflaged trapdoor, they seldom suspected the ploy.

Microterrain Can Have Strategic Significance

Wars are won by destroying the other side's strategic assets. Eastern armies do not always distribute their materiel through their chain of command. They sometimes stage it near the front and then issue it according to need. As a result, their soldiers have become expert at hiding supplies and equipment within the natural contours of the land. Asians prefer to do so below ground. It takes a skilled tracker to find the hidden entrance to one of their caches. U.S. forces found only a tiny percentage of the supplies that had been prestaged for enemy units in Vietnam.

Russians like to hide things above ground, but still within the natural flow the countryside. A contemporary Russian manual shows a truck in a covered gully, crates under the covered dip in a slope, and a tank beneath the extended overhang of a roof.[14] In each case, the area appears unchanged. Russia's surrogates follow the same method. Combining camouflage with natural contour must still be fairly effective, as most Serbian heavy weaponry escaped the U.S. bombing campaign in Kosovo.

> According to a suppressed Air Force report obtained by *Newsweek,* the number of targets verifiably destroyed was a tiny fraction of those claimed: 14 tanks, not 120; 18 armored personnel carriers, not 220; 20 artillery pieces, not 450. Out of 744 "confirmed" strikes by NATO pilots during the [78-day] war, the Air Force investigators, who spent weeks combing Kosovo by helicopter and by foot, found evidence of just 58.[15]
>
> — *Newsweek,* 15 May 2000

Harnessing the Senses

- *Can Eastern soldiers actually see better than Western?*
- *What is a person's "sixth sense"?*

NORTH KOREAN WARRANT OFFICER, 1952

(Source: Courtesy of Cassell PLC, from *World Army Uniforms since 1939,* © 1975, 1980, 1981, 1983 by Blandford Press Ltd., Part II, Plate 79; *FM 21-76* (1957), p. 194)

Where One's Natural Senses Are Better Respected

According to grandmaster Masaaki Hatsumi, "In *ninpo,* one's whole body must act as his eyes and ears."[1] An unexpected touch or odor may signal danger. Often operating during periods of reduced visibility would enhance the *ninja's* nonvisual senses.

To resist the Japanese takeover of Malaya, the British had formed a loose alliance with the predominantly Chinese Malaysian Communist Party.[2] From that alliance, the British gained some insight into the role senses play in close-quarter fighting.

Other Nations Use Privates as Observers

Sun Tzu valued spies, and Eastern armies prefer human "intelligence" to that which is electronic. As such, most systematically collect what frontline soldiers learn.

Every Soviet soldier in WWII had to be able to function as an observer. When so assigned, his duties involved watching the enemy, his own and neighboring units, and his commander. He was told to find a high location in the daytime, and a low one at night. The low one offered more horizons. At night, he was also encouraged to use his senses of smell and hearing.[3]

Seeing

It has been insinuated since WWII that the average U.S. soldier may not see as much as his Asian counterpart.

> There must be training in difficult observation, which is needed for the offense. It is my observation [on Guadalcanal] that only 5% of the men can really see while observing.[4]
>
> — Col. Merritt A. Edson after Guadalcanal
> *FMFRP 12-110*

There may be a simple explanation for the disparity. Like the *ninja,* the Asian soldier may do eye drills. During the day in the woods, one can force his eyes to focus on whatever is behind the nearest layer of foliage. As this is somewhat of an unnatural act, it takes practice. By alternately focusing on something far and near, one could build up the muscles for this technique and others.

After dark in rainy weather or noisy conditions, one must depend almost entirely on his night vision.[5] The *ninja* method of seeing through the darkness is called *ankokutoshijutsu.*[6] *Ankokutoshi no jutsu* involves the study of shadows.[7] Eastern night fighters remain a full 40 minutes in total darkness before a mission and then shut one eye to any light source.[8] The *ninja* not only has special nighttime seeing techniques, but he also remains alert to his quarry's ability to see. He remains in shadow or outside the sentry's field of vision. To pass the sentry, he has only to trick him into looking at a bright light.

During the Korean War, colorblind U.S. Marines were often

chosen for night patrol.[9] Perhaps they could more easily distinguish shades of gray. Or their well-exercised rods—the colorblind retinal receptor cells—may have drawn in more light.[10] Rods are responsible for both night and peripheral vision. For this reason squinting and then fully opening one's eyes will enhance the latter.[11] *Ninjas* heighten their peripheral vision through defocusing exercises: (1) narrowing the eyes, (2) loosening the intense focus on some object, and then (3) allowing the vision to blur.[12] Another way to see better after dark is to make a viewing tube with one's hand.[13] The tube lessens the light reflected off those things around the object being observed.

Even on a moonlit night, the Malaysian guerrillas of WWII needed some sort of light to follow a jungle path. They soon discovered that a few fire flies or luminous centipedes in the reflector of flashlight gave off enough light "to read a map, lay a charge, or follow a path."[14] In the primary jungle during the day, one can see farther at ground level.[15]

What There Is to See

Looked for most often on a battlefield are signs of enemy occupancy. They run the gamut from unnatural movement to wilted foliage, fresh dirt, linear shapes, and subtle glint. Yet, for the highly trained eye, the battlefield holds all sorts of signalling opportunities as well.

The enemy in Vietnam relied heavily on an intimate knowledge of each area and its occupants. They moved with ease along known trails or behind local guides. Water, camps, letter boxes, mines, caches, and safe routes were often marked with any number of signs.[16] After following U.S. patrols to their night ambush positions, the local VC probably marked the occupied trails.

Hearing

At night, it is hard to discern by sight, and judgments must be based on sound.[17] U.S. government manuals contain little, if any, listening techniques. A field expedient megaphone might work. There is a *ninja* technique that uses every cranial cavity to collect vibrations. One cups one's hands behind his ears, tilts his head

Figure 5.1: Putting One's Ear to the Ground
(Source: *FM 21-76* (1957), p. 38)

backwards a little, and opens his mouth.[18] While this may sound somewhat ludicrous, U.S. special forces do teach cupping one's hands behind the ears and turning one's head in the direction of the sound.[19] Through "hearing" battledrills, the pre-WWI Japanese prepared the individual soldier to gauge the enemy's direction and distance at night.[20]

Sound travels farther in the direction the wind is blowing. When there is no wind, conditions are also excellent for hearing. In the dead of night, surrounding noises are more pronounced than at dusk or dawn. In level open terrain, where there are no obstructions to sound waves, noises travel farther. Sound travels farther in the winter.[21] With one technique, a *ninja* can hear stealthy footsteps at a range of 30 meters.[22] To increase hearing ability, one can also hold his breath.

Certain substances carry sound. Others may merely give that impression. Heavily booted troops and any vehicle can sometimes be "heard" by putting one's ear to firmly packed ground (Figure 5.1). The ground, in effect, transmits their vibrations.

What There Is to Hear

Pre-WWI Japanese soldiers had to know the sounds of marching, digging, and artillery setup. They could also recognize sloshing canteen water, bayonet scabbard noise, and footsteps.[23] Those with a cough were left behind.[24]

More recently, Trinh Duc—a former VC—had no trouble differentiating the shot of an M-16 from an AK-47.[25] His comrades probably had little difficulty estimating the distance of an American sweep.

There are any number of noises that portend danger. A sudden flight of wild birds, a dog barking, or an agitated insect might all qualify. Then there is the rustling of bouncing gear, the "twang" of a grenade spoon, and the clack of a claymore plunger. Or, there may be no background noise whatsoever. After becoming familiar with an area's normal twitter, one would want to investigate any abnormality.

James Corbett—the legendary man-eating tiger hunter from India—listened to the "jungle folk" as he was growing up. In northwest Bengal, he identified a number of animals that would warn of danger—to include the approach of a weapon-toting human being.[26] Among them were birds, deer, monkeys, and squirrels.[27] In India, the best of the "talking" birds were drongos, red jungle fowl, and babblers.[28] Corbett claimed that certain jungle folk could understand and even imitate the languages of other species.[29] Excerpts from one of his books should illustrate the point. Those who, after reading it, still doubt Corbett's sanity should refrain from confronting any jungle inhabitant.

> Racket-tailed drongos are also found in association with cheetal [a barking deer], feeding on the grasshoppers and other winged insects disturbed by the deer. Having heard the cheetal give their alarm call when they see a leopard or a tiger, he learns the call and repeats it with great exactitude. I was present on one occasion when a leopard killed a yearling cheetal. Moving the leopard away for a few hundred yards, I returned to the kill, broke down a small bush, tied the kill to the stump, and, since there were no suitable trees near by, sat on the ground with my back to a bush and my movie camera [as opposed to a weapon] resting on my drawn-up knees. Presently a racket-tailed drongo arrived

> in company with a flock of white-throated laughing thrush. On catching sight of the kill, the drongo came close to have a better look at it, and in doing so, saw me. The kill was a natural sight to him but my presence appeared to puzzle him; after satisfying himself that I did not look dangerous, however, he flew back to the white-throats who were chattering noisily on the ground. The birds were on my left and I was expecting the leopard to appear from my right front when suddenly the drongo gave the alarm call of a cheetal, on hearing which the white-throats—some fifty in number—rose in a body and went screaming into the branches above them, from where they started giving their alarm call. By watching the drongo, I was able to follow every move of the unseen leopard who, annoyed by the baiting of the birds, worked round *[sic]* until he was immediately behind me.[30]
>
> — James Corbett, *Jungle Lore*

To match Corbett's achievement, one must learn to distinguish between nearly identical calls of the same animal. After watching a female deer tell her young to gather and another tell her herd to flee, Corbett knew the secret.

> The bark of the anxious mother recalling her young one[s] and the bark of the hind [three-year-old female] warning the herd of the presence of a human being [axe carrying woodcutter] had to my untrained ears, sounded exactly alike; only after I had gained experience did I detect that the difference in the call of animals and of birds was not to be found in the call itself but in the intonations of the call. A dog barks . . . to welcome its master . . . or with excitement on being taken for a run . . . or with frustration at a treed cat . . . or with anger at a stranger . . . or because it is chained up. In all these cases it is the intonation of the bark that enables the hearer to determine why the dog is barking.
>
> When I had absorbed sufficient knowledge to enable me to identify all the jungle folk by their calls, ascribe a reason for their call, and imitate many of them sufficiently well to get some birds and a few animals to come to me or to follow me, the jungles took on an added interest.[31]
>
> — James Corbett, *Jungle Lore*

Figure 5.2: Odor Can Carry a Long Way

(Source: *FM 7-70* (1986), p. 186)

Smelling

The *ninja* has a four-step smelling technique: (1) take a deep breath, (2) exhale it, (3) take shallow sniffs until the lungs are full, and (4) concentrate on the nostril linings at the very top of the nose. It enables him to detect standing water after dark.[32] Among his other smelling techniques may be wetting the ends of his nostrils. Just as wild animals depend on the power of smell to find their next meal, so do Eastern warriors depend on it to detect Western visitors. If so much as one visitor uses Ivory soap, the rest will not surprise an Eastern camp from upwind. (See Figure 5.2.)

> The ninja practices an "odor-free" discipline: he avoids garlic, leeks, all spicy foods, salty things, oily things, and so on. If his body gives off the smell of something he has eaten, for example, when in hiding, his adversary may become aware of him. On the other hand, if he does not eat such foods, he becomes far more sensitive to external stimuli.[33]
>
> — Dr. Masaaki Hatsumi

The Malaysian guerrillas of WWII gave up smoking to improve their powers of smell.[34] Against the partially *ninja*-trained Japanese, they had little choice.

What There Is to Smell

As early as 1913, Japanese troops were told to bury their excrement or urine and avoid smoking.[35]

To keep from being detected by Malayan rebels, those tracking them in the early 1960's stayed away from chewing gum, tobacco, toothpaste, hair tonic, insect spray, and soap.[36] Among the odors one might encounter on patrol are cigarette smoke, wood smoke, food cooking, fish stains, garlic sweat, and explosives.

Touching

The Asian soldier also relies on his sense of touch. As he works almost exclusively at night, he needs it to warn him of impending danger. To enhance this sense of touch, he dresses lightly. That way whomever he encounters clothed can be considered enemy. The Turks followed this procedure in Korea.[37] So did the Viet Cong in Vietnam.

> The reason the breaching team (the first sappers through the defender's minefields and barbed wire) wore only shorts, or went naked, was so that they had more skin exposed that could feel things, like wire, mines, or boobytraps.[38]

Exposed skin worked the best. Bare forearms helped the crawler to locate trip wires.

What There Is to Touch

In effect, the Asian night fighter attempts with his whole body to feel anything out of the ordinary. Such a feeling could involve a solitary object or entire medium. Something soft and warm would be the most unsettling, but metallic prongs and wires wouldn't be much better. Any indicator of another human being would qualify. Generally, the object's shape, moisture, temperature, and texture are assessed. Or the night fighter might sense a difference in the very medium through which he is moving—unnatural rippling of the water, a puff of air in an airtight house, crushed grass underfoot, etc.

The human skin is extremely sensitive to changes in temperature. Inside a house, residual heat on a chair or bed could spell trouble. Shell casings from silenced weapons might still be warm. When someone lies on the ground for a prolonged period of time, he leaves a momentary heat signature behind.[39] For reasons that are yet to be explained, a skilled *ninja* can often sense another person nearby without actually touching him.

Tasting

According to grandmaster Masaaki Hatsumi, a *ninja* develops the sensitivity of his taste buds.[40] After a while, he comes to recognize what seems out of place.

Experienced night fighters can discern odors through their tongues, or more precisely the soft membranes under their tongues.[41]

> To cultivate your ability to taste the air, purse your lips as if about to whistle and slowly draw air in. Allow the air to swirl around and under your tongue. Try to discern salty from sweet from oily. With practice you will be able to discern perfume or deodorant (a sweet or oily taste under your tongue) and residual tobacco (sharp, pungent) left in the air or on closing by a passing enemy.[42]

What There Is to Taste

A professional sailor can tell when he is near the ocean from the taste of salt in his mouth.[43] There may be other subtle odors that can only be detected through taste.

In war, poison is not an uncommon weapon. In nuclear, biological, and chemical (NBC) war, the only warning of an attack may be an unexpected taste.

That Sixth Sense

Any number of sources have been attributed to the mysterious sixth sense. Some claim that it is extrasensory perception (ESP). A man who has tracked African guerrillas for the better part of three decades claims it to be the product of the subconscious mind. He says it springs from one's instinctive comparison of subtle sensory input with deeply buried memories.[44] In other words, the inexperienced woodsman shouldn't count on having much. The reigning grandmaster of *ninpo* calls it the working knowledge of one's own subconscious level of thinking.[45] In effect, the accomplished *ninja* enjoys a broader perception of reality than the average person.

> It is of utmost importance to immerse and enjoy oneself in the world of nothingness. In this world of nothingness, one must see through to the essence of common sense, or knowledge, or divine consciousness, make a decision, and translate it into action. This is the one way to enlightenment. This is also the key to cultivating the sixth sense required of martial artists and ninjas.[46]
>
> — Dr. Masaaki Hatsumi

James Corbett attributed his sixth sense to his guardian angel.[47] With only two close calls out of 250 victories over something that can see after dark, Mr. Corbett's opinion must be respected.

There are scientific explanations as well. Researchers have found that the human brain contains magnetite (Fe_3O_4)—the same mineral that acts as the biological compass for birds and fish. They believe this mineral to be responsible for the electromagnetic field that surrounds and extends out from the human body. Night warriors try to enhance their "peripheral body sense" awareness.[48]

Staring at the back of someone's head will often cause him to turn around. Precisely why is not understood, but night stalkers routinely avoid looking directly at an intended victim.

Counterambushing Requires a Combination of Senses

To detect an impending ambush, the experienced point man must use of all his senses. While a single, isolated abnormality can often be attributed to coincidence, several at the same time can't.

> To avoid detection by an ambush post, the [North Korean infiltration] teams made frequent listening stops while crossing the DMZ. Reportedly, an ambush can often be detected by rattles, rustles, coughing, whispers made by its members and the smell of urine.[49]

It's Risky to Enhance One Sense at the Expense of Another

Putting anything over one's ear, eye, or skin severely curtails its overall function (Figure 5.3). Wearing anything with an odor will detract from one's ability to smell. To function at maximum efficiency, the body's built-in, early warning system must be unencumbered by paraphernalia.

Figure 5.3: Potential Threats to Sensory Balance

(Source: *FM 20-32* (1976), p. 60)

To stand the best chance of detecting a highly proficient opponent, one must have all of his senses working at maximum efficiency and in tandem. That way any one can trigger the alarm, while the rest are fully dedicated to the threat.

The Problem with Sensory-Enhancement Devices

Every electronic surveillance device has an inherent flaw. It must process everything it detects. As the Russians did throughout their amazing two-week victory over the Japanese in Manchuria in 1945, the attacker can come through the rain.[50] Heavy rain will not only erase sound but also virtually negate thermal imaging. To thwart night vision devices, an attacker has only to call for continuous illumination behind his penetration point.

The rural Easterner works harder to understand and comply with the laws of nature than does his industrialized Western counterpart. It is his knowledge of nature that permits him to outsmart Western technology. Until each U.S. private takes it upon himself to close this gap, he will continue to operate at a disadvantage in war.

> [B]e assured that if you are not interested, or if you have no desire to acquire knowledge, you will learn nothing from nature.[51]
>
> — James Corbett, *Jungle Lore*

6 Night Familiarity

- *Which part of the world produces the best night fighters?*
- *Why must a soldier learn to predict shadows?*

VIET CONG "SUBMACHINEGUNNER," 1960's-70's

(Source: Courtesy of Osprey Publishing Ltd., from *Armies of the Vietnam War 1962-75*, Men-at-Arms Series, ©1980 by Osprey, Plate C, No. 1; *FM 21-76* (1957), p. 54)

The American Aversion to Non-Illuminated Night Attacks

Touted as the world's "best night fighters" because of their night-vision equipment, U.S. troops have once again resorted to illuminating the darkness.

> Marines were fighting house to house in Nasiriyah. . . . A reporter for WTVD in Durham, NC, attached to the Camp Lejeune Marines . . . said Marines were using flares to light areas so they could see the enemy.[1]
>
> — Associated Press, 27 March 2003

One's equipment does not qualify him as a night fighter. Darkness provides tactical opportunities that no vision enhancement device can totally dispel. U.S. troops have been outclassed "skillwise" after dark for so long that their night fighting potential has until recently been officially dismissed. On Iwo Jima, 5th Amphibious Corps "decreed that all personnel moving after dark were to be considered as enemy. American troops were to hunker down in their foxholes at night and shoot anything moving!"[2] At first in Korea, Marine legend Wesley Fox wondered why his superiors bounced searchlights off the clouds on a dark night. Then, he realized how good the Communists were at non-illuminated night attacks.[3] Throughout the Vietnam era, the U.S. Marine Corps had taught only the illuminated variety.[4] Upon becoming a night attack instructor at The Basic School in Quantico, Fox discovered how little he had known about night attacks in Vietnam.[5]

As U.S. commanders try to capitalize on their edge in firepower, they are less likely to attack when they might have difficulty maintaining their unit's alignment. In other words, to risk fewer friendly-fire losses, they try not to attack in total darkness. As other nations have less ammunition, they have learned how to attack at night without shooting each other. They simply create stricter guidelines under which their assault troops can fire.

Asian Armies Prefer to Fight in the Dark

In WWII, the Japanese thought U.S. soldiers to be night-blind. Of course, their privates had been doing *ninja* night-vision-enhancement exercises since 1913.[6]

The Chinese in Korea thought U.S. troops "afraid of night and close-range engagement,"[7] and "not familiar with night fighting or hand-to-hand combat."[8] By U.S. admission, night training is of great importance to the Chinese army.[9]

Other Eastern Armies Also Prefer the Night

According to *TM 30-340,* the WWII Russians preferred night fighting when "terrain, minefields, and other obstacles—combined with enemy preparedness—eliminate the possibility of surprise and make heavy casualties in daytime operations a probability."[10] A U.S. Army study estimated that 40 percent of all Soviet attacks in 1944-45 were at night.[11]

> Soviet regulations before the outbreak of war in 1939 recognized that "night operations will be common under modern warfare conditions to exploit surprise, reduce losses, and disorganize the enemy."[12]
>
> — U.S. Army, *Leavenworth Papers No. 6*

> The Germans . . . admitted that their [Russian] opponents were better than they at fighting at night, in forest and swamps, at camouflage, and quick digging in.[13]
>
> — Republisher's introduction
> *Soviet Combat Regulations of November 1942*

> Although the Soviets at times suffered heavy casualties and even reverses at night . . . this was more the exception than the rule. Most senior German officers who fought the Soviets on the Eastern Front acknowledged their "natural superiority in fighting during night, fog, rain or snow," (Simon, in *Night Combat,* by Kesselring, 1952, 59) and especially their skill in night infiltration tactics, reconnaissance, and troop movements and concentrations (Kesselring, *Night Combat,* 1952, 2-5).[14]
>
> — U.S. Army, *Leavenworth Papers No. 6*

In Afghanistan, the Soviets routinely established their ambush and blocking positions before dawn.[15] However, their zeal for night fighting was starting to wane.

> The Soviet and Afghan government forces apparently did little to contest the *mujahideen* ownership of the night. Night patrols and ambushes were a singular planned event, not a routine mission. . . . Squad-sized ambushes were prohibited by 40th Army regulations. . . . The Soviets did not allow squad-sized ambushes in Afghanistan since their NCOs were not professional and perhaps not trusted.[16]
>
> — Editor's commentary
> DoD-published Soviet military academy study

In Afghanistan, motorized Russian infantry encountered an enemy who liked to fight at night and closer than they did. "Among the guerrilla forces' tactical strong suits were all types of night actions."[17]

> The *mujahideen* had . . . developed tactics and techniques to deal with the sweep. As a rule, they would lure us into predetermined areas and then open fire on us at a distance of no more than 50 to 100 meters. They would only fight in close contact to us since we could not use our artillery or aviation within 100 meters of friendly forces. The *mujahideen* knew the local area and local terrain features quite well and were thus able to outmaneuver the Soviet forces.[18]
>
> — DoD-published Soviet military academy study

Distinguishing Readiness from Weakness

The professional night fighter makes fear into a friend. Confident of his abilities, he continues on through threats that will often never materialize. To help himself do this, he practices relaxation, concentration, meditation, and visualization (imagining what he is about to do).[19] He knows that any type of tension will decrease the efficiency of his senses.[20]

What distinguishes the self-assurance of a *ninja* from that of a Western soldier is its source. Whereas the *ninja* has personally done the impossible, the Western soldier thinks he has magically inherited from his forefathers everything he might need in combat. By the time he discovers his error, it is often too late.

Not Being Seen

Secretly moving around at night takes staying off the horizon. (See Figure 6.1.) Unfortunately, the smallest of mounds can provide a natural horizon to a prone enemy soldier. Any light-colored surface will give him an "artificial" horizon.

While *ninjas* try not to reflect any light at all,[21] all night fighters move from shadow to shadow. The ever-shifting play of light and shadow has a name—*chiaroscuro*. The brighter the light is, the darker the shadows are. The shadow warrior plans every move so as to take full advantage of the *chiaroscuro*.[22] He is able to predict where shadows are likely to appear throughout the night. He can also predict their length. The *chiaroscuro* expert can also travel along the fleeting shadows of moving vehicles. He knows that a sentry is momentarily blinded every time a cloud drifts in front of the moon.

One of the best times to move is just before sunset or after sunrise. The human eye has trouble adjusting simultaneously to bright sky and dark ground.[23] To avoid being seen while moving, the soldier has only to keep dark objects behind him.[24]

Not Being Heard

Sound travels differently at night. Operating during periods of reduced visibility requires obeying those parameters. The WWII Japanese had strict rules during any night movement.

Figure 6.1: The Telltale Silhouette

(Source: *MCRP 3-02H* (1999), p. I-5)

> Soldiers will not load or fire without orders. . . . There will be no talking or whispering. Men who have a cough . . . will not be taken along.[25]
>
> — Japanese night-fighting manual, 1913

Night-Firing Techniques

The 1913 Japanese manual contained detailed instructions on night firing. Probably to preserve the element of surprise, the manual discouraged firing altogether. However, it also gave the following advice on how to do it at point-blank range: (1) so as not to shoot high, take care not to incline upper part of the body to the rear or raise the muzzle of the rifle above the horizontal; and (2) turn off the safety without fail.[26] Once through the exterior ring of Allied holes during WWII, Japanese assault troops may have been told to shoot downward to preclude fratricide. On Guadalcanal, they were given wooden bullets. It is the indiscriminate shooting of small arms that forces Western assault forces into exposed formations.

7 Nondetectable Movement

- *How can a ninja secretly cross different types of ground?*
- *How does an Eastern army hide its resupply route?*

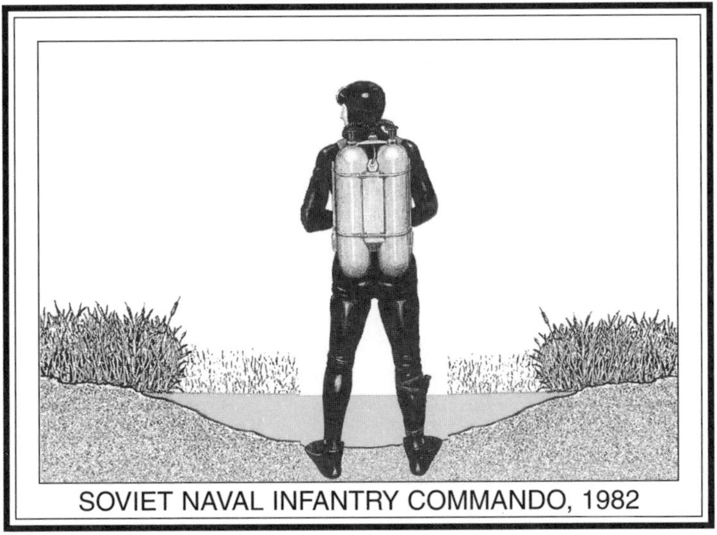

SOVIET NAVAL INFANTRY COMMANDO, 1982

(Source: Courtesy of Cassell PLC, from *Uniforms of the Elite Forces,* ©1982 by Blandford Press Ltd., Plate 17, No. 49; *MCRP 3-02H* (1999), p. VII-7)

A Long-Established Eastern Tradition

Of utmost importance in the East is remaining unseen in battle. Grandmaster Hatsumi calls *ninjutsu* the art of stealth, and *ninpo* the way of invisibility.[1] To practice them, one must learn the laws of nature—not only of vegetation and microterrain, but also of light, shadow, and sound. One will also need self-discipline and to pay close attention to what the enemy is doing. The *ninja* attempts totally to immerse himself in his mission.[2] According to Hatsumi, the *ninja* merges into his background and is acutely responsive to changing circumstances.[3] He creates no form or shape that is out of place.[4] He doesn't overreact to what is happening but simply flows with it,[5] and his techniques flow as well.[6] He has great forbearance and considers no obstacle insurmountable; he just keeps going.[7] He tries to bypass the enemy by studying his weaknesses.[8] He is well aware of his own weaknesses but will normally conceal them. He can do things that "big-picture thinkers" would deem impossible. Most distressing is the ease with which a *ninja* can capture or kill a Western scout. The *ninja* has only to hide behind a trail-side tree. As the scout passes the tree, the *ninja* slides along its opposite side and ends up behind him.[9]

According to the *ninpo* grandmaster, the *ninja* must revert to subconscious instinct to disappear.[10] If he shows nothing and thinks nothing, he is nothing.[11] His "benevolent heart" (or conscience) must stay intact.[12] How the *ninja* "empties himself" may give a clue to how he becomes less visible. A longtime student of psychology and Eastern religions provides an educated guess at the method.

> Psychologically, emptying oneself . . . refers to stopping the internal dialogue. You know, the one that . . . reminds you of what you failed to do, or should do, etc. . . .
>
> Once the internal dialogue stops, one can pay attention to what else [the] mind has up its sleeve. . . . This is the process of meditation. For an excellent description of what the Buddhists say about this, read *The Tibetan Book of Living and Dying* by Chogyal Rinpoche. Having "emptied" oneself, one can become attuned to what is.[13]
>
> — Dr. David H. Reinke

Entering a War Zone Undetected

There are many ways to sneak into a war zone. Among the best at deception and camouflage are the Chinese. During the Korean War, no fewer than 300,000 Chinese volunteers secretly journeyed hundreds of miles through territory under almost constant U.S. aerial surveillance. They traveled mostly at night or under the cover of forest-fire smoke and then hid during the day. Whole platoons would cram into a single-family dwelling or cover themselves with mud, straw, and other materials in a draw or gully. When forced to move during the day, they carried straw mats with which to cover themselves. By closing ranks and lying down in formation, they could simulate a freshly cut field. While dressed in white, or like refugees, or like ROK soldiers, some were able to move openly on the roads.[14]

> In rear and assembly areas, CCF [Chinese Communist Forces] troops dug two-man or squad-sized foxholes on reverse slopes. The soldiers carefully distributed the evacuated soil, covering it with branches and straw and replanting the turf. These foxholes were cunningly selected and camouflaged to blend in with the [surrounding] terrain; they were virtually undetectable except from close range (U.S. Army, *Engineer Intelligence Notes,* 1951, 8:4). The Chinese also selected bivouac sites in or near burned-out villages, often taking up residence among the rubble. The CCF avoided high ground, which characteristically attracted air strikes (U.S. Army, *Enemy Tactics, Techniques and Doctrine,* 1951, 5). In addition, the Chinese used natural materials almost exclusively for camouflage and concealment . . . UN patrols sometimes approached within killing range of the CCF, but the Chinese often did not fire, choosing to remain hidden.[15]
>
> — U.S. Army, *Leavenworth Research Survey No. 6*

A Russian Forerunner of the Ho Chi Minh Trail

According to *Leavenworth Papers No. 8,* the Soviet army broke through the Japanese Kwantung Army's Manchurian defenses in Manchuria in just two weeks in 1945.[16] In one heavily wooded sector, task-organized regimental forward detachments built their own road as they moved.[17] As the infantry cut the trees, the engineers worked them into a corduroy underlayer. In more open sectors, the Soviets hid their approach march in a different way.

> Along roads and tracks under Japanese observation, engineers erected vertical masking walls and camouflage overhead covers. The 5th Army zone contained eighteen kilometers of verticals walls and 1,515 overhead covers (editors of Voennoe Izdatel'stvo, *History of the Second World War: 1939-45,* 1980, 2:199-200).[18]
>
> — U.S. Army, *Leavenworth Papers No. 8*

In swampy sectors, the Soviets combined corduroy roads with screening fences and overhead vegetation.[19] The same techniques would be used eight years later—on a somewhat smaller scale—at the battle of Dien Bien Phu.

The Chinese Refinement

Much of what the Chinese needed to pursue their war aims in Korea was hand carried by conscript coolie and transport bicycle. As it came in along a myriad of routes, it was relatively immune to U.S. interdiction from the air.[20]

Though not commonly known to Americans, the Chinese also played a considerable role in the Viet Minh uprising a thousand miles to the south. They provided supplies,[21] volunteers,[22] advisors,[23] and even a covered supply road.[24] How this supply road was constructed may shed new light on its more famous descendent.

> The Red Highway [from China into North Vietnam] was a masterpiece of camouflage. It had been cut through the jungle without allowing as much as ten yards to be exposed to the skies. At a few less densely wooded sections, hundreds of trees had been roped together and drawn closer to one another with the aid of pulley-like contraptions. They had been fastened in such a way that their crown intertwined over the road. In the open ravines, networks of strong wire had been stretched between the slopes to support creepers, which had soon blotted out the road beneath.
>
> The jungle road included permanent bridges, twelve to fifteen feet wide, most of them constructed a few inches underwater to fool aerial observation. Difficult or swampy sections of the road had been "paved" either with stones or with logs leveled with gravel. Along the Red Highway were checkpoints where guerrilla MP's [military police] controlled transport or troop movements. Rest-houses and service stations where carts and bicycles could be repaired were also located on that incredible network of trails. Its very existence was ridiculed by some French statesmen. It was simply too incredible to believe. Yet it was there![25]

For the final battle of the war, the Viet Minh followed the Chinese example. In just three months, they built a 600-mile road from the Chinese resupply point to Dien Bien Phu.[26] Over that road, they would secretly move 100,000 fighters, 200 heavy guns, and 8000 tons of supplies.[27] While they had some Russian trucks, they relied primarily on 500-pound-carrying bicycles. Because the trail had been so well camouflaged, it was invisible to French observation planes. The Viet Minh had roped the treetops together to create a tunnel of vegetation.[28] Later, they would hide their heavy guns in bunkers overlooking the French-occupied valley.

The North Vietnamese Masterpiece

The Ho Chi Minh Trail of the U.S. Vietnam era provides one of the best clinics on nondetectable movement in history. The Truong Son mountain range is not tall but a full 1000 km long.[29] It runs almost the full length of Vietnam. Except from 1954-59, the Truong Son resupply line had been in continuous operation since 1946. Early builders of the trail avoided contact with local residents and kept out of sight of government troops. Some portions of the trail followed the Cam Lo River. Where that river met Route 9—the highway from Dong Ha to Laos—the trail workers crawled through culverts.[30] They moved when government outposts relaxed their vigilance and often right under their noses.[31] Morning fog or evening mist provided other opportunities.[32] They traveled quickly through open or heavily shelled areas.[33]

As time passed, the trail network was expanded toward the Laos and Cambodian borders. After the Americans arrived, the only portions the "trail" open to truck traffic in the daytime were the "K" roads. In all, they stretched 3140 km.[34] They were nothing more than narrow byways covered by foliage. Where those byways passed through thin foliage, the road workers joined tree branches by weighing them down with heavy rocks.

Not all of the north-south thoroughfares were at the edge of South Vietnam (Map 7.1). There was also a "strategic route" that by 1969 ran south out of the DMZ to a terminus near the Cam Lo Marine outpost (Maps 7.2 and 7.3). In October 1966, due north of Cam Lo and 2000 yards short of the DMZ, a platoon from Alpha Company, 1st Battalion, 4th Marines discovered a foliage-covered cobblestone road at Long Son (grid coordinates 149721).[35] Some 300 meters north of Route 605, its entrance was perfectly camou-

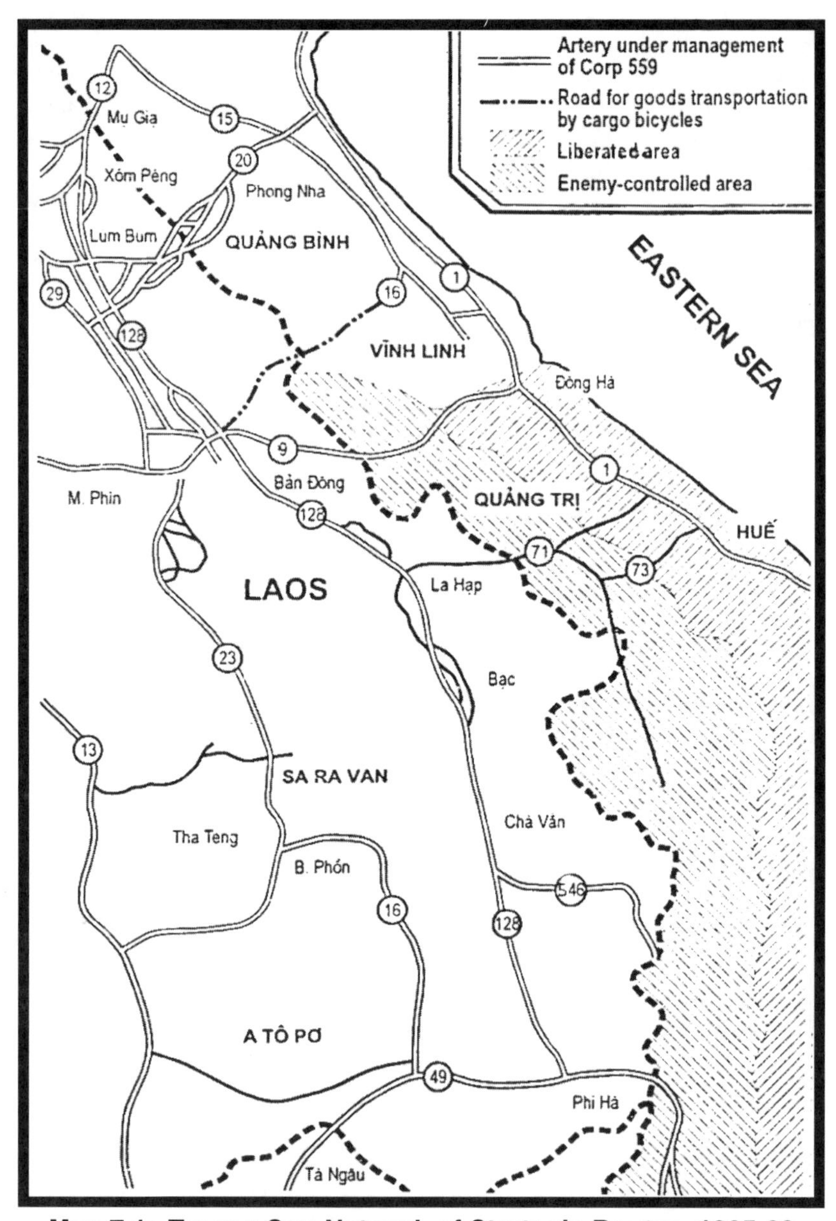

Map 7.1: Truong Son Network of Strategic Routes, 1965-68

(Source: Courtesy of The Gioi Publishers, from *The Ho Chi Minh Trail*, © 2001 by Hoang Khoi and The Gioi, p. 26)

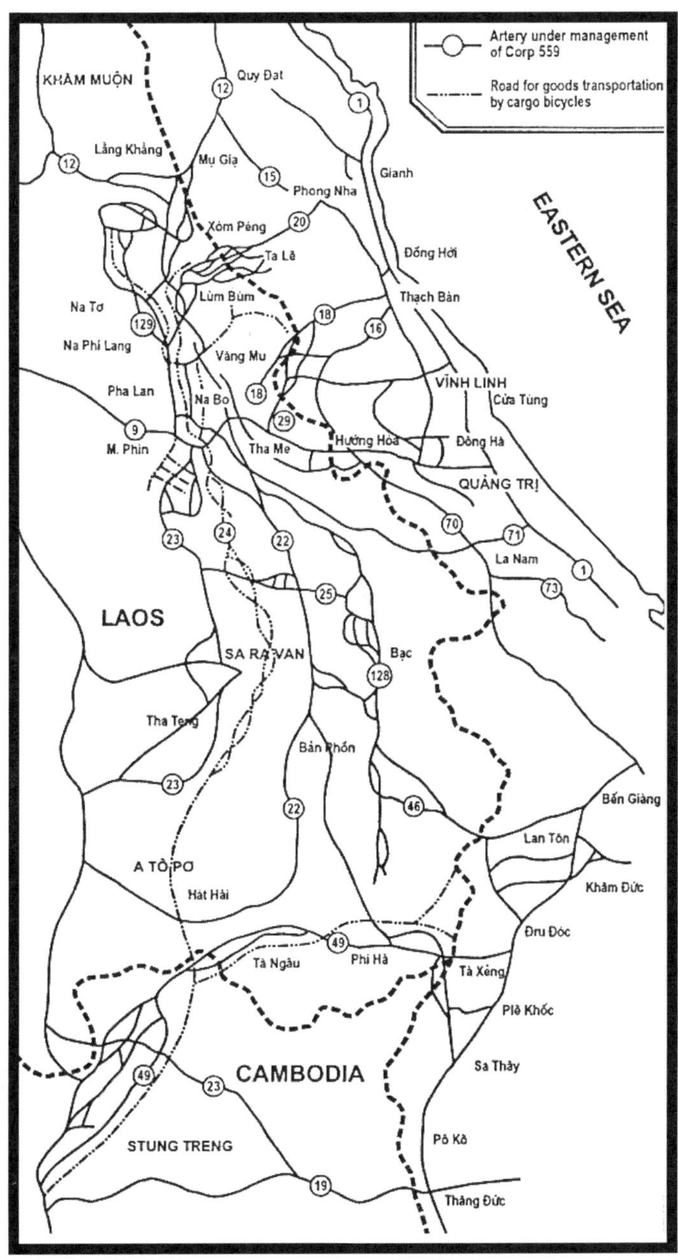

Map 7.2: Truong Son Network of Strategic Routes, 1969-73

(Source: Courtesy of The Gioi Publishers, from *The Ho Chi Minh Trail,* © 2001 by Hoang Khoi and The Gioi, p. 70)

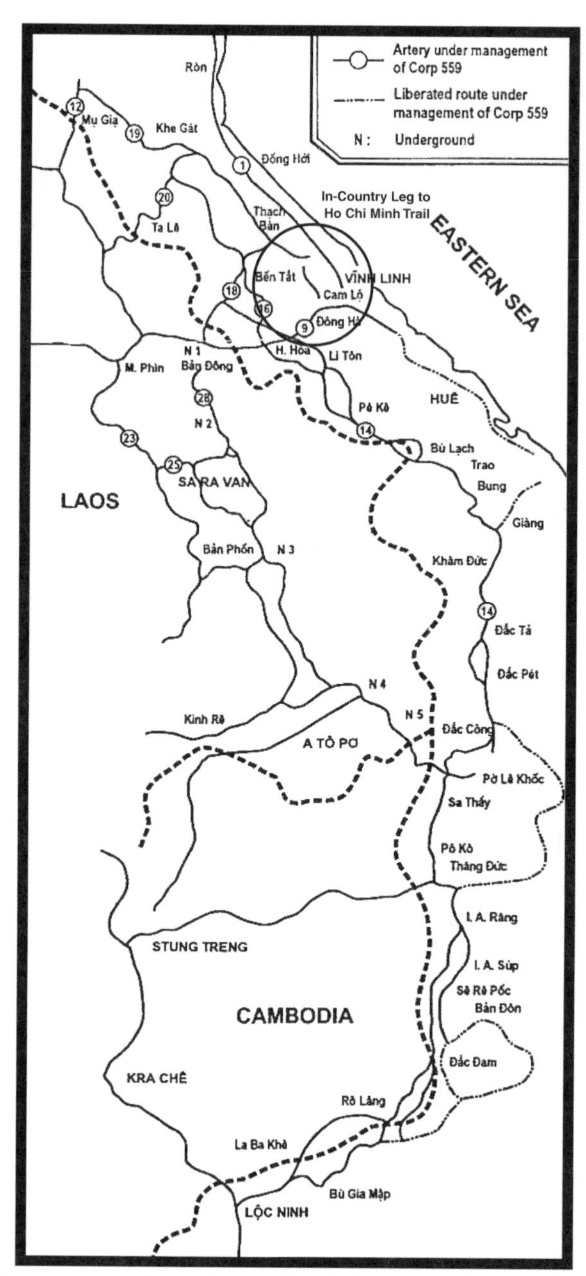

Map 7.3: Truong Son Network of Strategic Routes, 1973-75

(Source: Courtesy of The Gioi Publishers, from *The Ho Chi Minh Trail,* © 2001 by Hoang Khoi and The Gioi, p. 70)

flaged. This part of Route 605 had seen so much enemy activity as to be nicknamed—"Three Gateways to Hell" (Map 7.4). At this same location shortly after noon on Easter Sunday, 26 March 1967, Delta Company bumped into an NVA unit. It was dug in all along the trail. At about 4:00 P.M. on that same day from Hill 39 (some 900 yards to the southeast), Alpha Company spotted a northward-moving NVA column. As the enemy column emerged into the open at a point about 1000 yards farther to the southeast,[36] it exchanged 60 mm mortar fire with the Marines and then quickly disappeared. The leader of the patrol sent to investigate the sighting has just recently confirmed its precise location.

> [During the first sighting below Hill 39,] the NVA were 540 meters south of the stream [at grid coordinates 153707].[37]
> — CWO-4 "Tag" Guthrie USMC (Ret.)

In April 1967, Guthrie and two buddies—Perry and Kreh—were scouting ahead for Lt. Bill Roach's platoon when they discovered a covered trail between Long Son and Tan An.[38] He reported the sighting to Sgt. Parker, his platoon sergeant. (Please look at Map 7.4 more closely.)

> The Area . . . with the covered trail was in and around Tan An in the 150720 grid area. [During subsequent patrolling], the NVA were pulling us northwest [as a diversion] toward Gia Binh. The trail was moving west to east along the ridgeline and followed the streambed where the vegetation was thick.
>
> [It was] where the vegetation was not thick that the NVA used the large trails at night to move heavy equipment. Then they would move off the side trails into the harbor sites where they had small bunkers in the treelines that were all interconnected in this region.
>
> At night, the U.S. Army would fly the two-engine Mohawks which would take infrared pictures. . . . [This looked] like a large flash. . . . The planes would fly back to Phu Bai, and [technicians would] develop the pictures. One of the guys I used to work with as a government [intelligence] contractor . . . used to fly those things. . . . [He said] it was limited what those planes could do and see. The NVA took advantage of this by hiding in the treelines and cam-

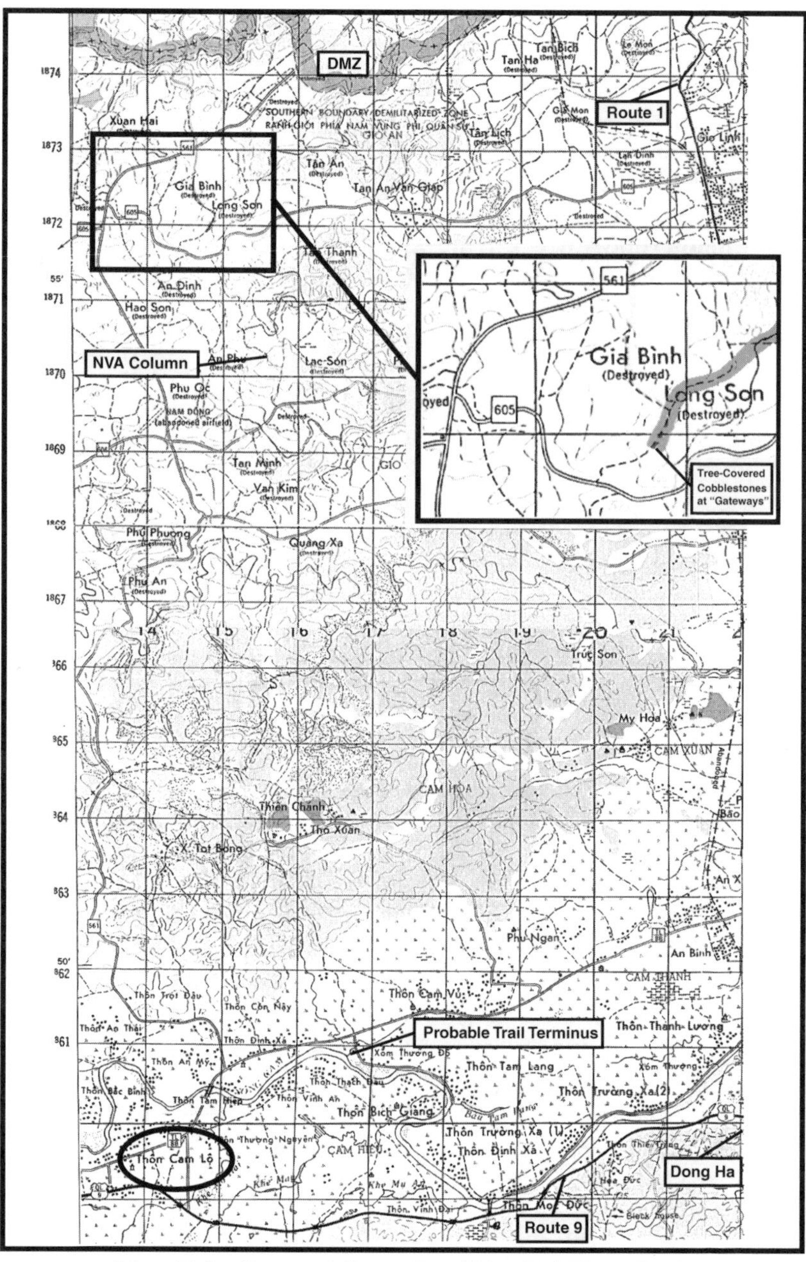

Map 7.4: Coastal Branch of Ho Chi Minh Trail

(Source: Quang Tri, Vietnam 1:50,000, Edition 3-TPC)

ouflaging the trails to conceal troop movements and harbor sites.[39]

— CWO-4 Charles "Tag" Guthrie USMC (Ret.)

Two separate sightings of a "cobblestone-paved tunnel" and two more of units moving toward it can mean only one thing. This was the "strategic route" shown on the North Vietnamese map. Guthrie thinks the NVA infiltration route may have followed the streambed that ran from Long Son north-northeast into the DMZ. This would make perfect sense from the standpoint of Oriental deception. The streambed would have been deeply recessed and heavily vegetated. U.S. aerial observers would have been watching the north-south trails and Route 561.

So detailed an analysis of a few acres of lightly vegetated terrain may seem unimportant to some. Yet to others, it shows why the U.S. intelligence community had failed to fulfill its mandate. To determine the whereabouts of the interior leg of the Ho Chi Minh trail at the time, that community would have needed a better understanding of Eastern tactics and history. Even if it had assimilated what U.S. patrol leaders saw, it would have still been unable to process the information. There is great value in discovering the exit point from the DMZ of the interior leg of the Ho Chi Minh Trail. That trail keeps going south. Simply by tracing a line along the side paths that paralleled 561, one can closely approximate where the trail may have emptied out into the Cam Lo River extension of its more famous western relative.[40]

There were also parallel road networks on either side of the Truong Son range. Of significant interest, the "N" markings on Map 7.3 reveal underground portions of the main Ho Chi Minh Trail. Whole sections were also devoted to cargo bicycle traffic (Map 7.2). Prior to 1968, the route running parallel to, and just north of, the DMZ was also consigned to cargo bicycles (Map 7.1). By 1975, the road network extended more than 20,000 km.[41]

There were scores of underwater bridges along the western segments of the Ho Chi Minh Trail as well.[42] However, they would also prove useful near the NVA's final objectives. A submerged bridge may have helped an entire NVA division to escape Hue City in 1968.[43] Further, the NVA and Viet Minh were by no means the first to use underwater bridges. The Russians had used them since 1939. By U.S. admission, their WWII manuals even showed how to build them.[44]

> The ingenious Soviet underwater bridges that plagued the Japanese at the Halha throughout the Nomonhan [Manchuria] affair took the Germans by surprise a few years later.[45]

By 1975, the North Vietnamese had constructed a vast network of improved roads and fuel pipelines right under the noses of the South Vietnamese. This network led right up to the outskirts of Saigon.

> The representative from the General Logistics and Technical Departments reported that thanks to the strengthening and extending of the main supply routes on the east and west of the Truong Son Range, the extending of the pipelines to the South and the improvement of all connecting highways, within a short period of time, 100,000 tonnes [tons] of military supplies had been sent from the North to the front. . . .
>
> In the Mekong River delta, eastern Nam Bo and even around Saigon, the entire population was mobilized to work for the front. The nearly 100,000 tonnes [tons] of war supplies accumulated on the spot were enough to meet more than half the needs of the front. Tens of thousands of supply workers, together with hundreds of boats and junks, were also brought into action.
>
> To this source was added considerable booty captured on various battlefronts in the south, Central Highlands and coastal provinces of central Vietnam.[46]
>
> — Gen. Hoang Van Thai

On Patrol

WWII Russian infantrymen were told, "Only the scouts move through small woods and brush, while the bulk of the squad goes around."[47]

Of course, the master of nondetectable movement on patrol was the Asian. As early as 1913, the Japanese manuals contained the following instructions for walking silently at night: (1) in tall grass keep feet low, (2) in short grass raise feet high, (3) in descending a hill plant the heel first, and (4) in climbing a hill plant the toe first.[48]

The Japanese were also given specific instructions on how to conduct a night patrol: (1) avoid entering woods or villages when possible; (2) keep a low posture in open terrain; (3) walk in the shadow along roads, treelines, and streets; (4) don't all start or stop at once on gravelly ground or a descending slope; and (5) lie down and listen at any suspicious indication.[49] (See Figure 7.1.) The Japanese patrol would always take a different route back to its base camp. However, it sometimes left telltale signs to help guide the next patrol: (1) pieces of white paper, (2) white powder, (3) broken tree limbs or small trees, (4) white rags tied onto branches, and (5) various types of road or trail markers.[50] The job of the Japanese "hidden patrol" was to hide near an enemy location or heavily traveled path. It would then collect intelligence through the capture or observation of enemy soldiers.[51]

The Malaysian guerrillas of WWII used a different technique to travel silently through the jungle. They walked heel first over hard ground and toe first over soft ground.[52] Several hundred miles to the north and years later, the Viet Minh found a way to remove any trace of their passage.

> Often the Viet Minh make no paths at all but use different routes between bases every day, allowing the grass to recover. Such trails cannot be detected.[53]

For the Approach March

WWI German stormtroopers secretly approached Allied lines through old communication trenches. According to *TM-E 30-451,* WWII Germans infiltrated to the line of departure in squad size, rendezvoused, and then hid for a day before attacking.[54]

Once again in Vietnam, the North Vietnamese would also infiltrate a platoon at a time to the point of attack. Their movement techniques were several and sophisticated. An enemy soldier later recalled this one:

> For trail camouflage, a unit would move in single file, each person in turn stepping in the footsteps of the man in front. The last man in the line would be responsible for eliminating whatever tracks there were.[55]
>
> — Hoang Tat Hong, NVA Sergeant

Where supporting arms threaten, speed acts as a type of security. Like the North Vietnamese,[56] the Chinese would often stay in single file or column and run in the approach march. Their scouts provided all the command and control that was necessary.[57]

The *mujahideen* in Afghanistan also had movement skills. "Among the guerrilla forces' tactical strong suits were . . . the ability to rapidly and clandestinely move in the mountains."[58]

Of the many ways to enhance surprise, keeping track of the wind direction is the most important. Easterners depend heavily on their natural senses. The wind can carry both noise and smell great distances. It is virtually impossible to sneak up on an Eastern unit from upwind.

During the Attack

Of all the world's armies, only the Western ones still attack in the daytime. Even during a nighttime assault, WWII Russians were instructed to stop and get down every time they were illuminated.[59] *TM 30-340* indicates that WWII Soviet troops jogged in the night attack.[60]

Figure 7.1: The Eastern Art of Shadow Walking

(Source: *MCRP 3-02H* (1999), p. I-7)

WWII Germans, on the other hand, were encouraged by their manual to crawl in the attack.[61] They sometimes used artillery to create enough shellholes to provide cover.

WWII Japanese enlisted men drilled on how to do the following quietly at night: (1) route march, (2) cross rough ground, and (3) attack.[62] During the crossing of rough ground, changing from line to column and back again was stressed. Also practiced was the changing in direction of a squad on line.[63] Without constant practice, tiny groups of soldiers cannot quickly execute even the simplest of maneuvers without becoming hopelessly separated.

The training the VC received on this type of thing was extraordinary. A captured D445 VC Local Force Infantry Battalion training schedule shows the extent to which the enemy prepared its soldiers for covert movement. During two weeks in the summer of 1970, the members of C2 Company received a full 36 hours of formal instruction on how to cross any combination of dry leaves, obstacles, and flares; 11 hours on how to cross a swamp under illumination; and 12 more hours on how to conduct sapper-style reconnaissance.[64] That the daily training only lasted 7 or 8 hours may have meant that the remaining time was dedicated to free-play (force-on-force) exercises.[65]

North Vietnamese sappers got training that was even more sophisticated. Here's what an NVA Master Sergeant recalls of the curriculum:

> The training was elaborate. We learned how to crouch while walking, how to crawl, how to move silently through mud and water, how to walk through dry leaves. We practiced different ways of stooping while we walked. In teams of seven men, we practiced moving in rhythm to avoid being spotted under searchlights, synchronizing our motions, stepping with toes first, then gradually lowering heels to the ground, very slowly, step by step.
>
> Wading through mud, we were taught to walk by lowering our toes first, then the rest of the foot. Picking our feet up, we would move them around gently [to break any suction], then slowly pull up the heels to avoid making noises. If you just pulled them up, without first moving them around gently, you'd make sounds. The same thing would happen if you didn't put your toes down first. We

used the same methods for walking through water. On dry leaves, we'd sling our weapons over our backs and move in a bent-over position using hands as well as feet. We were taught to move the dried leaves away with our hands, then pull our feet up underneath our palms so that we wouldn't step on the leaves. We'd keep moving that way until we neared the objective. Time made no difference. In training, it might take two or three hours to crawl like this through five fences of barbed wire.

. . . [To reconnoiter a position from the inside,] we would first have to crawl through the barbed wire, but without cutting the wire or removing the mines—we couldn't leave any traces. We were supposed to tie the wire up with string and mark the mines on the route we were following.[66]

— Nguyen Van Mo, 40th Mine Sapper Battalion

Eastern Soldiers Regularly Practice Movement Techniques

To survive a firefight, the U.S. private must have the strength and ability to move around below the bullets—i.e., to crawl. Yet, this ignominious skill is not stressed nearly as much in the West as it is in the East. Only on occasion is the American recruit exposed three methods: (1) on hands and knees, (2) fully prone with opposite hands and feet, and (3) slowly pushing the body forward after elevating it on toes and elbows.[67] He gets so little practice that he never really develops his crawling muscles.

The Eastern private routinely practices moving around on all fours. He has many techniques. (See Figure 7.2.) He even learns how to obscure his crawling. By U.S. admission, the WWII Soviets employed screening smoke even at individual level.[68]

Urban Movement

Through built-up terrain, any type of movement can be extremely dangerous. There are too many openings through which one can be seen or silhouetted.

According to *TM-E 30-451,* the WWII Germans moved through back yards and over rooftops, rather than up streets.[69] The WWII

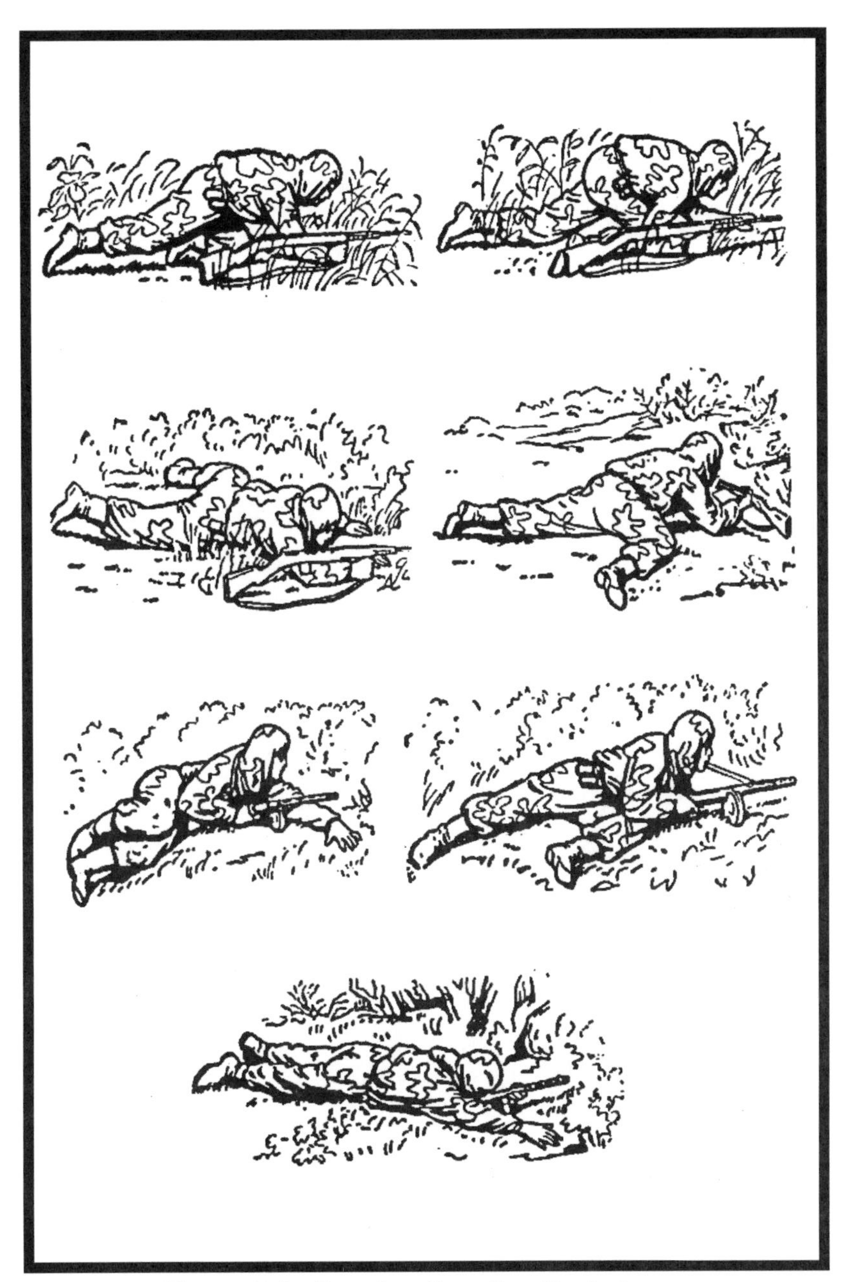

Figure 7.2: Russian Crawling Techniques

(Source: *Podgotovka Razvedcheka: Systema Speznaza Gru*, © 1998 by A.E.Taras and F.D. Zaruz, p. 150)

Russians also avoided moving up streets. Instead, they were instructed to move through yards, gardens, and holes in the walls.[70] Chapter 20 will discuss these issues in depth.

Guarded Communication

- *Do electronic communications have limitations?*
- *Would old-fashioned signaling sometimes work better?*

SOVIET "MOTOR RIFLES" PRIVATE, SUMMER 1975

(Source: Courtesy of Cassell PLC, from *World Army Uniforms since 1939,* © 1975, 1980, 1981, 1983 by Blandford Press Ltd., Part II, Plate 9; *FM 21-76* (1957), p. 110)

Electronic Warfare Has Its Deficiencies

Without radios, there is little need for encryption. In August 2002, some 13,500 U.S. military personnel completed Operation Millennium Challenge—a full scale Joint Forces Command wargame. In charge of the enemy forces was the previous head of the Marine Corps Combat Development Command. Though greatly constrained, the retired general "used motorcycle messengers to transmit orders, [thereby] negating the Blue Forces' high-tech eavesdropping capability."[1] He had once more demonstrated how easy it is to counter electronic-intelligence gathering. To confound electronic eavesdropping equipment, a "low-tech" army has only to revert to the nonelectronic ways of communication.

Over the years, the American intelligence community has focused almost entirely on the opponent's order of battle and equipment. With little personal infantry experience, its members could only gingerly assess the enemy's operational ability, and rarely discuss his tactics. Compulsorily loyal to the official record, they never suspected the validity of another interpretation. Oblivious to the ongoing evolution in small-unit tactical techniques, they assumed that all armies used the same ones. Unfortunately, Eastern armies thrive on deception and short-range combat. They routinely capture a Western adversary's attention with large units while destroying his strategic assets with small units. That adversary's advanced technology only makes the first step easier. Whereas before the Easterner had to construct elaborate dummy positions, he now has only to create a myriad of misleading electronic signatures.

The Asian Way of Doing Things

In the Far East (where Sun Tzu first expounded upon the advantages of surprise), anything overt is a ordinarily a feint. The same holds true for the region's tactical communications. Any signaling between tiny contingents is discreet (Figure 8.1). During a short-range encounter with an opposing force, the Eastern squad or platoon leader will seldom telegraph his intentions by shouting instructions or setting off strange-colored pyrotechnics. Only for large-unit operations are pyrotechnics normally employed. In August 1945, red "rockets" helped to coordinate the Soviet crossing of the Amur River in Manchuria.[2] Some 23 years later, attacks on Con Thien and Hue City were each coordinated by a single red star cluster.

The Oriental leader likes to stack the odds in his favor. He has his soldiers brainstorm and rehearse every engagement before it is fought. By the opening round, those soldiers have devised and practiced a response to every counterstroke. This gives them the option of using—as their signals—what a highly predictable enemy will invariably do under certain circumstances.

Early Eastern Signaling Codes

Since the invention of the machinegun, Eastern infantry units have preferred to operate silently whenever possible. Where their numbers were many or lighting conditions poor, they would rarely resort to audible signals. The Japanese night-fighting manual of 1913 talks of troops in column passing the word through hand slaps to their rifle butts.[3] As late as 1953, the Japanese, Chinese, and North Koreans were using whistles, bugles, and shouts to control their human-wave feints. They wanted the defender's undivided attention and may have believed the sounds unsettling. As Western firepower increased in subsequent years, their southern neighbors found less-obvious ways to coordinate their holding attack.

To the Western soldier, rifle shots are singular in purpose and commonplace in nature. He can't tell the difference between the report of an M-16 and an AK-47. However, the Eastern soldier can differentiate between the two sounds. To make matters worse, he protects the element of surprise by seldom firing his weapon. He instinctively knows that any rifle shot is either Western in origin or has special significance. In short, he has discovered that gunshots make good signals. So, to warn his unit, he will often use his rifle. As many a tree-climbing Japanese sentry found out the hard way during WWII, that kind of signaling has its down side around a sniper-oriented opponent. Still, Nipponese outposts would use numbers of shots to designate size or distance of an approaching American force. On Guadalcanal, Henderson Field infiltrators regularly signalled with their rifles when their radios failed. Most were probably artillery or naval-gunfire forward observers.[4]

> Japs who have infiltrated signal to each other with their rifles by the number of shots.[5]
>
> — Best NCO to Chesty Puller, *FMFRP 12-110*

Some 25 years later, the NVA would occasionally use another type of "bang" to signal an assault. According to POW Nguyen Van Mo, their lead sappers would explode a "signal [fire] cracker" once they had successfully infiltrated an Allied perimeter.[6]

Even more pronounced as a warning would be automatic-weapons fire. It seldom happens by accident. When the Soviet forward security posts fired their weapons on full automatic in

Afghanistan, they wanted the main-position defenders to man their posts. When they fired on semiautomatic, they were just alerting or probing.[7]

Of course, Eastern armies also have quieter ways of communicating. To signal to each other while walking single file through the blacked-out jungle, the Malaysian guerrillas of WWII made "a clicking noise between the upper teeth and side of the tongue—the sound to encourage a horse." It was a noise that would carry for a considerable distance without attracting undue enemy attention. One click meant "stop or danger," while two clicks meant "go on or OK." For a rallying signal, they used a non-indigenous bird call.[8]

The Eastern Way of Visually Signaling

For many of the same reasons, Easterners avoid ostentatious signaling. The 1913 Japanese night-fighting manual showed several ways in which patrol members could visually communicate after dark. One was by flag or light. Included were white cloth, matches, match-cord, and darkened lanterns. Of note, a physical object or human body had to be interposed between the light source and the enemy. Those light sources generally faced backwards.[9] So as not to miss a signal or lose sight of the man in front, the Japanese soldiers were taught never to hang their heads while moving.[10] For the Nipponese patrols, still another way of communicating was by example—everyone did exactly what the patrol leader did.[11]

As Easterners show their foe only what they want him to see, they like to use their own overt movements as signals. They then accomplish their follow-up secretive maneuvers by prerehearsed battledrill. That would make the completion of the previous step the signal for the next. For example, on the night of 27 November 1950 at the Chosin Reservoir, a row of specially trained crawling skirmishers preceded each of the Chinese assault columns.[12] Their job may have been to support the main attack with concussion grenades.[13] The assault force itself was composed of alternating rows of burpgunners and grenadiers.[14] If the initial exchange between opposing forces consisted of grenades only, the first wave would run in using their bayonets.[15] If there was any machinegun

Figure 8.1: There's No Talking on a "World-Class" Patrol

(Source: *FM 7-8* (1984), p. 5-22)

fire, the first wave would charge in using their burpguns.[16] It was as if the defenders' reaction signalled the assault force's next move. As Asians like to make a plan for every eventuality, they would be more likely to think of built-in signaling.

In Vietnam, other forms of inherent signaling were apparent. At the outset of hostilities at Ia Drang,[17] Operation Meade River,[18] and Operation Buffalo,[19] two enemy soldiers ran away from the American lead element—thereby suckering it into a trap. By the very act of running, those scouts could have signaled the ambushers that the Americans were in hot pursuit.

Eastern defenders can also easily pull back under fire. Their routes have been prerehearsed, and the cover party knows how many people to expect. Simply by counting noses, that cover party could tell when to fire at any subsequent movement. That's a type of battlefield communication.

Verbal Signaling

The 1913 Japanese night-fighting manual contained another way of communicating between patrol members after dark. It was by verbal relay—each successive man in column would simply tell the next man what to do.[20] Of course, he did so with a minimum of fanfare.

When Japanese skirmishers were discovered crawling up on a U.S. position at the beginning of WWII, they would shout "banzai" to signal an all-out assault. By the time Peleliu, Iwo Jima, and Okinawa rolled around, they had been pretty much weaned of the habit.

Late into the Vietnam War, North Vietnamese skirmishers would sometimes call out for a corpsman in English. This was very much like the Japanese ruse in the semidocumentary, *Sands of Iwo Jima*. On defense, the NVA conversed over telephone "land lines."[21]

Many of the Old Ways Are Once Again Viable

To wage electronic warfare, today's soldiers have only to do what their predecessors did. When worried about radio intercepts, they must practice radio silence and revert to more primitive methods. At the top of the list are communications wire and messengers. In Grozny, the Chechen rebels defeated electronic eavesdropping with messengers.[22] In Afghanistan, the al-Qaeda has recently done likewise.

> [The] al-Qaeda leaders greatly reduced their time on telephones and radios after realizing the United States' unmatched technical ability to monitor voice communications. During the summer, the military found a large cache of brand-new satellite phones—unused. This signaled that al-

Qaeda fighters have found other ways to talk without being detected, a Pentagon official said.[23]

— *The Washington Times,* 15 January 2003

Almost any orchestrated event can qualify as a signal for what has been planned next. During a firefight, the distant impact of a rifle grenade can serve as a signal to a friendly maneuver element or base of fire. If carefully placed, that same grenade will also distract the enemy.

Communications Wire Has Other Uses

During one WWI night attack, young Rommel had his lead elements unroll communication wire to help the rear elements find the objective. In the Soviets' strike into Manchuria in 1945, they did the same thing.[24] That very wire could have been used to transmit "tug" messages over a short distance. Of course, this plan has its downside, if the enemy has patrols out.

As will be seen in a later chapter, the North Vietnamese liked to defend in parallel lines instead of circles. With a compass and communication wire, they could have more easily established a hasty defense on a dark night. After a chance contact with an estimated NVA battalion in late 1966, D Company, 1st Battalion, 4th Marines encountered a long line of enemy holes connected by communication wire.[25]

More on Messengers

The Japanese night-fighting manual of 1913 also contained detailed instructions for messengers. One could infer that every Japanese private was formally trained in how to do the following: (1) map read, (2) land navigate through terrain association, (3) land navigate through celestial dead reckoning, (4) use an offset to avoid an obstacle, (5) use an aiming point to traverse a depression, (6) judge the tactical aspects of terrain, (7) cross danger areas, (8) memorize the physical objects along the route, (9) imagine how much shadowing would affect the route after dark, (10) leave discreet route markings, and (11) correctly relay a message.[26] By providing every private with such sophisticated land navigation skills, the Japanese army also garnered a world-class, forward-security-element capability.

In the Russian technique manuals of WWII, there are also instructions for messengers, albeit considerably less detailed than the Japanese. It would appear that the Eastern armies have better prepared their troops for nonelectronic signaling.

Discreet Force at Close Range

- *Do Eastern soldiers shoot their small arms less often?*
- *What are their weapons of choice at close range?*

VIET CONG "RPG" MAN, 1960'S-70'S

(Source: Courtesy of Osprey Publishing Ltd. from *Armies of the Vietnam War 1962-75*, Men-at-Arms Series, © 1980 by Osprey, Plate C, No. 2; *FM 21-76* (1957), p. 68)

Minimal Force Generates More Surprise

According to Sun Tzu, "The supreme art of war is to win without fighting."[1] Eastern operations and tactics are largely based on this paradox. Wars are won by destroying a foe's strategic assets and will to fight, not his people. That's why the *ninja* infiltrator avoids confrontation.[2] Any veteran of intense combat can confirm that his survival depended more on calm professionalism than impromptu machismo. Contemporary U.S. privates should not try to emulate the stand-up heroics of Hollywood films, nor be fooled by their martial-arts training. Most of the true heroes are dead.

Flexibility and naturalness are the basis of *ninjutsu,* not heightened aggressive drives or superhuman strength.[3]

The Local Population Plays a Pivotal Role in any Battle

A guerrilla-based army goes to considerable effort to enlist the support of the local populace. Not all of this effort is brutal. Lacking the money to rebuild the region's economic infrastructure, that army will often preserve it. Within a few days of taking Buon Me Thuot in 1975, the NVA had provided its residents with enough supplies, air raid shelters, and public order to reopen their shops and markets.[4] When the legitimate government then indiscriminately bombed the town, the people had more trouble distinguishing friend from foe. An iron fist is precisely what the Communists had counted on.

> In order to subdue the occupied territory, the [Western] enemy will have to become increasingly severe and oppressive *(Mao Tse-tung on Guerrilla Warfare,* 1961, 107). (Obviously this would play into the hands of the guerrilla because the people would then see him as their savior.)[5]
>
> — Maj. Frank D. Pelli USMC

Winning the Hearts and Minds of the People

The indigenous population of a region can be of great help to its occupiers. In Korea while Chinese commanders would regularly execute incumbent officials, the average Chinese soldier was taught to "care for and love the Korean people's interests in all respects."[6] That is not to say that atrocities did not happen.

Apparently, the NVA followed this same *modus operandi.* An enemy POW remembered the following instructions:

> Our motto is: The Army is welcomed wherever it goes, and loved wherever it stays. We will never touch even a piece of thread that belongs to the people. What belongs to the people remains theirs, and if by mistake it's damaged by us it will be compensated.[7]
>
> — an NVA prisoner

In Afghanistan, however, the Soviets were not as careful with community cohesion.

> The Soviets apparently showed little concern for the civilian population and started each sweep with an artillery bombardment. This did not win many hearts and minds for the Soviet forces. Often the Soviet effort seemed deliberately aimed at killing civilians and forcing them out of the rural areas.[8]
>
> — DoD-published Soviet military academy study

Standard procedure for a unit's headquarters will affect standard procedure for its lower echelons. If an infantryman sees his commander's supporting arms hurting noncombatants, he will be less discriminate with his small-arms fire. Not surprisingly, the soldier from a firepower-poor, surprise-oriented army more often shows mercy to any enemy picket.

Sentry Removal

Even before WWI, Asian armies preferred to capture, rather than shoot, an enemy sentry. They were simply trying to preserve the element of surprise.

> If the enemy's sentries are encountered, capture them (without firing) or kill them with the bayonet, but it must be done without noise.[9]
>
> — Japanese night-fighting manual, 1913

To acquire replacements and dispel rumors of barbarism during the Chinese Civil War in the 1930's, Mao Tse-tung encouraged the taking of POWs.[10] Of course, also central to his writings are the precepts of encirclement and annihilation.[11]

While attacking during WWII, the Russian infantryman was given the option of taking a defender captive.[12] He was also encouraged to do so on patrol.[13] Still somewhat puzzling to the U.S. editor of *The Bear Went over the Mountain* was why a Soviet platoon authorized to kill *mujahideen* in ambush would ask for their surrender.[14] He had apparently forgotten the rationale behind the WWII instructions.

> In the attack the rifle squad has the duty to . . . quickly close with the enemy, to attack him with hand grenades and in close combat destroy him or take him prisoner. . . .
>
> The platoon will get as close to the enemy as possible, assault without battle cries and using bayonets and grenades. . . .
>
> Enemy security posts and scouts are taken prisoner or destroyed with bayonets, without opening fire.[15]
>
> — *Soviet Combat Regulations of 1942*

Pictured in the Russian Spetznaz manual are a silenced semi-automatic rifle and crossbow with telescopic sight.[16] Also shown are a silenced automatic rifle and a crossbow with some type of payload at the end of its arrow. That payload could be either shape charge or fuel air explosive. As Eastern armies often detach commandos to line outfits, U.S. point elements should expect the worst.

Offensive Concussion Grenades

The next best thing to total silence on offense is a feigned mortar attack. At the outset, it drives defenders deep into their holes. The most plentiful weapon for such a ruse would be the grenade. Yet, not just any grenade would do. To keep from injuring the assault force, those grenades would have to be of the "concussion" variety. Only Eastern armies routinely issue grenades that don't throw shrapnel.

At the start of WWII, the Japanese had grenades with serrated casings (for enhanced fragmentation), grenades with smooth casings and wooden handles (presumably for better distance), and grenades with thin, smooth casings (probably for reduced fragmentation).[17]

Figure 9.1: Bayonets Work

(Source: *OPNAV P34-03* (1960), p. 406)

During the same period, the Germans used the famous "potato masher." It had a hollow wooden handle and thin, sheet metal head. To improve its antipersonnel effect, one had to add a fragmentation sleeve. In other words, the grenade could be used on either offense or defense. The Germans also had wooden and concrete "offensive-type" grenades.[18] "[The wooden] grenade is designed to produce a blast effect and may be used by troops advancing in the open."[19] Even the egg-shaped, high-explosive grenade had a thin, sheet metal casing.[20] It must have been intended for something other than killing.

Almost a decade later, the Chinese threw concussion grenades at the Chosin Reservoir.[21] As late as 1960, their manuals showed a grenade with a serrated-body for defense and one with a thin, smooth casing for offense.[22]

During the Vietnam War, U.S. fighting men thought their opponents ignorant for throwing satchel charges that were devoid of shrapnel.[23] Only a few of those Americans—decades later—would discover that their adversaries had been "fox-like" in their stupidity.

During the late 1980's, the Soviets were still using concussion grenades on offense in Afghanistan, just like their WWII predecessors had.[24] While entrenched, they would hunker down to escape the shrapnel from the defensive fragmentation grenade.

> Soviet grenades came in three basic varieties. The offensive grenades . . . were similar to the German "Potato Masher" types, and could be thrown further *[sic]* than they could themselves throw fragments back . . . , so you could throw the grenade and run right up behind it. The defensive grenade was egg-shaped and segmented, like the American "Pineapple," and could throw lethal fragments further *[sic]* than a soldier could throw the grenade. . . . Specialized grenades included smoke-producing and incendiary types, and most importantly: antitank grenades.[25]
>
> — *Soviet Combat Regulations of November 1942*

Currently displayed in Moscow's Armed Forces Museum is a Russian grenade holder (with its own throwing handle) into which five or six offensive grenades can be mounted.[26] Such a device would be of great use to a ground assault as it would make an explosion more like that of an artillery shell. Also on display was the fragmentation sleeve that can be attached to the offensive grenade. Even the Spetznaz manual shows the various throwing stances for this grenade.[27]

Minimal Shooting of Small Arms on the Assault

There are many ways to prevent fratricide during an assault. One can go through the complicated procedure of shifting the base of fire, aligning the troops, and illuminating the objective. Or one can simply disallow the use of small arms and fragmentation grenades altogether. If the assault troops have been restricted to concussion grenades and bayonets (Figure 9.1), the base of fire no longer needs to be shifted, the troops aligned, or the objective illuminated. With total surprise, one doesn't need a base of fire. With no bullets or shrapnel being generated, one doesn't need alignment or illumination. If the defenders happen to be American, British, or French, they will now be much easier to identify. They will be the only ones producing muzzle flashes and tracers.

As early as 1913, Japanese troops were taught to assault with bayonet alone at night so as not to compromise surprise.[28] In fact, they were often not even allowed to load their weapons. The emphasis was on hand-to-hand combat with bayonets.[29] Even if they were fired upon during that assault, the Japanese were told not to shoot back.[30] Sometimes non-infantry troops were tasked with throwing grenades at the moment of the charge.[31] The Asians were not alone in their common-sense approach to short-range combat.

During WWI, German "stormtroopers" withheld their small-arms fire in the assault. They used bayonets and grenades to further the bombardment ruse.[32] Their successors followed the same method. Many of the German divisions at the Bulge and Berlin were *volksgrenadier* or "people's infantry."[33] Throughout WWII, German squads preferred "cold steel" to shooting their weapons in the assault.[34] They also liked to throw grenades just before any assault on an entrenched enemy.[35]

The Russian and Japanese riflemen of that era were again told *not* to fire during a night assault.

> For a night attack (attack in smoke) the moment of surprise must be utilized to the fullest extent to destroy the enemy. Under conditions of total silence the squad quickly

> approaches the enemy, attacks him without opening fire and without battle cries and destroys him with bayonets or hand grenades.[36]
>
> — *Soviet Combat Regulations of November 1942*

> The infantry assault [during the night attack] is with the bayonet without firing.[37]
>
> — "Handbook on Japanese Military Forces"
> U.S. War Dept., *TM-E 30-480*

If the Chinese lead element at the Chosin Reservoir could get close to Marine lines, the first wave assaulted silently—without any shouting, shooting, or explosives.[38] Those that got in bayoneted rather than shot their immediate adversaries.[39] Only if the lead element could not get close, would the first wave come in throwing concussion grenades behind a rolling mortar barrage,[40] or shooting burpguns under overhead machinegun fire.[41]

Some 20 years later, NVA infantrymen were again instructed on how best to preserve the element of surprise during an assault.

> Success in the [NVA] attack is dependent on being able to breach the perimeter undetected. The assault is violent and invariably from more than one direction. It begins with a preparation, usually mortar and RPG fires. . . . Small arms are not employed except to cover the withdrawal in order to avoid disclosing the location of attacking forces. Once defending troops are forced into the bunkers, penetration of the perimeter is effected. Mortars cease firing, but the illusion of incoming fire is maintained through the use of RPG's, grenades, and satchel charges.[42]
>
> — *NVA-VC Small Unit Tactics & Techniques Study*
> U.S. Department of Defense

Not firing one's small arms can help in many ways. One has more maneuverability and less of a logistical burden. One can also tell who and where the enemy is. He is the only one shooting.

A Greater Emphasis on Close Combat

Eastern armies define "close combat" differently than their Western counterparts. For the Soviets, it implies hand-to-hand or bayonet fighting.[43] While the Western soldier can shoot his opponent at up to 500 yards, the Eastern soldier is ready to get more personal.

In the assault, the WWII German *gewehrschutzen* (or riflemen) were to use their bayonets.[44] After the initial penetration, they were expected to engage in hand-to-hand fighting.[45]

The Japanese also preferred close contact. In the dense jungle on Guadalcanal, grenade-throwing ability was at a premium.

> Most of the fighting here has been carried out at extremely close range, and there has been as much throwing of grenades as . . . firing a weapon.[46]
>
> — Guadalcanal veteran, *FMFRP 12-110*

Once inside an objective, the Chinese also preferred grenades, bayonets, and hand-to-hand fighting.[47]

Eastern Assault Forces Bring Along Their Fire Superiority

During an assault, a Western platoon will routinely position its machineguns outside the objective as a base of fire. Its Eastern counterpart is less likely to leave them behind. During WWII, each German squad got as close as it could to its objective without shooting. It then carried its light machinegun on the assault.[48]

Paradoxically, the Soviets and Chinese give their assault troops more automatic weapons. In WWII, Soviet submachinegunners would penetrate enemy lines (by infiltration or atop a tank) and then attack those lines from the rear. Small-arms fire from the rear would do less to compromise surprise. Every other row of Chinese at the Chosin Reservoir may have had burpguns to establish fire superiority if the grenadier rows were prematurely discovered.[49]

While the Soviet AK-47 has a selector that goes from "safe" to "full automatic" to "semi-automatic,"[50] the American M-16 will only permit three-round bursts. Once defenders had begun to fire, assault troops would have less reason to limit their small-arms fire.

> The main tasks of the submachinegunner:
>
> - to operate on the flanks and rear of the enemy . . .
> - to infiltrate through enemy lines.[51]
>
> — *Soviet Combat Regulations of November 1942*

Weapons of Choice during Chance Contact

During a chance encounter with an opposing force, one would prefer to remain hidden. As the trajectory of grenades can be difficult to ascertain, Easterners rely on them heavily.

> Troops must receive a high degree of individual [skill] training. . . . Individuals must have thorough practice in throwing hand grenades in woods.[52]
>
> — Guadalcanal veteran, *FMFRP 12-110*

The Eastern unit also has another solution. It will use automatic-weapons fire as a diversion when performing a double envelopment of a smaller opponent. Then while closing the noose, its members will only take random, well-aimed shots. That way, the encircled troops have great difficulty pinpointing the source of the fire. The 1913 Japanese night-fighting manual tells the troops to avoid continued individual fire and shooting all at once for long. The idea was to keep the enemy from pinpointing any particular fire source.[53]

The Same Principle Applies to Ambushing

The Soviets did not use the "claymore and grenade" ambush in Afghanistan.[54] However, U.S. veterans of WWII, Korea, and Vietnam, did encounter command-detonated mines and unexpected hand grenades.

The whole idea in modern war is to neutralize one's immediate adversary without attracting the attention of his buddies. While it may be more expeditious to shoot him, one should never lose sight of the fact that he may have a thousand comrades over the next rise. Sun Tzu may have preached minimal force for this reason alone.

The Moral Issue

On every battlefield, one must win the hearts and minds of the local inhabitants. They provide the best intelligence. As an Eastern army routinely fluctuates between guerrilla and mobile warfare, its Western opponent will always be tempted to consider every inhabitant as hostile. Undertrained Americans had little choice but to fire indiscriminately to extract themselves from Mogadishu in 1993.

> Late in the afternoon of Sunday, Oct. 3, 1993, attack helicopters dropped about 120 elite American soldiers into a busy neighborhood in the heart of Mogadishu, Somalia. Their mission was to abduct several top lieutenants of Somali warlord Mohammed Farrah Aidid and return to base. It was supposed to take about an hour. . . .
>
> Instead, two of their high-tech UH-60 Blackhawk attack helicopters were shot down. The men were pinned down through a long and terrible night in a hostile city, fighting for their lives. When they emerged the following morning, 18 Americans were dead and 73 wounded. . . .
>
> Carefully defined rules of engagement, calling for soldiers to fire only on Somalis who aimed weapons at them, were quickly discarded in the heat of the fight. Most soldiers interviewed said that through most of the fight they fired on crowds and eventually at anyone and anything they saw. . . .
>
> Official U.S. estimates of Somalian casualties at the time numbered 350 dead and 500 injured [many of which were women and children]. . . . The Task Force Ranger commander, Maj.Gen. William F. Garrison, testifying before the Senate, said that if his men had put any more ammunition into the city "we would have sunk it."[55]
>
> —*Philadelphia [Inquirer] On Line,* November 1997

More recently in Iraq, many U.S. forces "shot at anything that moved" during the initial onslaught.[56] Throughout the drive on Baghdad, each evening newscast brought more pictures of burned-out cars. Surely, not all could have contained irregular troops or suicide bombers. No one stopped to verify the combatant status of the 2000-3000 Iraqis killed during the first American "thunder run" into Baghdad on 5 April 2003.[57] This may be the first war in which U.S. infantrymen sighted in their rifles every time they changed location. While good for a contested assault, this technique may overfacilitate deadly force at other times. With more skill comes better target selection and ways to win without shooting.

10 Combat Deception

- *How do Eastern troops appear from nowhere?*
- *Why can't U.S. intelligence detect them?*

SOVIET SENIOR SERGEANT, SUMMER 1965-79

(Source: Courtesy of Cassell PLC, from *World Army Uniforms since 1939*, © 1975, 1980, 1981, 1983 by Blandford Press Ltd., Part II, Plate 2; *FM 21-76* (1957), p. 88)

Overpredictability Has Its Price

According to Sun Tzu, "All warfare is based on deception."[1] Yet, battles can occasionally be "won" without it. An affluent army will often depend more on firepower than surprise. While U.S. forces have deception manuals, they are far more secretive before each battle than during its composite engagements. They try harder to mask the main landing point than any subsequent small-unit maneuvers. Unfortunately, those small-unit maneuvers cannot generate much surprise without differing from the published example. Too much initiative on the part of a low-level commander might jeopardize the overall plan. So U.S. infantry platoons and squads have become almost totally predictable. The only ruse built into their standardized attack is placing the base of fire to one side of the avenue of advance. Even that is disallowed after dark (when the base of fire must shoot directly over the heads of the assaulting troops).

At the division, regimental, or battalion level, this inherent lack of surprise seems easily remedied by more firepower. However, at the platoon level and below, it has dire consequences. In essence, American squads, fire teams, and riflemen have been "bum-rushing" alert and intact enemy machinegunners since 1918. Those machinegunners were alert because of the preparatory fire that precedes every U.S. ground assault. They were intact because of the dummy positions and bombproof shelters that protected them from the shelling.

Less Powerful Nations Practice More Small-Unit Deception

That U.S. forces consider bottom-echelon deception disruptive does not mean that all armies do. Some of the fiercest fighting of World War II occurred along the Eastern Front. Every Russian infantry leader (down to platoon level) was required to have a deception plan.[2]

Among the oldest tricks is masking one's own intentions while waiting for a foe to make the first move. Another is to allow that foe to think he has won; for it is then that he will let down his guard. Both are included in the ancient Oriental "36 Strategies for Deception,"[3] and both are practiced by modern-day *ninjas*.[4]

There's a Style of War That Fools the Foe at Every Level

From Asia has come a way of fighting based almost totally on deception. Much of this deception occurs at the lower echelons, where Western-style forces would least expect initiative. Communist writers, in particular, have closely protected its details. Still, some of those details have already been assimilated by other armies. By the outbreak of WWII, German advisors had served with the Japanese Army (off and on) since 1872.[5] The Russians have their own Oriental heritage. Both nations have come to understand the Asian style of war. America never has. More disturbingly, this way of beating a more powerful opponent has both evolved and become common knowledge throughout the Eastern World.

By overemphasizing large-unit, long-range combat, the U.S. military may have unintentionally lost its proficiency at small-unit, short-range fighting. This shortfall has not gone unnoticed by America's traditional foes. They may have arrived at their squad techniques through *yin / yang* antithesis. By design, many of those techniques may be the opposite of what U.S. forces would expect. For example, ambushers may assault when expected to pull back, or defenders may pull back when expected to hold the line. Published as guidelines rather than doctrine, these alternative techniques produce small-unit tactics that are difficult for Westerners to predict. If the enemy headquarters subsequently decentralizes control over training and operations, each of its subordinate units will apply those guidelines in a slightly different way. What then occurs across the board will appear to be "asymmetric" warfare—something for which overwhelming force might seem the only solution.

Central to Asian thought is holding an opponent's attention with "ordinary forces" while beating him with "extraordinary forces."[6] He routinely practices the "false face" and "art of delay."[7] This is true at both big- and small-unit levels. For example, the battle for Khe Sanh was nothing more than a feint.[8] To undermine the U.S. claim that it was winning the war, Hanoi had only to attack the cities in force. Issued by Hanoi in November 1967 and captured by the U.S. Army later that month near Dac To was a "plan of false battles."[9] In this plan, Hanoi appeared to be countering U.S. pacification efforts in outlying areas. It knew that American commanders would try hard to avoid the political embarrassment of another Dien Bien Phu. It further recognized that a "false attack" at Khe Sanh would lure U.S. forces away from the district capitals and most prestigious northern city—Hue. While historians have speculated for years about Khe Sanh being a feint, recently acquired North Vietnamese documents confirm it.[10] There was some tunneling beneath the airstrip's protective barbed wire.[11] The Communist campaign maps for Khe Sanh don't reveal any nearby troop movements,[12] like the maps for Dien Bien Phu did.[13] At Khe Sanh, the close combat occurred around outlying camps early in the campaign.[14] At the nearby Lang Vei Special Forces Camp, the enemy used two companies of light amphibious tanks (for the first time in the war) to heighten U.S. concern.[15] As at Dien Bien Phu, many shells hit the Khe Sanh airstrip. However, in 1968, their launching tubes were probably better dug into the surrounding mountainsides. In 60 days, they would have to endure 78,000 tons of U.S. bombs.[16] The North Vietnamese readily admit to losing half of one infantry regiment (roughly 500 men) to U.S. bombs during this period.[17] That they did not lose more is a credit to their ingenuity.

Easterners Rely on More Deception at the Tactical Level

The enemy in Vietnam required as much, if not more, secrecy on the easily accessible coastal plain. His attack was by the "three-pronged close-combat method (political, military, and work on U.S. troops)."[18] Most difficult to accept is how he was able to demoralize those troops. Under ultimate attack was the Americans' moral compass and self esteem. By having to rely more on bombardment than participation, U.S. forces alienated the very population they had come to save. To make matters worse, they were constantly reminded of their relative lack of skill at close range.

> [North] Vietnam's military art [is] to enhance the human factor so that a small number of troops can fight bigger forces, higher quality combatants may cope with numerical superiority, Vietnamese intelligence may overwhelm the weaponry and intelligence of the U.S. war-conducting machinery.[19]
>
> — *Tet Mau Than The 1968 Event in South Vietnam*
> The Gioi Publishers, Hanoi

For tactics, the enemy used a variation of the Russian double encirclement.[20] They drew U.S. maneuver forces to the outer circle to give themselves more freedom of action along the inner circle.[21] As former Marine Raider Samuel B. Griffith put it, "The enemy's rear is the guerrilla's front."[22] The ultimate goal of the NVA/VC was to destroy the strategic assets within U.S. camps. To their advantage would be doing so without appearing to attack. By keeping their activities secret, they could reuse their techniques over and over. Of interest would be which of the forces along the inner circle might be considered "extraordinary."

> For the inner circle, task forces units, commandos, specialized mortar gunners, guerilla *[sic]* forces on the "anti-U.S. snipers rings" attacked the U.S. forces wherever and whenever it was possible to attack.[23]
>
> — *The Tet Mau Than 1968 Event in South Vietnam*
> The Gioi Publishers, Hanoi

The "task forces units" were local contingents of regional forces or their "special-assignment-squad" attachments.[24] The VC had three kinds of troops: main force, regional force, and local militia.[25] The first two were primarily offensive in nature; and the third, defensive. According to Mao, guerrillas are formed into two types of units: combatant and self-defense. Combatant units are organized from platoons to regiments. At the company level and above, each has a military and political hierarchy.[26] Embedded—each within the next—were the following tactical areas of operation: (1) hamlets, (2) communes, (3) districts, (4) Bases, and (5) Zones.[27] (Notice the capitalization of the last two terms.) While each area recruited and controlled its own unit, those units were sometimes required to participate in a distant parent-unit operation. At the "Base" level may have existed a regimental hierarchy; and at the "Zone" level, a brigade.[28] Communes generally had 30-100 fighters.[29] The self-defense forces provided local security, intelligence collection, and policing. They may have also provided combat service support to combatant units passing through.[30] Within their respective areas, they monitored enemy activities, guided friendly forces, and guarded infiltration routes.[31]

Enemy "commandos" were those with the skill to secretly penetrate U.S. lines. As assault squads, they could enter under an indirect-fire deception. As sappers,[32] they could sneak through the barbed wire. They could also function as part of a "spearhead battalion" (roughly 350 men)—one that would enter the inner circle to attack a well-insulated target.[33] Local commandos routinely preceded spearhead battalions in the assault.[34]

The "specialized mortar crews" were those that could place pinpoint fire into a U.S. camp. To hit a target of strategic import, they would use surveying or preregistration. To walk rounds along a linear portion of U.S. lines, they had only to "direct-lay" their mortars from a defiladed position at one end of those lines. Their ability to precisely target U.S. fighting holes made possible a variation of WWI German stormtrooper assault technique. The 1917 method included shelling diversions to either side of, and then a timed barrage directly on, the planned point of penetration.

"Anti-U.S. sniper rings" were cells of skilled snipers. Their precise roles can only be guessed. As with the sniping of the U.S. 10th Army commander on Okinawa in 1945,[35] they could certainly disrupt the U.S. command and control apparatus. They could counter U.S. snipers. They could defend a small piece of real estate from afar. However, they could also cause a U.S. maneuver element to slow down or change direction. The North Vietnamese battle chronicle for the Tay Nguyen Campaign of 1975 brags about the ability to "attract the enemy forces in the direction of our choice, keep them moving according to our will and hold them back."[36] Why couldn't have VC snipers done the same thing on a smaller scale? By knowing the locations of guerrilla staging areas and main-force caches, the snipers could have helped to defend them. They could have shadowed U.S. maneuver elements or simply prepositioned themselves around any location of strategic significance.

More about the Inner Circle

U.S. installations were surrounded by an inner circle of "fighting villages, political pockets, guerrilla zones, and "anti-U.S. snipers' rings."[37] In concert, these four formations could cause U.S. forces to assume a defensive, passive posture.[38] They did so by giving U.S. "search-and-destroy" sweeps little to show for their offensive effort. Some facilitated attacks on U.S. installations, while others drew U.S. security patrols away from the staging areas for such attacks. Many of those staging areas must have been hidden within plain sight of (or right next to) U.S. perimeters. While adjacent-hamlet residents were thought to be there for protection, some probably had tunnels beneath the U.S. wire. The more distant staging areas were probably in hamlets inhabited by sympathizers, protected by boobytraps, and surrounded by snipers and killer teams. To find an active antagonist, U.S. patrols would have avoided any secure or mined area.

The "fighting villages" were hamlets that were more distant, heavily fortified, and ostensibly deserted. While concealed by vegetation, their peripheral trenchlines were so long as to protect the discreet occupant from aerial bombardment. As with today's North Vietnamese "Military Fortress" barrier along the Chinese border, these fortified villages must have formed a "soft, strongpoint defense." They were, after all, mutually supporting bastions that were little effected by losing a few. As with the "combat villages" of the modern Military Fortress, the "fighting villages" of the 1960's may have all had their own "vanish underground" capability.[39]

> The [Duong Minh Chau] Base [northwest of Saigon] was divided into 13 districts, each district was divided into communes and hamlets having their own fortifications, tunnels, and trenches; they formed a closed front line able to help one another when necessary.[40]
>
> — *Tet Mau Than 1968 Event in South Vietnam*
> The Gioi Publishers, Hanoi

Mao had written about the flatland guerrilla needing shifting, yet relatively secure, "base areas."[41] Note the lower case and what may very well be a misleading inference of dimension. Mao was talking about tiny impenetrable enclaves. His "base area" differs from the North Vietnamese "Base." While the latter may infer consolidation, it actually applied to a large region. It would appear that Communists have protected their battlefield methodologies through convoluted writing.

> To start a people's war with commune bases; the core forces are in the first place the communical guerilla *[sic]* and self-defense forces with a broad and direct participation of the people.[42]
>
> — 1961 Resolution
> Communist Central Board

> Guerilla *[sic]* bases and the system of fighting villages were increasingly expanded.[43]
>
> — *Tet Mau Than 1968 Event in South Vietnam*
> The Gioi Publishers, Hanoi

> [During the 1966-1967 dry season, the [Communists] . . . controlled . . . 700 communes, and 6,750 hamlets. On the other hand, the Saigon Administration controlled (to vari-

> ous degrees) only 5,400 hamlets out of a total of 16,293 hamlets in all South Vietnam.[44]
>
> — *Tet Mau Than 1968 Event in South Vietnam*
> The Gioi Publishers, Hanoi

The guerrilla base area for each commune must have rotated among (and existed beneath) that commune's ten or so fighting villages.[45] This was after all, the same way that maneuver elements operated.

> Few permanent camps [existed], especially within South Vietnam itself. In Laos, North Vietnam, and Cambodia, there were permanent camps, although as the war went on even these camps in "safe areas" tended to be underground. Within South Vietnam, each Viet Cong or North Vietnamese unit had several entrenched camps. Communist units would constantly move around, rarely staying in one place more than a few weeks, and usually never more than a few days.[46]

However, these flatland base areas could not have survived without some sort of district, Base, and Zonal infrastructure. Their support establishment was generally in the less-accessible forests and mountains. Yet, northwest of Saigon, it was below ground.

> [Within the Duong Minh Chau Base] were concentrated officials, personnel in government services, schools, factories, hospitals, storehouses, broadcast stations. . . . These centers [were] built along dried streams, below the forest foliage, and at half-underground level.[47]
>
> — *Tet Mau Than 1968 Event in South Vietnam*
> The Gioi Publishers, Hanoi

Early in the war by Communist admission, these centers were located within a vast network of communications trenches—to obscure their exact whereabouts.[48] Later in the war, many were connected by tunnel.

The "political pockets" were cadres secretly operating within Saigon-controlled areas.[49] The most useful members of these cadres owned property inside the cities or next to Allied perimeters.

Their homes provided the perfect staging area for covert attacks. By the end of 1967, there were 19 inner-city political pockets composed of 325 families—most near targets of strategic importance. In addition, there were some 400 hiding places for soldiers and armaments.[50]

"Guerrilla zones" (different from Zones) were parcels of real estate patrolled by "professional 'hunting teams' specialized in the fight against aircraft, tanks, armored vehicles, and GI's."[51] Guerrilla zones often contained these killer teams and very little else. Only seeming to hold something of strategic value, they helped to lure U.S. forces away from the guerrilla base camps. As when a few settlers moved between the firing ports of an undermanned stockade, guerrilla zone inhabitants tried to look more numerous than they really were. With few instructions and many places to hide, they could have easily captured the attention of a much more powerful opponent.

This Formation Has a Name

As shown in Map 10.1, the interspersion of fortified villages and guerrilla zones resembled a leopard's skin. Not surprisingly, that's what the North Vietnamese called the formation.[52] At the center of every ten or so spots was an underground guerrilla base camp.

> We controlled many areas in the mountains and plains alike, and had established our bases in the towns.[53]
>
> — Gen. Hoang Van Thai

The Asian Sucker Stance

The Asian will only allow his opponent to see what he wants him to see and then wait. After that opponent acts hastily on partial information, the Asian will complete his secret maneuver and invariably capture the momentum. This misleading posture is displayed at every echelon. The NVA created a diversion on the outer circle at Khe Sanh so they could more easily attack the inner circle at Hue. Once the inner circle had drawn sufficient Allied interest, gains were again possible along the outer circle. Khe Sanh was abandoned in July of 1968. On the outer circle, mobile war was possible; on the inner circle, guerrilla war was the norm.[54] However, as the first derives from the second, both remain mutually compatible. The Tet attacks were both from the inside out and from the outside in.[55]

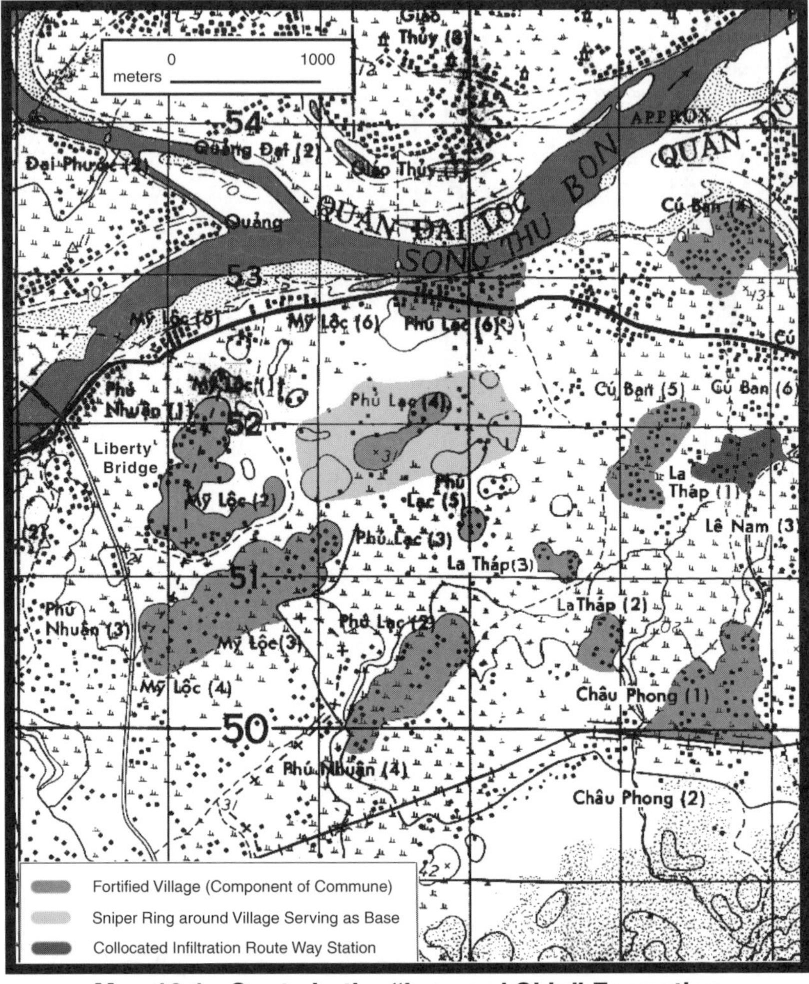

Map 10.1: Spots in the "Leopard Skin" Formation

(Source: 1:50,000 map of area northeast of An Hoa Marine Base)

While constant paradoxes make perfect sense to a Taoist, they confuse a Westerner. To fight an Eastern contingent of any size, that Westerner must come to know the secret maneuver associated with each sucker stance.

There is No Lack of Deception at the Eastern Squad Level

To whatever extent Americans are allowed to practice deception at the squad level, they must look to the East for inspiration. Concealment, by itself, does not qualify. True deception includes a misleading action or signal.

Most of what is written about war lacks tactical detail. It contains little if any individual or small-unit technique. To learn how an Eastern soldier fights, one must study his thought processes and recognizable actions, and then guess at his real intentions. To discover the inner workings of his ruses, one must work from the bottom up. Each bullet, explosion, or movement must be analyzed from the standpoint of deception. Then, the soldier ruses will predict the squad ruses and so forth up the chain. Of interest will be how and why each engagement was initiated.

Once again, the major Eastern armies—Japanese,[56] Germans,[57] Russians,[58] Chinese,[59] and North Vietnamese[60]—all discourage the shooting of small arms during a reduced-visibility assault (their most prevalent type). This helps them to alleviate the age-old problem of telling friend from foe after dark. Just by obeying their doctrine at night, Western riflemen compromise their safety. To deceive a firepower-dependent adversary, one used shots and explosions as ruses. Each category of enemy contact would have its own possibilities. Perhaps most often discussed is the "sniping incident."

The Sniping Deception

Because U.S. maneuver elements lack the skill to capture an objective quietly, they have unwittingly acquired a battlefield signature. To America's adversaries, the sound of shooting means the presence of U.S. ground troops. With one shot, a tree-climbing sentry could concurrently announce a U.S. unit and remove its primary decision maker. To the leaders of the American unit, one misdirected shot would connote accidental discharge or wildlife incident. On Guadalcanal, Chesty Puller's NCOs finally determined why so few Marines were being killed by the scores of Japanese snipers within the Henderson Field perimeter. Many of those "snipers" were supporting-arms forward observers. They were using shots to adjust fire when their radios failed.[61]

In late 1966 on Operation Deckhouse VI south of Chu Lai, a Marine rifleman got sniped from afar as a platoon from A Company, 1st Battalion, 4th Marines was about to enter a fortified but apparently abandoned village.[62] Why was that rifleman sniped instead of his clearly visible platoon commander? To "medevac" the man, the platoon had to cut short its sweep. Perhaps that rifleman was poking around something of importance.

While moving north toward the DMZ on Operation Buffalo in July 1967, the first sign of trouble for B Company, 1st Battalion, 9th Marines was the removal by a lone bullet of the lead platoon commander's trigger finger.[63] Promptly relieving himself of command, he walked to the rear of the company column. Thinking this the work of a nearsighted peasant, his platoon fanned out to check the area. As it neared the source of the shot, it was cut down by machinegun fire from both flanks. Inappropriate to the enemy's plan would have been a helicopter medevac. One "poor" shot had been enough to remove the primary decision maker and lure the unit into a trap.

On 18 August 1968, in what would become the final engagement of the third Tet Offensive, the commander of G Company, 2d Battalion, 5th Marines was barely missed by a small-arms round from the rear while his unit assaulted a contested fortified village.[64] Believing the round to have come from a lone enemy soldier just seen exiting the objective, the commander continued with his airstrike. Sunset was only an hour away. That NVA might have been trying to gain valuable time for his compatriots. After dark, they could slip undetected into a nearby stream.

If snipers had—from afar—kept an eye on infiltration waystations and guerrilla base areas, they could have discouraged GI interest in the entrances to underground facilities.

The Boobytrap Deception

To the well-supplied American, a mine is an instrument of killing, and a minefield is a place to be avoided. Not since the Revolutionary War has he had to flee a more powerful opponent. Several hundred yards north of the An Hoa airstrip lay a hamlet so heavily boobytrapped as to be routinely avoided by Marine patrols.[65] It has just recently been discovered that its residents had secretly tunneled right up to the edge of An Hoa's protective barbed wire.[66] A boobytrap maze will discourage either pursuit or further movement in the same direction.

How the Temporary Base Areas Were Protected

While functioning as temporary base areas, the fortified villages may have been surrounded by boobytraps and defended from afar by preregistered mortar or skilled sniper. Such snipers would have had a fourfold purpose: (1) watch for U.S. snipers, (2) sound the alarm should U.S. troops approach, (3) try to get those troops to slow down or change direction, and (4) protect from a distance the hidden entrances to the underground facilities.

When U.S. forces entered a heavily boobytrapped area in Vietnam, they were nearing something of strategic import to the enemy. When they followed "the sound of the guns," they were moving away from anything of strategic importance.

A Temporary Base Area Was Tough to Find

According to the Communist regime in Hanoi, American forces completed a 50-day operation (presumably northwest of Saigon) in May 1967 without finding a single guerrilla base area. With little to show for their effort, the U.S. forces purportedly traveled 15,000 kilometers.[67]

Should the occupants of a temporary base area be surrounded, they would man the surface fortifications until dark and then mysteriously disappear. The German brigade of the French Foreign Legion may have discovered their escape tactic in the early 1950's. The underground-hide facility in one village had two underwater portals.

> The tunnel had several exits but the guerrillas used an opening placed half way down the well. There was also an underwater exit into the river for emergency use. . . .

When the smoke cleared, Sergeant Krebitz surveyed the tunnels, altogether two hundred yards long. It was a well-built complex with several large chambers for stores and sleeping quarters. Krebitz counted three hundred bunks.
Three of the chambers were loaded to capacity with weapons and ammunition.[68]

The Larger Underground Facilities Had a Dual Role

At Tham Than Khe just south of the DMZ in December of 1967, L Company, 3rd Battalion, 1st Marines learned a hard lesson. On its second visit to the same village in 24 hours, the company ran into a fully deployed enemy battalion.[69] That battalion had been occupying an underground facility. It had no trouble escaping the subsequent U.S. encirclement. As the Japanese had on Okinawa,[70] the North Vietnamese established semipermanent transplacement routes with underground way stations. For such a way station, any of the larger fortified villages would have qualified.

All along these semipermanent "infiltration" routes were hidden supplies. They had come from Zonal, Base, and district "logistic storehouses."[71] In people's war, supplies are prestaged at the front and then assigned as needed. This alternative way of provisioning an army had facilitated the North Korean stalemate of 1953. Operating within South Vietnam by the end of 1967 were five full logistics regiments. Armament transportation was largely accomplished by volunteers. Around Saigon alone, tens of thousands of persons transported materials from the logistics bases to the suburbs.[72] Only while in use were these big caches actively defended.

The Elusive Base Area Served Another Purpose

By forcing a powerful adversary to cordon off a huge area to discover a single camp, one causes that adversary to widely disperse. He then becomes more susceptible to piecemeal destruction from the rear. As early as the 1930's, Mao Tse-tung would flood a region with tiny, loosely controlled patrols.[73] As soon as those patrols heard shooting, they knew where the opposition was. They could either attack it from unexpected directions or ambush the relief column.

The enemy in Vietnam may have used the same thought process to take tactical advantage of the Allied quest for guerrilla base camps. Both Mao and Giap had said, "He [the Communist] never allows the enemy to mass his forces against a lucrative guerrilla target."[74] In all probability, guerrilla zone occupants and anti-U.S. sniper rings needed no guidance from above to mount a concerted effort. They had only to do something to keep U.S. forces away from guerrilla camps and installations. In this way, the enemy fighters gained direction without forfeiting initiative.

The Ultimate Deception

As yet unrecognized in the West is the tactical strength of people's war. The Communists will run an occasional mass assault just to keep their opponent honest. That assault draws his attention away from the sappers and stormtrooper squads that are beating him. Their so-called "suicide" assaults aren't what they seem. During the "human-wave" assaults at the Chosin Reservoir on the night of 27 November 1950, the Chinese put roughly four times as many people at risk as the Americans,[75] and came away with only twice as many casualties.[76]

Paradoxically, a tiny people's war contingent can successfully engage a much larger and better equipped opponent. By U.S. estimate in February of 1968,[77] roughly 80,000 enemy fighters managed to attack—with varying degrees of success—almost all of the major Allied nerve centers in South Vietnam. Those were installations being defended by upwards of a million men. While suffering roughly the same number of casualties as the Allies (40,000), those badly outnumbered enemy forces created the illusion that they cannot be beaten. They can, of course, but not until U.S. forces decipher their deceptions and acquire commensurate small-unit skill. (See Appendix A.)

Frontline Deception Is Proportional to Allowed Initiative

There are many ways to fool an antagonist in any scenario. One of the more psychologically draining of the Easterner's tricks is to withdraw tactically at the last moment—thus leaving the bloodied Westerner with little to show for his efforts. Infantry veterans of

WWII, Korea, or Vietnam can remember fighting hard for a bunker, only to find it vacant. Normally, the occupant's withdrawal is rearward through trenches and tunnels, but not always.

> One time in Afghanistan, the *mujahideen* withdrew parallel to his forward positions instead of straight back.[78]
> — DoD-published Soviet military academy study

The Easterner has little trouble escaping an encirclement. As will be shown in later chapters, his method of fighting creates the expectation that he will occasionally be surrounded. Among his exit tricks are diversions, exfiltration, and tunneling. The diversion will often take the form of a mortar or artillery barrage. By U.S. determination, a Soviet barrage would often precede a grenade and bayonet attack in a different sector during WWII.[79] The Asian rear guard demonstrates with automatic-weapons fire or firecrackers at the fake exit point.

Many Westerners have yet to realize the Easterners' skill at exfiltration. According to *TM 30-340,* exfiltration was established procedure for the Soviets when all else had failed in 1943.[80] Squads or fire teams would escape one at a time through a previously discovered gap in enemy lines. In this fashion in 1968, the remnants of an entire NVA division used tunnels, rivers, sewage trenches, and VC-controlled corridors to escape—primarily to the south—the Hue City Citadel.[81]

> From 23-25 February, the main forces, the commandos, the local troops, the militia, and the wounded gradually and secretly withdrew [from the Citadel] to the liberated areas and the zonal bases in the Tri Thien mountain ranges.[82]
> — *Tet Mau Than 1968 Event in South Vietnam*
> The Gioi Publishers, Hanoi

Making Good One's Escape

Important to any breakout is getting away. Besides rearguard action, pursuit can be discouraged by various ruses. Among the world's best at such ruses are the North Koreans.

> When pursued by U.S. or ROK soldiers, [North Korean]

> agents withdrew in unison to the DMZ as fast as possible. They had been instructed to escape through high mountains and to step on rocks and grass to leave no footprints. To mislead pursuers, tree branches or small bushes were broken in a direction other than the escape route. A false trail was sometimes made with footprints and discarded items.[83]

Only an affluent Westerner fights upright at ground level. The undergunned Easterner will either climb, crawl, or dig. At Iwo Jima, he fought almost entirely below ground.[84] To do so, he must hide the entrances to his subterranean thoroughfares. At Khe Sanh, he spread out the digging spoils to look like freshly plowed fields.[85] More recently in Korea, he has caused them to look like the tailings from road construction.

> The road work and construction of fortification detected along the northern edge of the DMZ were apparently a part of a KPA deception plan to ensure that the huge quantities of spoil produced by these tunnels would not be spotted by ROK/U.S. reconnaissance.[86]

Squad-Level Defensive Ruses

Most productive of the defensive deceptions is hiding one's assets. Camouflage is only one way to do so. It can also be accomplished through dummy positions and covert maneuver. Other nations have had to rely more on these alternative means.

The undergunned Easterner has made an art form out of fooling aerial observers. He can make a pastoral scene look like a target and a target look like pastoral scene. The WWII Russian combat engineers were often tasked with building dummy positions for infantry units.[87] The contemporary North Korean gets double duty out of his equipment mock-ups.

> Pieces of dummy equipment are generally used [initially] for training [in North Korea]. They normally include replicas of tanks, field artillery, antiaircraft artillery, missiles, and air frames. Tank mock-ups are used as training devices for antitank training. . . . [T]roops practice demolitions or close-in approach and simulate destruction on the

> mock-ups. . . . The dummy runway probably serves as a decoy to attacking air crews . . . or a rehearsal area for demolition training.[88]

Among the defensive maneuvers is moving just to the rear of the original line of resistance. There are many variations. One is to shift to alternate positions or nearby shellholes after dark (as the Germans did midway through WWI).[89] Shoddily constructed obstacles, grass-stuffed uniforms,[90] and unmanned radio "squawk-boxes" would further the deception. (See Figure 10.1.) During WWII, the Germans often kept a machinegun behind their main lines that would not participate in the initial defense.[91] Asians prefer to hide below ground and come up behind the assault force. They also revel in confounding opposition close-air-support pilots. They do so by duplicating—at the wrong time and place—either the target designator round or friendly position marker. They need not determine the marker's color through radio intercept, just watch. In early 1968 while still an advisor to the South Vietnamese, Wesley Fox watched an NVA/VC unit do precisely that.[92]

Squad-Level Offensive Ruses

The most useful of the offensive ruses is to attack without appearing to attack. This is generally done by covertly entering the enemy position by stealth or during a bombardment. Easiest of the latter is the "dry airstrike." While nowhere in published doctrine, Americans have copied it since Korea.[93]

If the foe can't be secretly assaulted, then he must be confused as to the direction of the attack. In Afghanistan, the *mujahideen* attacked one morning with the rising sun at their backs.[94] While one would think Russian Spetznaz to be the equivalent of U.S. Special Forces, the Russian literature describes them as "diversionary" troops.[95] Perhaps a few are attached to each line outfit.

To attack without risk, one would trick the foe into surrendering prematurely. Staunch Anglophiles might be surprised to learn that the surrenderors of Singapore outnumbered their attackers by two to one.[96] During the drive down the Malay Peninsula in December of 1941, Japanese infantrymen would often melt into the jungle and subsequently "demoralize" an enemy roadblock into submission. They would sneak behind it and use fireworks and other

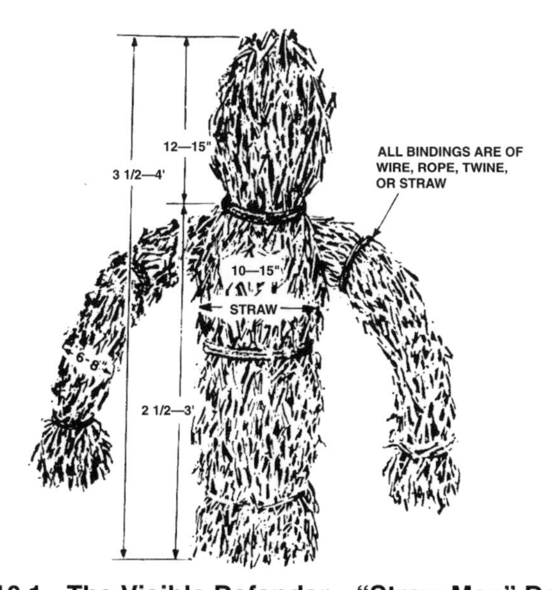

Figure 10.1: The Visible Defender—"Straw-Man" Decoy

(Source: Excerpt from *Chinese Manual on Field Fortifications*, in *A Historical Perspective on Light Infantry*, U.S. Army Combat Studies Inst. Research Survey No. 6, p. 72)

noisemakers to convince its defenders that they were surrounded by superior forces. In the resulting confusion, their tanks would overrun the British from the front, sometimes catching reinforcements on the road.[97]

Urban Ruses

As urban terrain differs from any other, it generates an unusual assortment of tricks. The greater profusion of confined spaces makes some quite lethal.

Much of what succeeds on offense in the city has been thoroughly reconnoitered ahead of time. By U.S. admission, the Russians have deployed urban infiltrators ahead of their main force since WWII.[98] The art of bypassing/silencing sentries has been studied and refined throughout the Eastern world since the time of Christ. Most easily accessible are the *ninja* techniques. They will be covered in more depth in Chapters 13 and 24. Where possible, there should be no grenade throwing or "pieing-off" through the same opening one plans to transit. To a seasoned defender, even a breaching blast will sufficiently telegraph the point of entry. The idea in urban fighting is to do the unexpected without fanfare.

While defending a built-up area, one's firing ports must look no different than any other opening. Further, they can't disclose any firing signature. WWII Germans would open all the windows in a building so as not to reveal which were being used. According to *TM-E 30-451,* they would also fire from deep inside a room or frequently change location.[99] At Hue City, the North Vietnamese did the same thing. Each square yard of one street in the Citadel was covered by thin bans of fire from recessed positions.[100] It was also covered by fire from beneath the bushes between buildings. The accomplished urban defender can employ so many ruses, mirrors, and hidden explosives as to make his building virtually impregnable. This subject will be revisited in Chapter 21.

11 One-on-One Tactical Decision Making

- *What does it take to win an unexpected encounter?*
- *Do Eastern privates get more decision-making practice?*

SOVIET MOTOR RIFLES PRIVATE, WINTER 1945-70

(Source: Courtesy of Cassell PLC, from *World Army Uniforms since 1939,* © 1975, 1980, 1981, 1983 by Blandford Press Ltd., Part II, Plate 4; *FM 21-76* (1957), p. 95)

The Rumor Is Wrong

Asian riflemen may normally lack facial expression and occasionally participate in a "human-wave" feint, but they are not mindless automatons—nor treated as such. That is the discriminatory vestige of U.S. wartime propaganda, and Eastern armies are in no hurry to dispel the myth. With fewer tanks and planes of their own, those Eastern armies must rely on highly opportunistic ground units. So as to be ready to make a strategic contribution to the war effort, each Eastern private is continually apprised of his commander's intent. Chinese "mobile" warfare grew out of, and easily transitions back into, "guerrilla" warfare.[1] That takes squads with enough skill to operate alone against a much larger opposing force.

> Those who understand big and small units will be victorious.[2]
> — Sun Tzu

To the extent that Asian entry-level training has been based on the *ninja* tradition, it encourages initiative among privates. (See Appendix B.) The *ninja* leaves what is known and journeys into what is unknown.[3] To do so, he must make spontaneous decisions.[4] This is his military heritage.

> Do not repeat tactics which gained you victory in the past, but let your tactics be molded by . . . circumstances.[5]
> — Sun Tzu

> Sun Zi [Sun Tzu] . . . stressed human beings' dynamic initiative in changing the disadvantaged into the advantaged, and the passive into the active.[6]
> — *The Strategic Advantage*
> New World Press, Beijing

It is more likely that the rank-conscious and firepower-dependent American military has caused its lowest grades to feel like easily replaceable trigger pullers.

> We [Americans] tend to think of our roles in the scheme of things as insignificant. . . . In the present nuclear age many people have the feeling that nothing can be done, and this leads to apathy and depression. What *ninjutsu* represents is a means of making even small actions significant. However, to do so means to understand the forces one is up against. But by gathering information and coming to terms with one's foes, one begins to see that like giving a small shove to an off-balance boulder to create a landslide, one person's action can have dramatic and far-reaching effects. This aspect of the *ninja* is probably one of their most appealing and important sides, and the reason many people have become attracted to their ancient, but effective, methods.[7]

For the American Private to Stand a Better Chance

Safely to operate in an Eastern war zone, the U.S. private must know more than just how to use his (or her) equipment and follow orders. When his woods-wise opponent suddenly appears at close range out of nowhere, he must decide quickly how best to fire and move across unfamiliar ground. In essence, he must be as good at tactical decision making over the next 50 yards as his company commander is over the next 1000.

Yet, to this day, U.S. privates are not trained in tactical decision making. As novices, they are thought to be better off following their leaders' orders than their own instincts. Unfortunately, that leader must base his decisions on overall impressions, whereas the private has situational specifics. Until this "not-so-slight" oversight is corrected, each American private must take it upon himself to discover the following: (1) his opponent's *modus operandi,* (2) when personally to take cover, (3) his opponent's weaknesses, and (4) how personally to capitalize on those weaknesses. He must learn how to avoid the enemy's initial "blow" and then land his own. Paradoxically, his opponent has been given more opportunity to achieve his full warrior potential. While Eastern armies may value the life of each soldier less, they depend on his strategic contribution more. As a result, Eastern privates are generally better prepared to survive.

> Whether in guerrilla . . . or limited regular warfare, waged artfully, it [armed struggle] is fully capable of . . . getting the better of a modern army like the U.S. Army. . . . This is a development of the . . . military art, the main content of which is to rely chiefly on [each] man, on his patriotism and . . . spirit, to bring into full play [through him] all weapons and techniques available to defeat an enemy with up-to-date weapons and equipment.[8]
>
> — Gen. Vo Nguyen Giap
> *The Military Art of People's War*

Things Are Not Always As They Seem

As U.S. citizens enjoy more individual freedom than anyone else, they naturally assume that their recently enlisted offspring will be able to fend for themselves better than those from Germany, Russia, or Asia. Those American recruits who live through their first few firefights probably can, but their Eastern counterparts will get more tactical-decision-making practice before being committed.

Contrary to popular opinion, German, Russian, and Asian "nonrates" have been granted more influence over their commanders' decisions than their American counterparts. Through collective opinions and command criticism,[9] Eastern privates get more opportunity to exercise their minds. Though serving a less democratic regime, they have, in effect, been allowed a more democratic process within their units.

> Apart from the role played by the Party, the reason why the Red Army can sustain itself [during the second revolutionary war] without collapse in spite of such a poor standard of material life and such incessant engagements, is its practice of democracy. The officers do not beat the men; officers and men receive equal treatment; soldiers enjoy freedom of assembly and speech; cumbersome formalities and ceremonies are done away with; and the account books are open to the inspection of all.[10]
>
> — Mao Tse-tung in 1928

> In *Kiem thao* sessions, the [NVA] soldiers offered judgments of their comrades [and commanders] and listened to evaluations of their own performances. The meetings sometimes featured discussions of tactics from the unit's recent engagements or *suggestions* [italics added] . . . from the army command.[11]

> Each soldier and officer criticized his own actions and the other members of the company regardless of rank. After each confession or criticism, a general [group] discussion ensued. . . . [I]t gave the soldiers a sense of participation in the unit's decision-making process. They viewed themselves, therefore, not as witless cannon fodder, but as thinking members of a team.[12]

Whereas a U.S. private often gets punished for showing too much initiative, a German private was recently punished for not showing enough.[13] This will come as no surprise to armchair historians. By the end of 1917, the Germans had not only put their NCOs in charge of ground attack spearheads,[14] but also defensive-matrix forts.[15] As will become evident from Chapters 12 and 19, the Japanese, Russians, Chinese, North Koreans, and North Vietnamese would soon do the same thing. Unfortunately, for cultural and political reasons, the United States has not followed suit.

The Extent of the U.S. Shortfall

That is not to say that U.S. troops have never shown any initiative (Figure 11.1), or that their leaders have never allowed them to make decisions. Without enlisted common sense, WWII might have turned out quite differently. Wesley Fox created squad-sized Tactical Areas of Operation (TAORs) in Vietnam.[16] Yet sadly, it has only been the great American commanders who have realized the direct correlation between a private's initiative and survival.

Figure 11.1: What Is Often the Best Decision
(Source: *FM 5-12B3* (1977), p. 2-308)

> If I had to train my regiment over again, I would stress small group training and the training of the individual. . . .
> . . . With proper training, our Americans are better, as our people can think better as individuals. Encourage your individuals and bring them out.[17]
> — Col. Merritt A. Edson after Guadalcanal
> *FMFRP 12-110*

The Germans

German appreciation for delegating authority predates WWI. It goes far beyond "empowering" NCOs.

> It is no less important to educate the soldier to think and act for himself. His self-reliance and sense of honor will then induce him to do his duty even when he is no longer under the eye of his commanding officer.[18]
> — *German Field Service Regulations of 1908*

By switching to NCO-led assault elements and defensive forts late in WWI, the Germans evolved tactically, but too late to secure overall victory. Nevertheless, the die was cast for giving privates more leeway.

> "The position of the NCO as group (squad) leader," wrote Ludendorff in his memoirs, "thus became more important. Tactics became more and more individualized."[19]

By U.S. admission during WWII, "German tactical doctrines stress the responsibility and the initiative of subordinates."[20] The German squad leader could still launch an assault and penetration on his own initiative.[21] He was further allowed to advance beyond enemy lines (sometimes bypassing enemy pockets of resistance) until his platoon leader told him to stop.[22]

The Japanese

After the Nomonhan [Manchurian] Incident of 1939, Zhukov remarked on how good the Japanese were at close-quarter combat.[23]

> Japanese strength . . . lay in small units and the epitome of Japanese doctrine was embodied in small-unit tactics. Night attacks . . . and the willingness to engage in hand-to-hand combat were the hallmarks of the Japanese infantryman. Indeed such tactics were very successful against Soviet infantry [in 1939].[24]
>
> — U.S. Army, *Leavenworth Papers No. 2*

As early as 1920, the Japanese had realized the need for decentralizing control over tactical operations. In essence, they were giving their lowest enlisted ranks more decision-making leeway.

> By 1920 IJA [Imperial Japanese Army] tacticians realized the need to disperse infantry formations in order to reduce losses when attacking a defender who possessed the lethal firepower of modern weapons.[25]
>
> These [Japanese changes] included . . . increased reliance on the independent decision-making ability of . . . *noncommissioned officers* [italics added].[26]
>
> — U.S. Army, *Leavenworth Papers No. 2*

The Russians

Every WWII Russian rifleman was required to know what role he, his squad, and his platoon were to play in the overall battle.[27] If he had been the mindless automaton of Western propaganda, he wouldn't have needed the extra information. In truth, he had to look for and report opportunities.[28] He also had to outwit enemy armor.

> The soldier must know how he (personally) can fight tanks.[29]
>
> — *Soviet Combat Regulations of November 1942*

As World War II progressed, the common Soviet rifleman—like his squad leader—may have been encouraged to display initiative in the absence of orders. In a top-down hierarchy, this is rare.

> In the case of a sudden change in the situation when it is not possible to wait for new orders, the leader [down to squad level] must handle the situation independently.
>
> The absence of orders can never be used by the leader as an excuse for inactivity in battle.[30]
>
> — *Soviet Combat Regulations of November 1942*

The Russians have repeatedly demonstrated a preference for armored thrusts. Lesser known is their dismounted tactical proficiency at the point of the spear. Even more obscure is how they secure—in advance with infiltrators—the armor's route of advance.

As Stalingrad is generally recognized as the turning point of WWII, one could say that the Russians are also adept at urban defense. As will be shown in Chapter 21, their holding out along the banks of the Volga set something of a tactical precedent. Semi-independent snipers, killer teams, and assault squads conducted what has subsequently been called an "urban-swarm" defense. Without enlisted thinkers at Stalingrad, WWII might have taken much longer.

The Chinese

The Maoist "mobile offense" has undoubtedly sprung from the ancient "cloud" battle array.[31] With it, a large unit could rapidly disperse and reform. That would explain why—at the outset of a chance contact—American units have invariably noticed enemy activity at their flanks and rear. No wonder watching the closest enemy concentration has done so little good. Tiny groups of loosely controlled and highly opportunistic personnel were conducting the main attack. Because of their guerrilla heritage, those personnel have had little trouble working semi-independently. With instruction on self-discipline and knowledge of the commander's intent, each could not only think for himself, but also make a strategic contribution. Initially, there were no commissioned or enlisted ranks in the Chinese "people's army." Every soldier was respected for his part of the whole. Officers were called "comrade leader" and had to endure their men's hardships. In combat, they maintained loose control by keeping every echelon informed of any changes to the unit's tactical or strategic missions.[32]

Similarly, the Maoist "mobile defense" has descended from the ancient procedure of "feigned flight." As will be shown by the next chapter, that defense is nothing other than a gradually receding belt of tiny forts. It's almost identical to what the defenders of Iwo Jima and Okinawa called "retreat combat."[33] This process cannot be successfully orchestrated from above, it must be left to the discretion of the NCO in charge of each squad and its organic machinegun.

The North Vietnamese

Simply put, guerrilla war and its inseparable "mobile" variant require and encourage more individual skill and initiative than the traditional U.S. style of fighting.

> When comparing the two sides, their strength and weakness should be considered in specific conditions, chiefly the actual effectiveness of strength in action and in the confrontation. A strength in action should be considered in . . . human dynamism (resolution, capacity, resourcefulness, creativeness) which are very decisive factors.[34]
>
> — The National Political Publishing House, Hanoi

> Combatants who fight well as guerrillas adapt to changes when introduced into regular units.[35]
>
> — Gen. Vo Nguyen Giap

To produce privates of commensurate skill and flexibility, U.S. commanders would have to decentralize control over company-level training and operations. Sadly, too few are willing to demand the time to delegate authority; and too few are willing to run the political risk of enlisted mistakes.

How the U.S. Private Can Help Himself

Regardless of occupational specialty, every U.S. private must be prepared to outsmart the most accomplished of Eastern infantrymen. He may stumble upon one of the *ninjas* who routinely penetrate U.S. bases. Or he may become a rifle company replacement. That happened to Marine cooks and bakers on Iwo Jima, and Army support personnel at the Battle of the Bulge.

The U.S. private can help himself by learning—through trial and error—how to present less of a target. He must do so within established unit parameters. Through all types of terrain, he must practice denying the adversary a full three-second sight picture of an upright human being. Simply by lowering his stance, he can halve his exposure to fire. By crawling, he can virtually eliminate it.

Part Three

What the "Eastern" Soldier Does

All warfare is based on deception. — Sun Tzu

(Source: *Sun Tzu's Art of War,* by Gen. Tao Hanzhang, © 1987 by Sterling Publishing Co., p. 95)

12 When Told to Hold

- *Would Eastern soldiers have defensive "lines"?*
- *To what extent do they employ "firesacks"?*

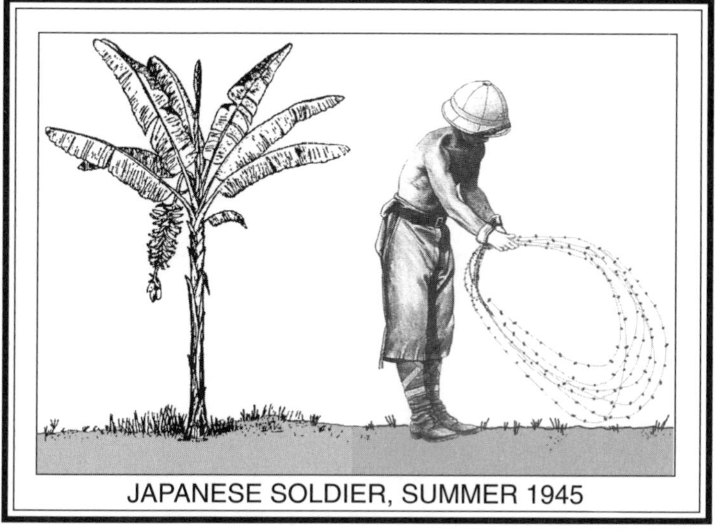

JAPANESE SOLDIER, SUMMER 1945

(Source: Courtesy of Cassell PLC, from *World Army Uniforms since 1939,* © 1975, 1980, 1981, 1983 by Blandford Press Ltd., Part I, Plate 201; *FM 21-76* (1992), p. 6-7)

WWI Produced the "Elastic Defense in Depth"

At first in WWI, the Germans protected their riflemen with trenches and their machinegunners with deep pits. Then they noticed that too many riflemen were succumbing to bombardment,[1] and too many machinegunners to capture. During a pre-assault barrage, both were better off scattering into nearby shellholes. From there, they could deliver unexpected fire into the advancing waves of Doughboys. As each shellhole accommodated roughly a squad, the role of the riflemen became one of protecting the crew-served weapon.[2] From a hastily fortified pit,[3] the machinegunner could

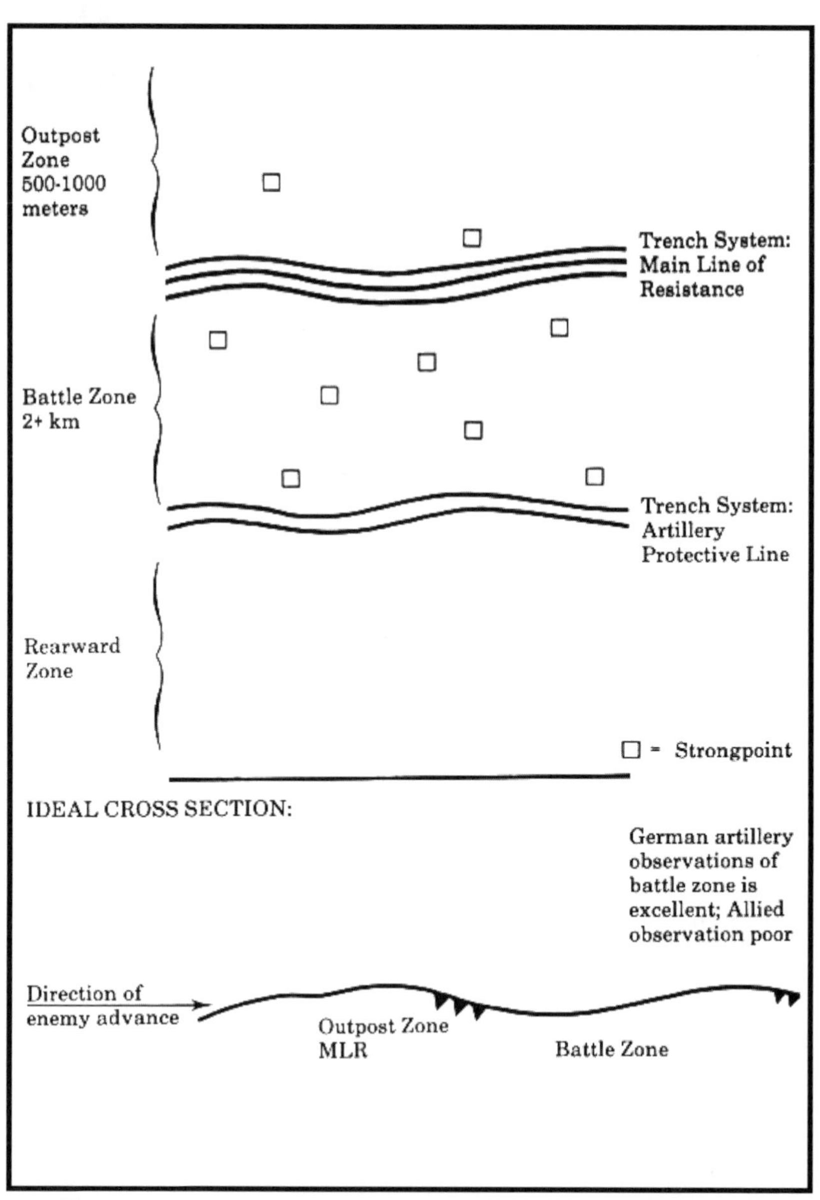

Figure 12.1: WWI Germans Depart from Linear Defense

(Source: *The Changes in German Tactical Doctrine during the First World War*, U.S. Army Combat Studies Inst., Leavenworth Paper No. 4 (1981), p. 14)

more easily avoid capture. In fact, squads were encouraged to move slightly forward, sideways, or backward as the situation dictated.[4] The loose conglomeration of positions was to bend without breaking. By April 1917, what had started out as a randomly chosen series of shellholes became a staggered matrix of carefully located bunkers (Figure 12.1).[5] The Germans had begun the logical transition from a linear defense in depth to a matrix of semi-independent strongpoints.[6]

Soon the Germans realized that on a reverse slope, these strongpoints would be out of sight of Allied gunners.[7] If camouflaged, they could elude aerial observers as well.[8] When staggered, each could be protected by the interlocking machinegun fire from those on either side behind it. This veritable barrier of hot lead would not only make the bastion harder to assault, but also easier to reinforce, resupply, or abandon. Even behind the main line of resistance (MLR), each fort would provide its own 360° defense. Its occupants were to hold out until relieved by counterattack.

Only after the war was it realized that a strongpoint matrix is almost as durable when forward-echelon personnel are allowed to establish new rear-echelon positions. A gradually receding defensive belt would need no coordinator, as long as the NCO in charge of each fort was allowed to decide when to pull back.[9]

What in WWII Only Looked Like a Linear Defense

Then came WWII. The Germans were still deploying squad-sized machinegun nests in echelon.[10] However, these nests were now surrounded by barbed wire and mine belts (Figure 12.2). According to *TM-E 30-451,* these tiny bastions could be located on either forward or reverse slopes.[11] Unlike their WWI predecessors, each contained a zigzag piece of trench with an indentation at its center. Their decidedly linear shape made them difficult to pinpoint from the air—hard to distinguish from the spider web of communications trenches. Their location could also be easily shifted. Three squad-sized strongpoints comprised one of platoon size, three platoons one of company size, and so on up the chain of command. At every echelon within this matrix of embedded strongpoints, the formation was "two up and one back."

Squad strongpoints normally are incorporated into platoon

> strongpoints, and the latter into company strongpoints.[12]
> — "Handbook for German Military Forces"
> U.S. War Dept., *TM-E 30-451*

To this basic design, the WWII Germans had the authority to add dummy positions, covered dugouts, breaks in the mine belts, gates in the wire, and communication trenches (for reinforcement or withdrawal).

Aerial photographs of the WWII Japanese defenses reveal no consistent pattern, but they do show partial circles within a patchwork of trenchline.[13] While most of this trenchline may have been for aerial deception or ground communication, the partial circles were probably the barrier outlines of squad- or platoon-sized strongpoints.

By WWII, the Russians had also replaced the linear defense with the strongpoint variety. Again the bastions were arrayed in "V" formation. For each Russian battalion, a partially manned dummy position was added between the strongpoints at the top of the "V" to conceal the firesack. Others were added on either side of the strongpoint at the bottom of the "V" to hide its exact location.[14] Each company bastion had real and fake positions in the same configuration (Figure 12.3).[15] That their platoon-sized strongpoints were themselves comprised of squad strongpoints is evident from the tactics manual for the period.

> The position of a squad set up in the depth of the main battle position must have a good field of fire . . . into the intervals between the squads set in front and on their flanks, and the alternate position must have a field of fire to the rear.[16]
> — "The Rifle Squad on Defense"
> *Soviet Combat Regulations of November 1942*

Within each platoon's intermittent circle of barbed wire and mine belts, lay three linear squad formations in the same "two-up-and-one-back" pattern.[17] (See Figure 12.4.) Left largely unmanned was the center and periphery of what from the air might have resembled a perimeter.

By 1986, the Soviets were calling this same formation a "firesack defense."[18] It had been widely adopted by their client states. It was also well known to their adversaries.

> Just as the [Soviet] platoon started to move . . . the *mujahideen* opened fire (from hills #1, 2 & 3).[19]
> — DoD-published Soviet military academy study

> In the spring of 1987, intelligence reports stated that the *mujahideen* had constructed a fortified region near Jalalabad in Nangarhar Province. The so-called "Melava" fortified region contained huge stores of weapons, ammunition, medicine and foodstuffs which had been brought from Pakistan. The dominant heights in this area were well fortified with dense minefields and with deep trenches and dugouts dug into the rocky strata. Each mountain had been turned into a self-sufficient strong point, prepared for defense in all directions.[20]
> — DoD-published Soviet military academy study

The Cross-Compartment Variant

According to *TM-E 30-451,* the WWII German Westwall system resembled the elastic defense of WWI (Figure 12.1) except for "concealed" reverse-slope strongpoints.[21] Their invisibility enhanced more than just durability. One wonders how they might have cooperated with those on the forward slope of the next ridge. After all, fire from every direction is hard to dodge.

Not all of the Soviet Union was flat. It had uneven terrain, and its armies often defended valleys from elevated bastions during WWII.

> Positions on different peaks are not held by occupying the areas between them, but with advanced fire nests firing between them (on several locations).
>
> The entrances and exits from the strongpoints must [be] covered with cross fire, the dead angles and hollows by mortar bombardment.[22]
> — "The Rifle Platoon in the Defense"
> *Soviet Combat Regulations of November 1942*

The Soviets occupied ridgelines on both forward and reverse slopes as well. By U.S. admission, their firing positions were echeloned vertically as well as in depth.[23] There were several tiers of squad-

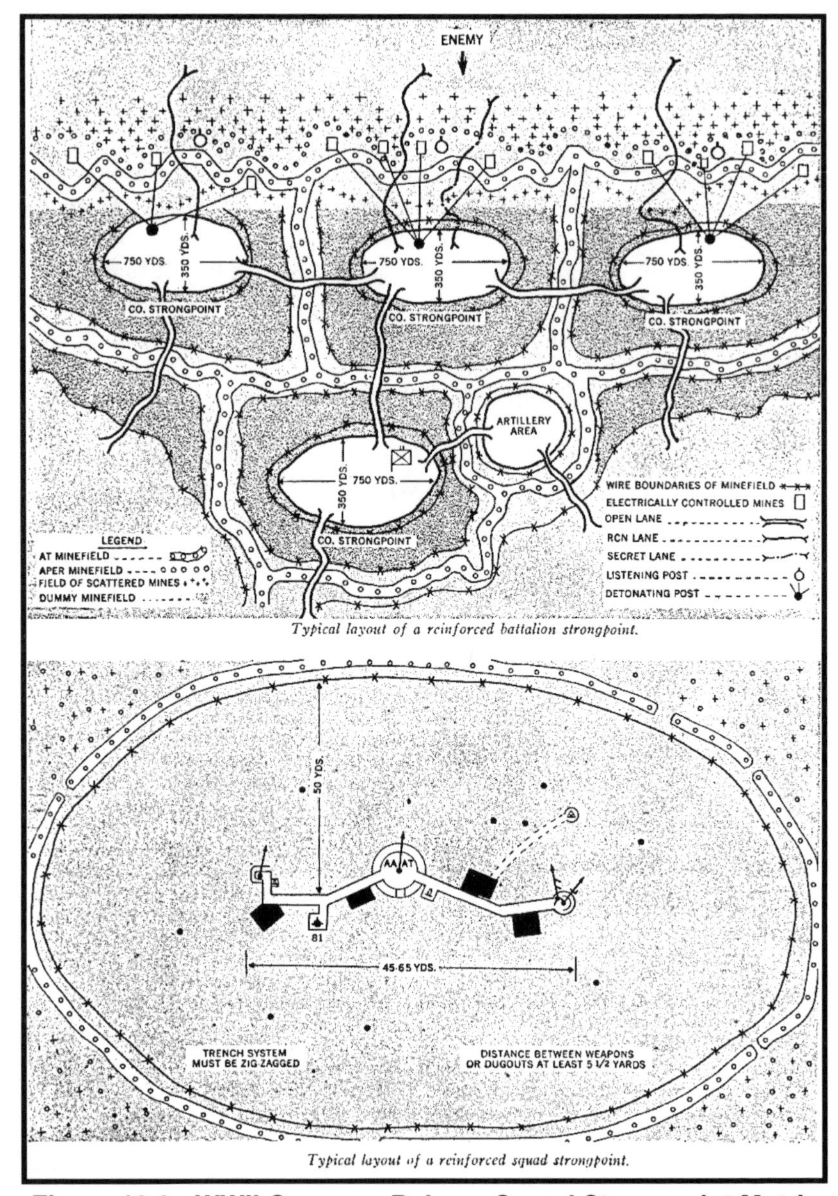

Typical layout of a reinforced battalion strongpoint.

Typical layout of a reinforced squad strongpoint.

Figure 12.2: WWII Germans Rely on Squad Strongpoint Matrix

(Source: *Handbook on German Military Forces,* TM-E 30-451 (1945), p. 230)

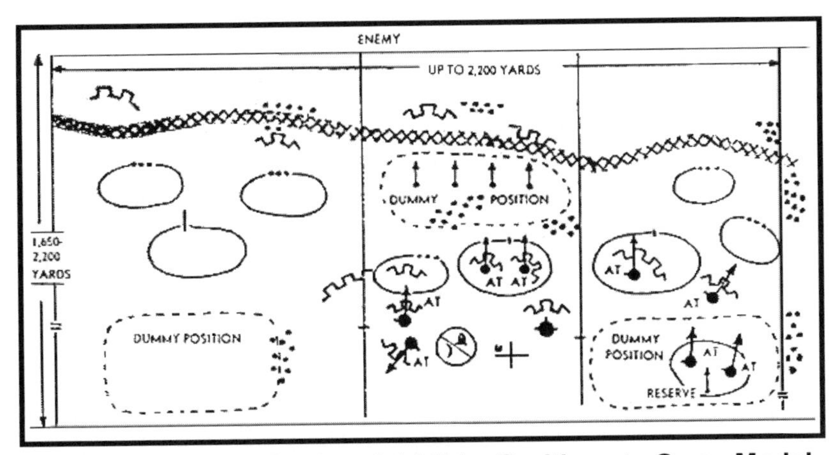

Figure 12.3: WWII Soviets Add Fake Positions to Same Model

(Source: *Handbook for U.S.S.R. Military Forces*, TM 30-340 (1945), p. V-47)

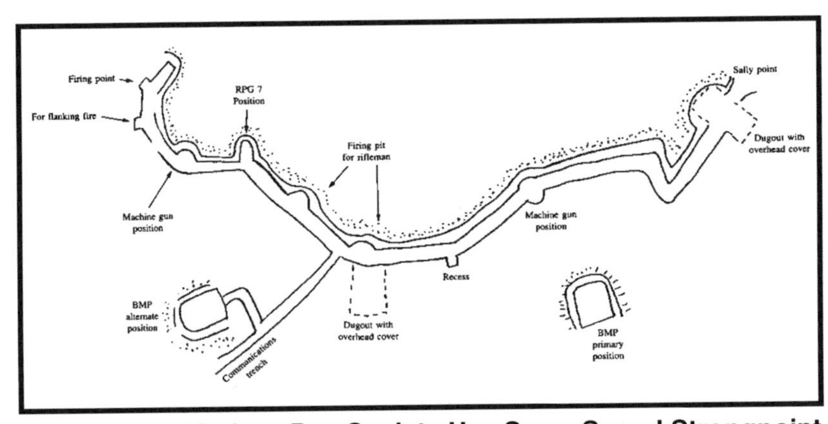

Figure 12.4: Modern-Day Soviets Use Same Squad Strongpoint

(Source: Courtesy of Presidio Press, from *Soviet Airland Battle Tactics*, by William P. Baxter, © 1986 by Presidio Press, p. 137)

sized perimeters on both front and back slopes. It was not long before someone thought of linking the ones in front and back by tunnel.

Across Iwo Jima's northern half stretched three separate bands of strongholds.[24] Each had, as its building block, a squad-sized position with crew-served weapon and a dozen or so riflemen. At the center of the squad formation was a cave or tunnel into which everyone could escape bombardment or fall back to the reverse slope. Defended in the same fashion on Okinawa were the east-west ridgelines just north of Shuri.

> [A Japanese] POW [prisoner of war] can recall only two specific tactical recommendations made in the [32nd Army] Battle Lesson(s): one that the standard A/T [antitank] satchel charge be increased in weight . . . and the other that they increase the distance between the so-called "octopus pots" *(KAKUTSUBO)* foxholes in preparing defensive positions in front of caves and similar entrenchments.[25]
>
> — Col. Hiromichi Yahara, IJA survivor of Okinawa

The Okinawan terrain was different. Whereas the ridgelines on Iwo had been closely bunched, those on Okinawa could be taken under direct, long-range fire. For this reason, Okinawa's defenders placed observation posts on forward slopes and main forces on reverse slopes (underground). Doing so may have served a dual purpose. Central to the Asian style of war are encirclement and close combat.[26] By letting an attacker pass over the first string of reverse-slope strongpoints, an Imperial Japanese Army (IJA) unit could automatically envelop and encircle him. Nothing accomplishes annihilation like multidirectional crossfire.

The Chinese did the same thing in Korea.[27] (See Figure 12.10.) By placing their positions on reverse slopes, they could cover them by fire from the rear. They could also escape low-trajectory bombardment. Undeniably, the Chinese were using the same strongpoint system as the Germans, Russians, and Japanese had before them.

> CCF defensive works exploited the terrain and followed an irregular shape, often triangular or ladderlike, so that rearward positions could fire in the gaps between the forward positions.[28]
>
> — U.S. Army, *Leavenworth Research Survey No. 6*

> During the last two years of the [Korean] war, the Chinese defense assumed a positional character of remarkable strength. By the end of 1951, the extensive trench network ran fourteen miles in depth (attributed to Marshal). As time passed, the works became more and more impregnable. By hand labor, using ordinary tools, CCF troops fortified the reverse slopes of hills and dug tunnels all the way through to the forward slopes for observation (attributed to Ridgway).[29]
>
> — U.S. Army, *Leavenworth Research Survey No. 6*

Everyone within Crawling Distance of a Bombproof Shelter

The Germans and Japanese were not alone in discovering how to evade pre-assault shelling. Only required are dummy positions and bombproof shelters.

The WWII Russian squad regularly dug a "combat trench (man-deep), shell-proof covered position, communications trench, and alternate positions."[30] Its members routinely retreated into that shell-proof position during bombardment.

> On the opening of enemy fire and specifically on the beginning of bombardment by enemy artillery, the platoon must go into fortifications or cover; in each squad position and the headquarters position, the platoon leader sets up observation posts.
>
> . . . [T]he platoon leader must bring the soldiers and the heavy weapons into firing position under cover as soon as the enemy gets within range.[31]
>
> — "The Rifle Platoon in the Defense"
> *Soviet Combat Regulations of November 1942*

In Afghanistan, those fighting the Russians would often have enough time to move the bulk of their forces into caves while maintaining observers in firing positions.[32]

The Defensive Ambush

As the prepared strongpoint made the logical transition from reverse slope to underground, it became virtually invisible. The WWII Germans disallowed unnecessary movement in their defensive formations.[33] Years later, the Soviets used hidden tanglefoot (low-strung barbed wire) in Afghanistan.[34] The *mujahideen* could often spot Soviet movement into an area in sufficient time to mine the target valley's entrance and passes.[35]

To escape an attacker's supporting arms, all but the most naive army would withhold their direct fire until the last moment. As early as 1913, Japanese defenders were instructed to shoot assault troops only at short range.[36] During WWII, even the Russians exercised common sense.

> Far from opening fire at 400 to 800 meters, numerous German accounts relate that Soviet infantry would wait until the Germans got within 100 meters or even 50 meters or less, and then suddenly open fire from camouflaged positions, forcing the German infantry to "go to ground" or retreat.[37]
>
> — *Soviet Combat Regulations of November 1942*

An Orderly Withdrawal to Prepared Fallback Positions

By 1917, the Germans had abandoned the belief that defended ground must remain inviolate.[38] If that ground were penetrated, the battlefield's depth would weaken the attacker, preserve the defender, and enhance that defender's chances of retaliation.[39] However, the WWI German squad could only move a short distance to the rear on its own initiative.

According to *TM-E 30-451,* a WWII German unit (of indeterminate size) could pull back when continuing to fight lacked promise or defeat was imminent.[40] It could also use mutually supporting strongpoints in conjunction with a delaying action.[41] Implicit in this instruction was the shifting of forward units into fallback positions. When possible, such a withdrawal was accomplished at night.

Meanwhile, at the MLR, the WWII Russian squad could only pull back after being ordered to do so by its platoon leader. However, the Russian manual for the period provides detailed procedures on how to accomplish the maneuver.[42] Often, the light machinegun would initially cover the withdrawal of the rest of the squad.[43]

The rearward-moving belt of manned positions really came into its own on the other side of the world. In fact, late in WWII, the Japanese were practicing what they called "retreat combat."

> They [the Japanese] called it [being on defense] "retreat combat." . . . Initially, Japanese firepower had been concentrated on the perimeter but subsequently the Army showed considerable skill at defense in depth. The offensive aspect of defense in depth was preserved by the use of underground passages from which Japanese soldiers would emerge when the ground had been overrun. Somewhat disconcertingly for their opponents, the exits were so well camouflaged and concealed that they were not easily detected. Japanese skill lay not so much in their use of nets and paint as in the use they made of [natural] materials. In some areas the exits would be covered with vegetation transplanted with such skill that it was still growing; in more barren areas the rocks and stones would have been so arranged that the eye would be led away from the vital area and so fail to detect it. Dummy defenses, often amazingly realistic even at close quarters, were used to shepherd attackers into positions well prepared to receive them.[44]

> Our general retreat policy [on Okinawa], following the principle of regular retreat operations, was aimed at a total retreat toward fortifications at Kiyan. Our war objective, however, remained a war of attrition, looking forward to a decisive battle in mainland Japan. We intended to carry out a German-army-style, local prolonged resistance, taking advantage of the rugged terrain and numerous caves along the twelve kilometers between the Shuri Line and the new front line.[45]
>
> — Col. Hiromichi Yahara, IJA survivor of Okinawa

During the initial stages of the Korean War, both the North Koreans and Chinese readily moved rearwards. While officially a "mobile defense," it was still accomplished through a succession of hastily fortified positions. Then in 1953, the lines stiffened and a below-ground version of the elastic, positional defense was born.

By 1994, the North Korean Army (NKA) was still using three defensive options—two of which involved moving backwards:

> NKA [tactical] doctrine calls for three types of defense: position[al] defense, mobile defense, and retrograde operations. . . . The mobile defense would be used to gain time, exact losses on . . . [enemy] forces, and preserve combat strength *while losing ground.* NKA retrograde (or disengagement) operations would be used to gain time to plan for the next operation or to restore combat capability [italics added].[46]
>
> — "North Korea Handbook"
> U.S. Dept. of Defense, *PC-2600-6421-94*

The North Korean positional option looks almost exactly like the "two-up-and-one-back" German and Russian elastic model (Figure 12.5). It provides all-around protection and mutual support.[47]

While capable of both mobile and positional defenses, the Chinese prefer the former. Even the latter makes provision for disappearing below the ground.

> The Chinese Communist concept of defense includes . . . trading space for time; complete abandonment . . . of areas of little importance; concentration of defensive strength in critical areas; conduct of extensive guerrilla activities in hostile rear areas, and . . . psychological warfare. . . .
>
> The Chinese Communists employ both mobile defense and positional defense, but favor the mobile defense since it is best suited to the concept of mobile warfare.
>
> . . . Mobile defense tactics are frequently employed to lure an extended hostile force into an . . . ambush.
>
> Positional defense is employed only when absolutely necessary. . . . Where a prepared position has been overrun by the enemy, the defending forces will fall back into bunkers, call artillery fire and tactical air support onto their own position, and wait to join with counterattacking forces.[48]
>
> — "Handbook on Chinese Communist Army"
> U.S. Army, *PAM 30-51*

In the outpost and forward-defense zones, modern-day Russians also systematically withdraw to fallback positions. They use deeply echeloned defenses with interlocking fires and obstacles to sap an attacker's strength.[49] Their forward detachments conduct a mobile defense in the security zone. To preclude encirclement, they would need withdrawal routes and exfiltration technique. In fact, to the modern-day Russian, retreat and withdrawal are as much a part of the defense as holding ground. He sees the defense not as a positional tool but as a temporary condition on a dynamic battlefield.[50] Even behind the MLR, he prefers a defense in depth.

> The main defensive position comprises a series of defensive belts arranged in depth (Ionin, "Modern Defense," *Voennyi Vestnik,* 1981, 4:16).
> . . . Defensive belts are organized along defensible terrain and comprise a series of mutually supporting defensive positions or strongpoints reinforced by obstacles and supported by fire.
> . . . A motorized rifle company may occupy a strongpoint 1000 meters wide and 500 meters deep, with 300 meters between platoons. Platoon positions are self-contained and are organized so that the company and the platoon can fight in any direction. . . . [Second] echelon units may be placed to provide defense in depth in gaps between first echelon units. . . .
> . . . A platoon strongpoint is up to 400 meters in frontage and 300 meters in depth and comprises squad positions (Merzylak, "IFV's on the Defensive," *Soviet Military Review,* 1983, 8:20). . . .
> . . . The final step is to prepare communications trenches for supply and evacuation and to permit covered movement within the platoon strongpoint.[51]

Within the main defensive position, the Russian relies on maneuver as well as dominant terrain and firepower.[52] He pulls back to avoid destruction by superior enemy forces, to trade space for time, or to redeploy forces to another axis.[53] If the plan calls for withdrawal, he does so on his own initiative; if it does not, he needs the approval of his commander two echelons up the chain of command.[54] To this day, the Soviet defense is predicated on strongpoints and firesacks.[55] To disguise a firesack, the center portion must be lightly manned and then evacuated. The Soviet defender has, as his first priority, to construct obstacles and barriers. He does so to canalize his foe. Then, he prepares strongpoints with the emphasis on deception and digging.[56] Finally, he works on "communication" routes for easy access and egress.[57] His squad strongpoint is of a design that can be easily hidden and shifted within the communication routes (Figure 12.4).

When the Russians deployed to Afghanistan, they realized that even nonaligned political movements knew the value of tactical withdrawal. The *mujahideen* routinely withdrew from almost every contact with the Soviet army. They normally did so over prerehearsed routes,[58] or through fog and darkness.[59]

> The Afghan *mujahideen* practically never conducted positional warfare and, when threatened with encirclement, would abandon their positions.[60]
>
> — DoD-published Soviet military academy study

How Can the Easterner Dig In So Quickly?

In the battle of Hue City, an NVA unit completely fortified a street (Phase Line Green) within two days of it being abandoned by South Vietnamese forces. How was this possible?

The 1913 Japanese manual contains several battledrills for night entrenchment in different types of terrain.[61] The North Vietnamese must have also practiced defensive battledrills.

How the Occasional Encirclement Is Handled

During WWII, both the Germans and Russians expected some strongpoints to get cut off and surrounded. Unless specifically told to wait for a counterattack, their occupants had the authority to fight their way out. As an alternative, the WWII Soviets permitted exfiltration by small units.[62]

Asians routinely slip through enemy lines after getting cornered. They also hide below ground until their attacker gets tired of occupying the surface. An entire NVA division secretly escaped the U.S. cordon of Hue City.[63]

The Easterner Uses the Same Method in Urban Terrain

During WWII, the Russians converted houses into strongpoints and connected them by trenches. Those without basements had positions, troop shelters, and escape routes dug under the floors.[64] The modern-day Russian still defends a built-up area with strongpoints arrayed in depth.[65] They are normally blocks and major buildings. Each is organized to resist attack from all sides and to fight independently. The areas between the strongholds are swept by fire from heavy weapons. The sewers function as "communications" routes.[66]

Hue City was defended in much the same way. Every other street in the southeast corner of the ancient Citadel was studded with fortified blocks. The intermediate blocks were contested by roving teams of snipers. Chapter 21 will discuss this in more depth.

What May Be the Mobile-Defense Formation

During first part of the Korean War, the Chinese mobile defense resembled a hastily constructed positional defense that moved rearward. The standard "one-up-and-two-back" formation had the potential for shifting quickly over to the attack.

> Before defensive lines became fixed in Korea, the CCF did not employ . . . position defense. Instead, they employed a basic defensive scheme of "one up and two back." In this scheme the "up" group operated as a screening and delaying force. The two "back" units . . . [prepared] for a counter-offensive or the defense. . . .
>
> A number of defensive principles characterized Chinese operations during this time:
>
> - Defensive units, disposed in great depth, deployed along a narrow front.
> - Forward elements playing purely delaying roles to time while the remaining units prepared a second line of defense.
> - Troops built defensive positions strong enough to afford protection from air and artillery attack.
> - Soldiers established dummy positions and gun emplacement for the deception of the enemy.
> - The Chinese placed light automatic weapons well forward, with the heavy weapons disposed in depth. . . .

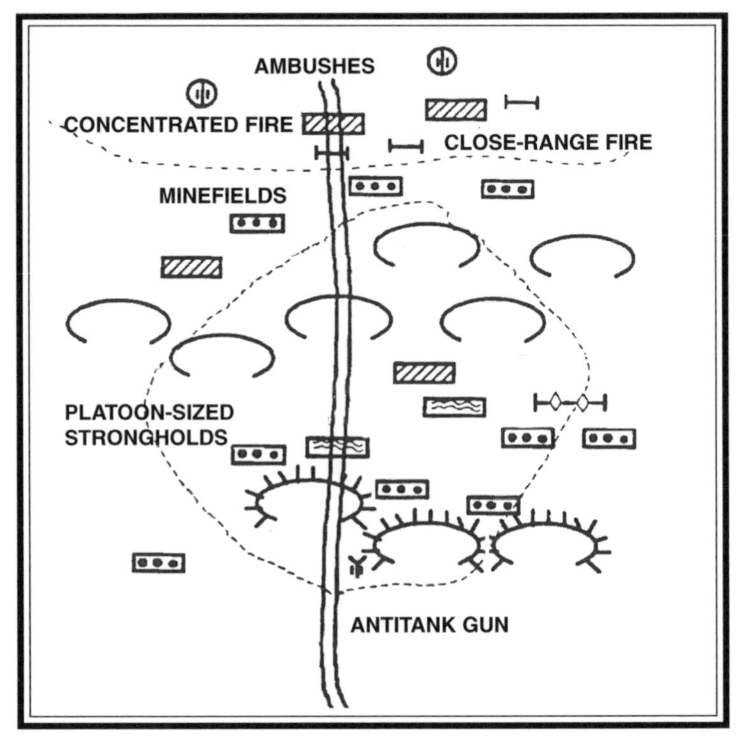

Figure 12.5: Current North Korean "Defensive Stronghold"

(Source: *North Korea Handbook,* DoD PC-2600-6421-94 (1994), p. 3-99)

Figure 12.6: North Vietnamese "Double-Line" Defense

- Defensive forces were withdrawn to successive defensive positions.[67]

— U.S. Army, *Leavenworth Research Survey No. 6*

By 1986, the Soviet motorized rifle battalion was still using this same formation, but with dummy positions on either side of the lead unit.[68]

The North Vietnamese Hasty Defense Looks Familiar

VC/NVA hasty defenses did not resemble Western blocking positions or perimeters. Designed to withstand bombardment, they ordinarily consisted of two roughly parallel lines of fighting holes (Figure 12.6).

> Under the dense canopy of vegetation, two lines or belts of fortifications were constructed fifty to two hundred meters apart. . . . These belts of defensive positions followed the outline of an L, U, or V so as to offer the possibility of a crossfire.[69]

That the lines had a built-in firesack may mean that one's center was only partially manned. If so, the formation would be identical to the earlier German and Russian models.

Even the Squad Strongpoint Has "Firesacks"

The Asian has a fondness for the V-shaped formation. It provides him with a ready-made firesack into which his pursuers can wander. When the firing starts, they have nowhere to hide. In the Pacific during WWII, the Japanese liked to protect their machinegun bunker with rifle pits connected by trenches that formed an "X." (See Figure 12.7.)

The opposition in Vietnam had a small-unit defensive array that also utilized a combination of "V's." It was configured like the spokes of a wheel. With the fields of fire marked, each element could easily cover (by fire) the fronts of the other two. (See Figure 12.8.)

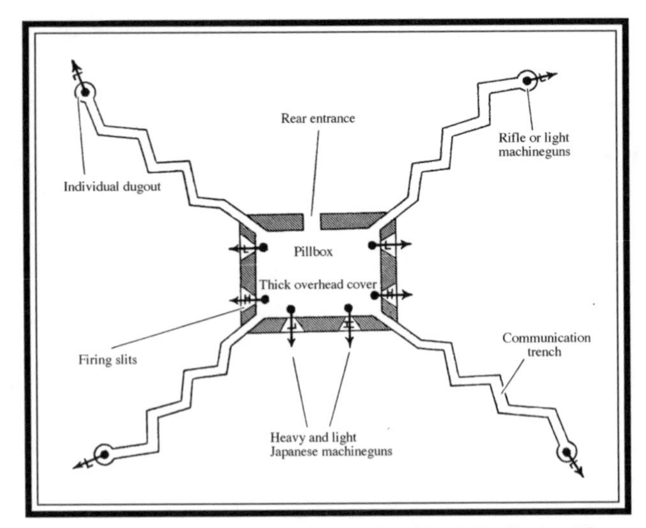

Figure 12.7: Japanese Strongpoint Had Built-In Firesacks

(Source: *TM-E 30-480* (1944), p. 160)

Figure 12.8: Viet Cong "Wagon Wheel" Outpost

The Rearward-Moving Asian Embrace

Against a supporting-arms-abundant foe, it makes little sense to totally withdraw. One pulls away just far enough to stay within danger-close range. That's what the Japanese did in WWII and what the NVA did in Vietnam.

> One tactic developed by the VC/NVA to counter this "surround and pound with fire support" technique was to "hug the enemy" rather than fall back to the second belt of defenses. This involved following the attackers as they pulled back to put in artillery and air.[70]

Another way to leave is to move into a secret subterranean chamber. With enough preparation, a sizable force could wait for its high-tempo attacker to move on.

Every Mountain a Strongpoint Defense

In a precipitous landscape, the positional elastic defense became more formidable as it moved below ground. Its inflicted great suffering on Iwo Jima in 1945 and prevented U.S. forces from reoccupying North Korea in 1953. As illustrated in Figures 12.9 and 12.10, its building block was a machinegun bunker surrounded by rifle pits.

> During that time [just preceding the 1953 Korean hill battles], anyone who happened to enter the mountainous area of our positions could hear the sound of earthwork. . . . Gradually, a defense system took shape. It was backed up by supporting [strong] points with tunnel fortifications as main structures. . . . Here one could find all kinds of facilities . . . as well as defended fortifications, communications entrenchments, and crisscrossing main tunnels and branch lines. . . . Even though the enemy continued dropping thousands of bombs that exploded on top of the mountains, our commanders were able to sit down and peacefully read books and newspapers or play cards and chess.[71]
>
> — *Mao's Generals Remember Korea*

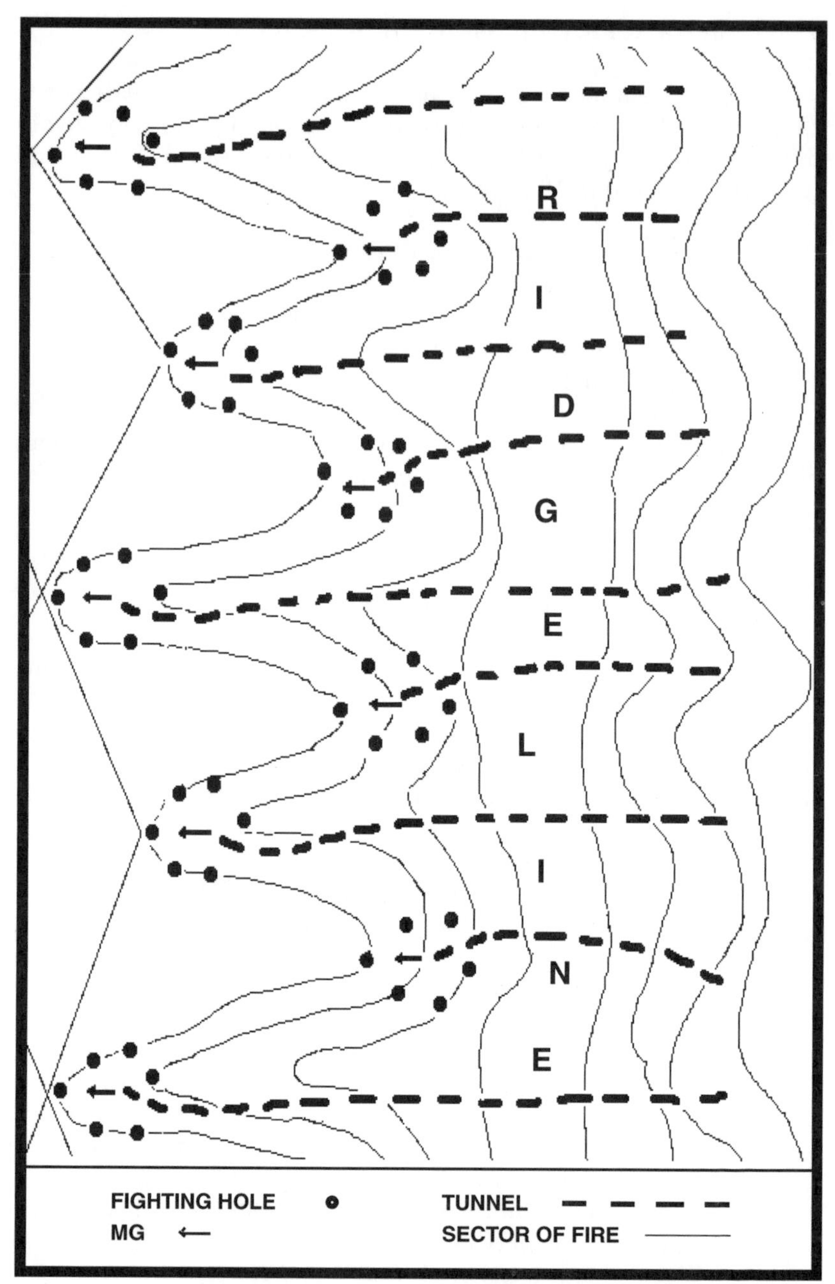

Figure 12.9: WWII Japanese Link Squad Forts to Reverse Slope

Under Russian control since WWII, North Korea may have developed a slightly different approach to positional defense. At least, that's what a recently acquired Russian history book implies. The North Koreans did some tunneling between forward and reverse slopes (Figure 12.11), but they may have relied more heavily on covered trenchlines than the Chinese and Japanese (Figure 12.12). Trenches would have been easier to dig. As in the Russian model, the manned portions of those trenches would have then become their strongpoints (Figures 12.13 and 12.14). Notice that the machinegun bunkers had rifle positions to their front and sides. A machinegun bunker would have been easy to add to a covered trench (Figure 12.15). Open gun pits may have had trapdoors and were connected by lateral (Figure 12.16). Figure 12.17 may show why U.S. troops found so many hard-won positions empty. The Reds were not only shifting to the reverse slope, but also forward (beneath their attackers). By so doing, they contributed to the evolution of the offense—more on that in Chapter 19.

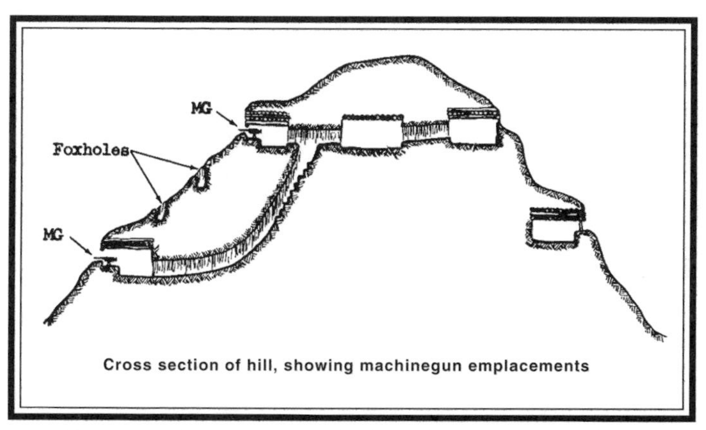

Figure 12.10: Chinese Defend Ridgelines Same Way in Korea

(Source: Excerpt from *Chinese Manual on Field Fortifications*, in *A Historical Perspective on Light Infantry*, U.S. Army Combat Studies Inst. Research Survey No. 6, p. 88)

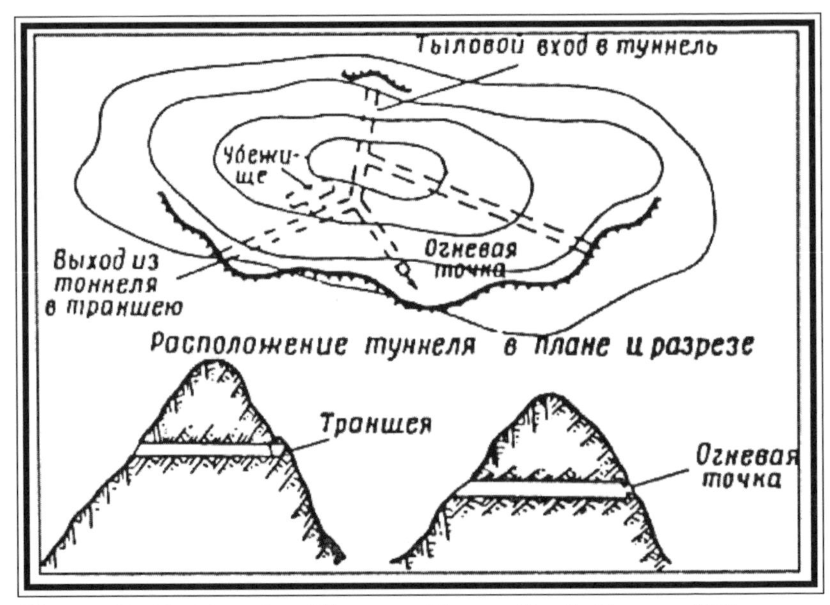

Figure 12.11: Russian Diagram Shows North Korean Difference

(Source: *Voina v Koree: 1950-1953,* by A.A. Kuryacheba, © 2002 by Polygon Publishers, p. 601)

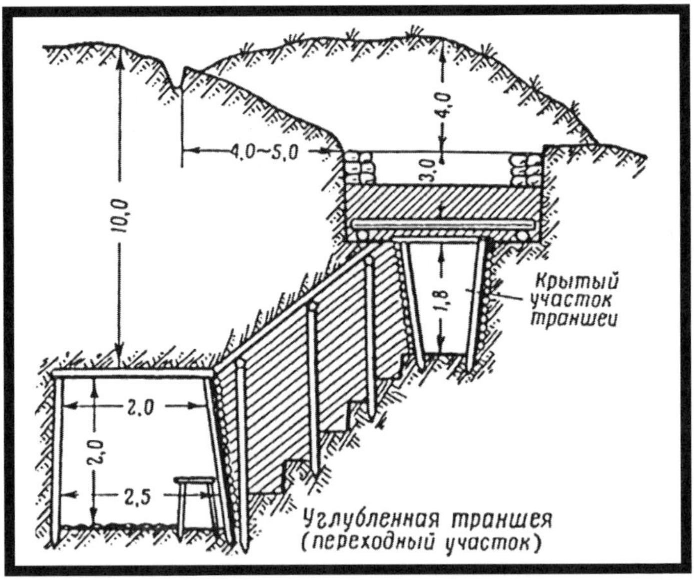

Figure 12.12: Many Trenchlines Were Covered

(Source: *Voina v Koree: 1950-1953,* by A.A. Kuryacheba, © 2002 by Polygon Publishers, p. 613)

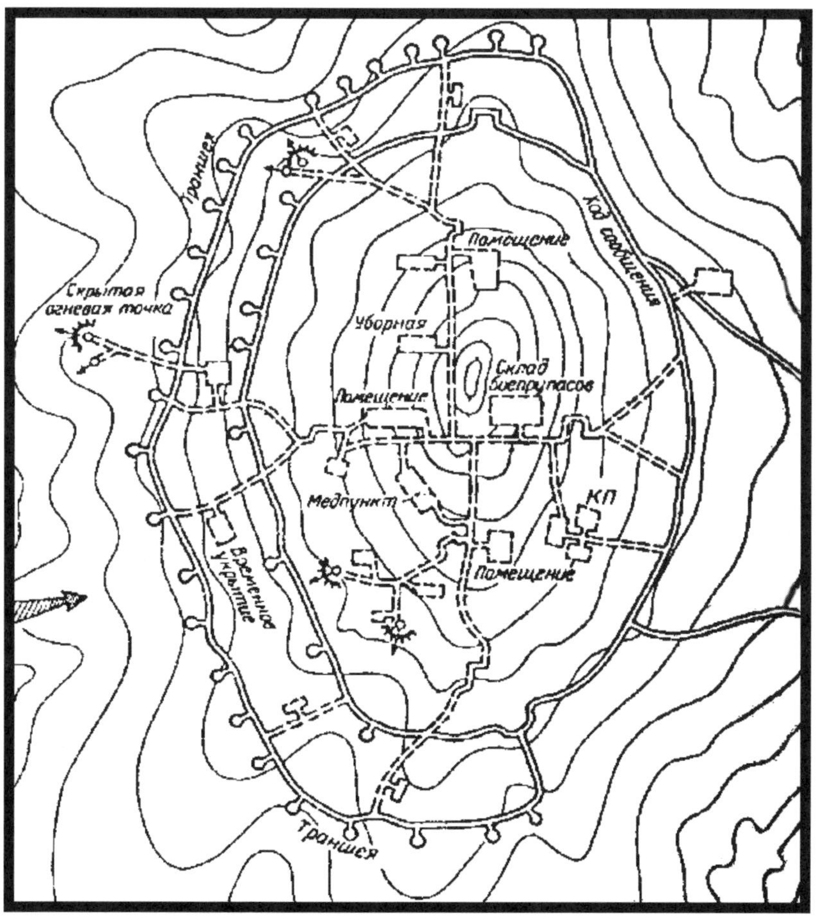

Figure 12.13: One Escape Tunnel for Every Few Positions

(Source: *Voina v Koree: 1950-1953*, by A.A. Kuryacheba, © 2002 by Polygon Publishers, p. 582)

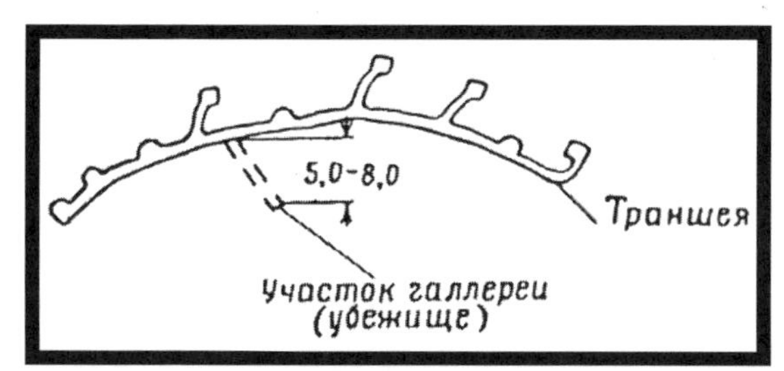

Figure 12.14: Manned Sections Became Strongpoints

(Source: *Voina v Koree: 1950-1953*, by A.A. Kuryacheba, © 2002 by Polygon Publishers, p. 613)

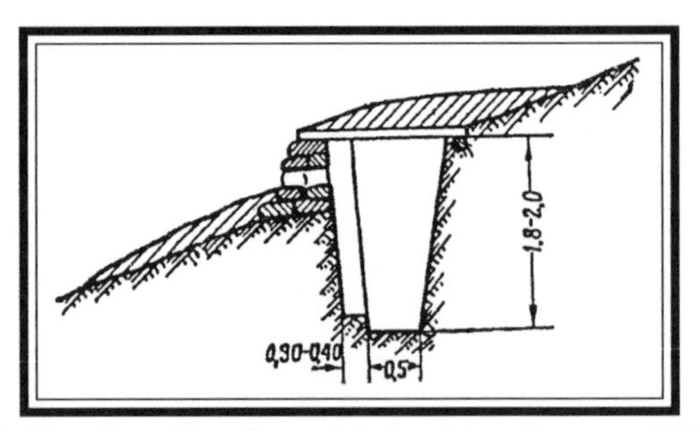

Figure 12.15: Gun Bunkers Linked by Covered Trenchline

(Source: *Voina v Koree: 1950-1953,* by A.A. Kuryacheba, © 2002 by Polygon Publishers, p. 585)

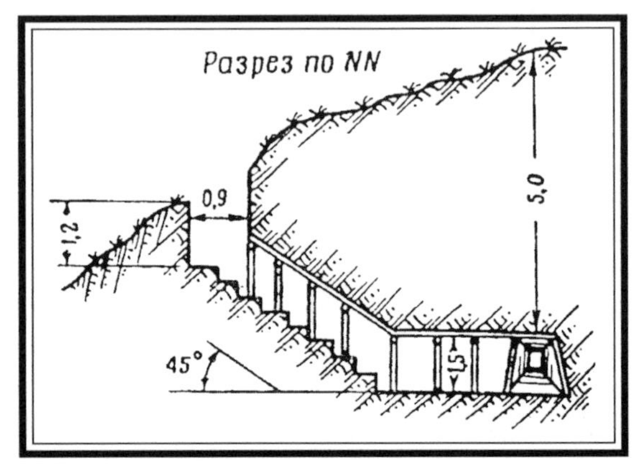

Figure 12.16: Gun Pits Linked by Lateral Tunnel

(Source: *Voina v Koree: 1950-1953,* by A.A. Kuryacheba, © 2002 by Polygon Publishers, p. 590)

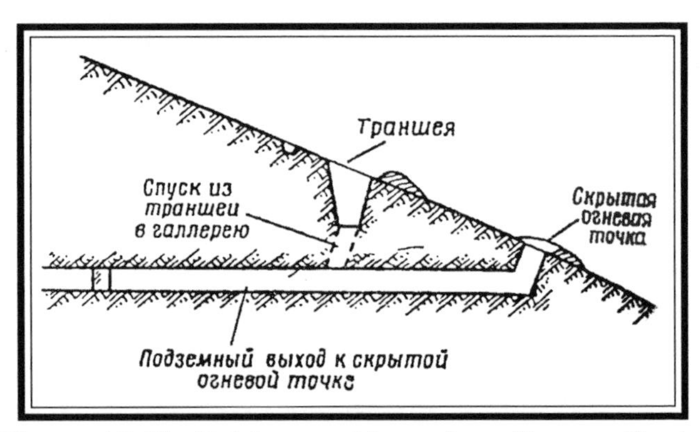

Figure 12.17: Evidence of Ingenious Escape Route

(Source: *Voina v Koree: 1950-1953,* by A.A. Kuryacheba, © 2002 by Polygon Publishers, p. 614)

Of late, the same type of defense has been spotted in Southwest Asia.

> In the spring of 1987, intelligence reports stated that the *mujahideen* had constructed a fortified region near Jahalabad in Nangarhar Province. The so-called "Melava" fortified region contained huge stores of weapons, ammunition, medicine and foodstuffs which had been brought over from Pakistan. The dominant heights in this area were well fortified with dense minefields and with deep trenches and dugouts dug into the rocky strata. Each mountain had been turned into a self-sufficient strong point, prepared for defense in all directions.[72]
>
> — DoD-published Soviet military academy study

The Soft, Underground Strongpoint Masterpiece

On a highly lethal battlefield, the strongpoints would have to be moved below ground. In 1953, the North Koreans and Chinese were able—with their much-vaunted hill defense network—to keep a much stronger U.N. army from reoccupying North Korea.

The Cu Chi tunnel complex northwest of Saigon may have been much more than an interconnected series of bunkers and storage rooms. From the 124 miles of Cu Chi district excavations,[73] the enemy could counterattack—from the inside—the night encampments of U.S. maneuver forces.[74] One NVA soldier remembers how easily his comrades moved considerable distances below ground.

> [In 1965] the Americans were carrying out military operation in Cu Chi. We were moving almost a thousand men from village to village through underground tunnels.[75]
>
> — Col. Huong Van Ba, NVA artillery officer

The Cu Chi tunnel complex had all the attributes of a soft, underground strongpoint defense. Here's what a former VC recalled.

> [About 1965] we took a bus to Cu Chi, where I was led into an ordinary village house and found myself being conducted through a maze of tunnels that started from the covered hole in the house's bunker.

> The main tunnel in this system was called Tong Nhut Road. It connected village to village and in some places had entrances from almost every house. In those places, if there was an enemy attack, an entire hamlet might disappear.
>
> ... The Cu Chi tunnels served as a defense, a trap for Americans or ARVN troops, a supply area, and a stop for new recruits coming out from Saigon into the jungle.[76]
>
> — Trinh Duc, VC village chieftain

Then, in February of 1979, the full significance of this new way to defend emerged. While North Vietnam's regulars were off fighting the Khmer Rouge in Cambodia,[77] its home guard was able to defend Vietnam's northern border.[78] Its reservists kept 17 Chinese divisions—backed by one-fifth the Chinese air force[79]—from penetrating more than 30 miles into its border region.[80] When Hanoi regulars finally advanced toward the last border town in contention, the Chinese declared victory and pulled out.[81] By China's own estimate, it had taken 20,000 casualties in 16 days.[82] Counting civilians, the Chinese also estimated the North Vietnamese had lost about 10,000.[83] (See Maps 12.1 and 12.2.)

> In the predawn darkness of ... February 17, when thick mist blanketed the forested hills on the Sino-Vietnamese border, the Chinese People's Liberation Army unleashed its fury. ... In an awesome display of firepower, hundreds of 130mm and 122mm long-range guns and 140mm multiple-rocket launchers poured shells into Vietnam at a rate of almost one a second. ...
>
> Then, like floodwater bursting through a dam, some eighty-five thousand Chinese soldiers, supported by armor, streamed into Vietnam through twenty-six points along the border. While the main thrust soon narrowed to five main entries leading to provincial capitals, the invading troops were thrown in a wide net to destroy Vietnamese outposts. With the celebrated "human-wave" tactic, used by the PLA [People's Liberation Army] during the Korean War, thousands of soldiers attempted to dislodge Vietnamese militia and border guards entrenched on hilltops and precipices. As Peking would later admit, the tactic proved a disaster.

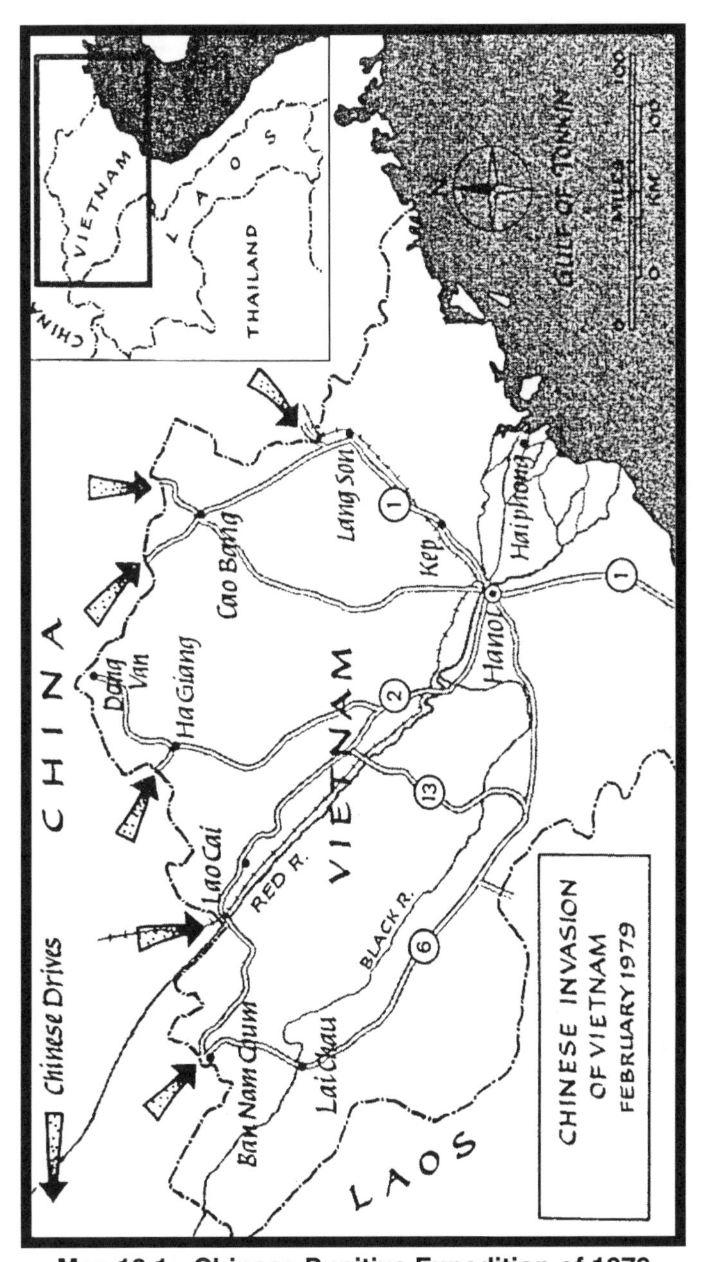

Map 12.1: Chinese Punitive Expedition of 1979

(Source: Courtesy of The Wylie Agency, Inc., from *Brother Enemy: The War after the War,* © 1986 by Nyan Chanda, p. 337)

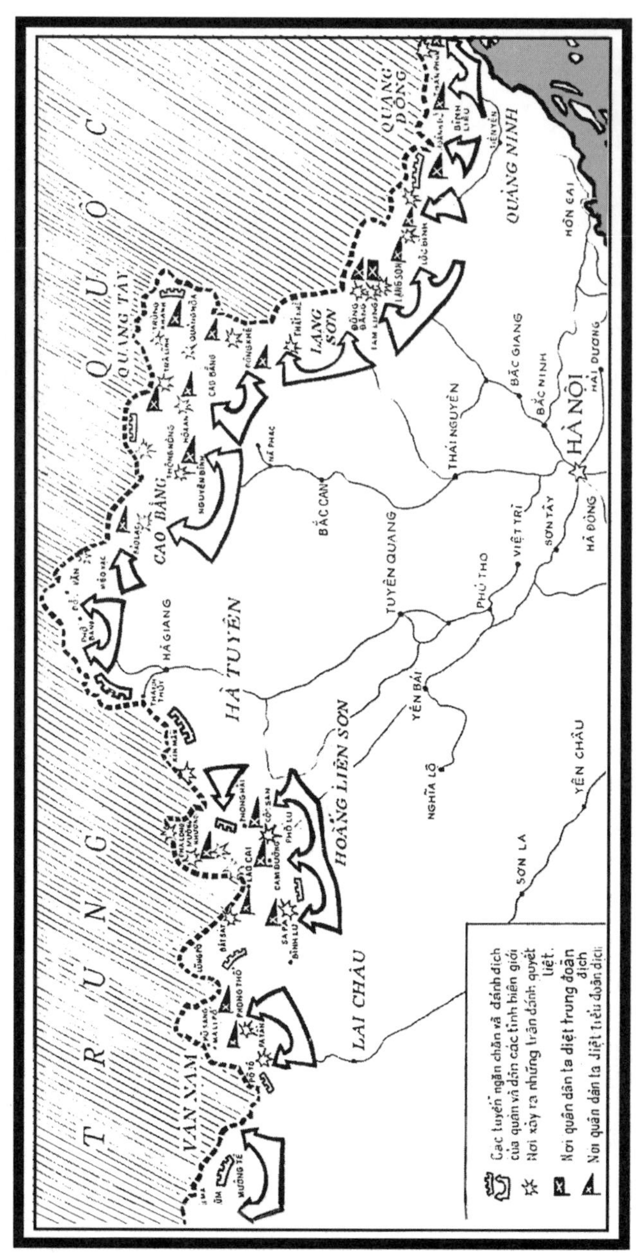

Map 12.2: Extent of the Chinese Penetration

(Source: *The War 1858-1975 in Vietnam*, by Nguyen Khac Can, Phan Viet Thuc, and Nguyen Ngoc Diep, Nha Xuat Ban Van Hoa Dan Toc Publishers, Hanoi, Figure 754)

> The Chinese had not foreseen the kind of traps the Vietnamese had laid and the maze of tunnels and bunkers they had constructed in the border region. In the first three days of fighting, the Chinese suffered heavy casualties as thousand of soldiers were cut down by machinegun fire from fortified positions, blown up by mines, and maimed by boobytraps. . .
>
> . . . [A replacement Chinese general] quickly abandoned the human-wave tactic and ordered a more discriminating attack in coordination with artillery and tank support. In ten days of fighting, armor-led Chinese . . . had slowly advanced between twenty and thirty miles inside the border, and after heavy fighting captured four [Vietnamese] provincial capitals. . . . With an intense artillery barrage followed by a tank assault, the Chinese had begun on February 27 the invasion of Lang Son—the only remaining provincial capital in the border region. . . . On March 5 [after continuous fighting] the Chinese had finally established control over Lang Son. . . . The same day they began withdrawing to Chinese territory.[84]

While North Vietnam may have achieved local air superiority during the latter stages of the battle, its soft, underground strongpoint defense is what scored the decisive victory over the Communist Chinese. From that and the U.S. experience, Vietnam has developed a veritable "Military Fortress" to blunt the next invasion from the north. (For its probable strongpoint design, see Figure 12.15.)

> A new concept has been devised to meet this contingency [another Chinese invasion]—called the Military Fortress. While new and innovative, it does have roots in the "combat village" of the Vietnam War and the still earlier "fortified village" of the Viet Minh War.
>
> The Military Fortress concept presently involves some two dozen [North Vietnamese] districts that abut on China—an inaccessible region of mountain, jungle, and Montagnard—that are to be welded into one contiguous defensive structure. Each village of the district is to become a "combat village," linked in tactical planning terms

to neighboring villages; the entire district thus becomes a single strategic entity, and all the districts together become a grand Military Fortress. Villagers are armed, and all have combat duties. . . . Each villager spends part of each day training and working on fortifications, for which he gets extra rations. The work includes digging the usual combat trench foxhole, trench, bunker, underground food and weapons storeroom, and the *ever-present "vanish underground" installation,* the hidden tunnel complex [italics added]. These are within the village. Some distance out,

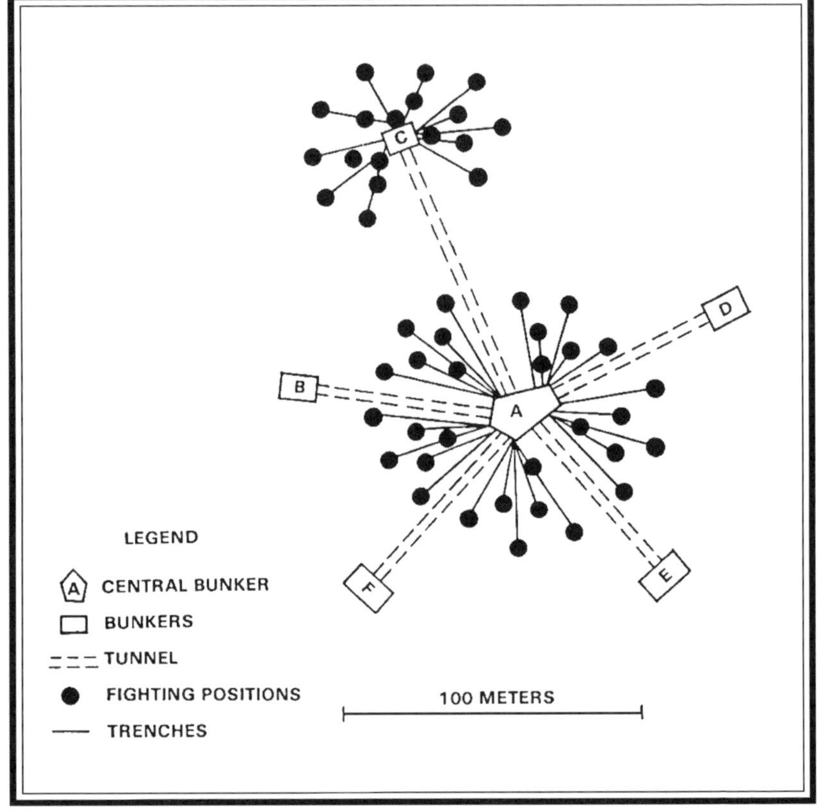

Figure 12.18: Possible Vietnamese "Military-Fortress" Segment
(Source: *Counterguerrilla Operations,* FM 90-8 (1986), p. A-6)

Figure 12.19: The Eastern Way to Kill a Tank

(Source: Courtesy of Michael Leahy, from preliminary drawing for *Phantom Soldier; FM 20-32* (1976), p. 57)

> usually two to three kilometers, is what is called the "distant fortification," a second string of interlocked trenches, ambush bunkers, manned by well-equipped paramilitary troops serving full time. Several villages (usually about five) are tied together by communication systems and fields of fire into "combat clusters" (about seven per district), and the whole becomes a single strategic entity.[85]

For tourists, the Vietnamese now tout the Binh Moc tunnels at the northern, ocean end of the old DMZ between North and South Vietnam as "bigger than Cu Chi."[86] They must have had either a "Military Fortress" prototype across their southern border during the Vietnam War, or an infiltration route under the DMZ, or both. There is certainly evidence of both. U.S. intelligence has long known of vast tunnels and caves just across the DMZ northwest of Con Thien.[87] There are also smaller, lesser-advertised excavations at the Cua Viet River mouth just south of the DMZ.[88]

Every Below-Ground Defender a Tank Killer?

Since WWI, most of America's adversaries have had personnel who specialized in fighting tanks "with their bare hands." They often shared their amazing techniques with the common soldier.

When overrun by armor during WWII, every German infantryman was supposed to hold his ground, blind the armor, and kill the follow-on infantry.[89] Every WWII Soviet soldier was expected to stop any tank that entered his area. He even had special antitank grenades with which to do so.[90] (See Figure 12.16.)

> The [Soviet] soldier must know he (personally) can fight tanks.
>
> If the tanks attack without infantry, they are to be attacked with antitank grenades and flame bottles, the vision blocks (slits) taken under fire, ball charges and antitank mines thrown under the tracks, crews exiting from a damaged tank will be destroyed by fire. If the tanks get close, the soldier conceals himself in a trench or fold in the ground.
>
> If the tanks attack with infantry, only designated soldiers will fight the tanks while all the rest attack the infantry.[91]
>
> — *Soviet Combat Regulations of November 1942*

On Iwo Jima, Japanese spider hole occupants had several ways of attacking nearby armor. One entailed pulling an antitank mine on a string under the tracks of a tank.[92] Another involved creating screening smoke, closing with the tank, and then attaching an explosive charge.

> Japanese sappers darted from bypassed spider traps with sputtering demolition charges hurled against tanks.[93]

> A Japanese squad rushed one tank with hand grenades that exploded in roars of green smoke. The Sherman was undamaged, but in moving to safety it plunged into a ravine and threw a tread.[94]

Random heroics can not stop a determined tank assault, but a 1945 battle in Manchuria demonstrated the latter technique's potential.

> Each individual of the main body of the 1st Company of the [Japanese] Raiding Battalion armed himself with explosives and rushed the enemy tanks. Although minor damage was

> inflicted on a majority of the them [the tanks], the explosives were not of sufficient strength (three to seven kilograms) to halt the tanks.[95]
>
> — US. Armed Forces Far East
> Military History Section

In essence, the Japanese needed larger or armor-piercing shape charges.[96] When they tried bigger charges at Mitanchiang, they succeeded.

> Soviet tanks also reached 126th [Japanese] Infantry Division headquarters, where "a squad of firemen from the transport unit, each armed with a 15 kilogram explosive, attacked the leading five tanks in a suicide charge, one tank per man, and successfully demolished all five tanks.[97]
>
> — US. Armed Forces Far East
> Military History Section

During the Korean War, rank-and-file Chinese infantrymen may have received training in some of the same techniques.

> All [Chinese] armies [entering Korea] offered [to the common soldier] classes in antitank training.[98]

While approaching a firefight near the DMZ in Vietnam, Marine tanks would often be disabled by mines.[99] Their nemesis was probably a spider hole occupant with the "mine-on-a-string" trick.[100] In the din of battle, he could have launched an RPG without anyone noticing it.

In an urban setting, it has been said that every man is a tank killer. During the 1956 Budapest Uprising, scores of Russian tanks were destroyed by hand-delivered molotov cocktails.[101] Multistory buildings along narrow streets made perfect launching platforms for fire bombs. Sewers provided the routes of access and egress. Would not a dense forest above honeycombed ground provide the same opportunities? In essence, the Asians have given rural terrain many of the defensive characteristics of urban terrain. The sooner the West realizes the implications the better. To this day, North Korean privates practice killing tank mock-ups at close range.[102]

Next Time Around

Many of this chapter's illustrations come from German, Russian, Chinese, North Korean, and North Vietnamese literature. Blatantly obvious are the trends in how Eastern armies fend off attack. One could reasonably conclude that future U.S. adversaries will do something similar. They will have two ways of defending close terrain. For a mobile defense, they will rely on a rearward-moving matrix of successively occupied fallback positions. It will closely resemble what maneuver warfare proponents call an "ambush in series." America's firepower will be largely ineffective against lightly manned forward slopes and recently abandoned positions.

For a positional defense, America's foes will use some variant of the below-ground elastic model. It is extremely doubtful that U.S. spy satellites, planes, or drones will ever be able to see below ground. If the enemy can't adequately bury or camouflage his positions, he will simply hide them among any number of dummy emplacements. Allowing one's opponent to pass overhead is a foolproof way to envelop him. As with the WWI Germans, the distinction between defense and offense will have blurred. This is a good indicator of tactical evolution.[103]

13 At the Listening Post

- *Is a nighttime observation post in much danger?*
- *What can a ninja-trained infiltrator do to it?*

CHINESE INFANTRYMAN, WINTER 1950

(Source: Courtesy of Osprey Publishing Ltd., from *The Korean War 1950-53*, Men-at-Arms Series, © 1986 by Osprey, Plate H, No. 1; *FM 90-10-1* (1982), p. B-43)

Herein Lies Great Danger

The nighttime listening post is the private's most dangerous assignment on defense. Without a good entrance and exit strategy, great location, and special training, he and his buddy will be at the skilled stalker's mercy. On the night of 24 October 1942, Sgt. Mitchell Page heard Japanese chatter where his three-man listening post had been and plastered the area with grenades.[1] By this time up the line, Marine legend Chesty Puller had already resorted to a 46-man outpost.[2] Clearly, the Japanese ability to spot U.S. skirmishers had exceeded the skirmishers' ability to hide.

To have the best chance of survival, the two-man observation post must have—in some fashion—done the following. So as not to be seen entering its lair, it was dropped from the end of a perimeter patrol. To discourage infiltrators, it has an unobstructed view to beyond grenade range on all sides. To prevent being rushed, it has natural obstacles to the front and sides. To disrupt enemy momentum, it has a preregistered mortar concentration and claymore daisy chain to its front. To keep prowling scouts from learning its precise location, it is not linked to its headquarters by wire. To return to its parent unit under fire, it has a covered route. To avoid being killed on the way in, it has a way of discreetly sharing intentions with the current shift of watchstanders. As Eastern armies tend to have a greater appreciation for microterrain and enlisted advice, they already practice many of these procedures. Any trends would be particularly noteworthy.

Germans

The WWII Germans used two-man "reconnaissance and visiting" patrols within the line of sentry posts. Their job at night and in broken terrain was to observe the intervening areas not occupied by posts, and perform liaison.[3]

To penetrate such a screen, an enemy scout would have a much harder time. While stalking a sentry in his lane, he would still be visible to a roving sentry.

Japanese

The 1913 Japanese night-fighting manual lists the following conditions for a good listening post: (1) broad and unobstructed field of view, (2) no obstructions to hearing like wind or gurgling stream, (3) in the shadow with the moon behind, (4) with a visible skyline to the front, (5) not known to the local populace, (6) with enemy-assault-stopping obstacles to the front and flanks, and (7) not stationary.[4] The sentries were also told to stay off any skyline and lower than the objects behind them.[5] To do so, they would have had to find good places to observe their sectors from low ground.

As the night sentinel often had to shift location, he was further trained in night reconnaissance. For that, he was given additional advice: (1) to remain unnoticed, follow along the base of physical objects and terrain features; (2) to hear better, hold breath; (3) to spot enemy more easily, look for objects on the skyline; (4) to avoid being surprised, watch for enemy in the shadows; and (5) to be forewarned, form judgments from the sounds heard.[6] If the Japanese sentinel had to fire, he did so quickly and from a protected position. When fighting the Russians, Japanese sentinels were often approached by people wearing their own uniforms and speaking their own language.[7] Another Russian trick was to create a diversion out front of the Japanese listening post while a few flankers crawled to within assaulting distance.[8]

Unique to the Eastern listening post is its potential to shift location. At one point during the Nomonhan Incident of 1939, Japanese defensive scouts were able to shadow Russian infiltrators.[9] While on the move, night sentinels were taught not to do the following: (1) make noise or cast shadows, (2) move quickly, (3) go in the wrong direction, (4) approach another sentinel unannounced, or (5) compromise a passing patrol.[10] In fact, the patrols and entire sentinel line were required to share with one another everything they had learned about the enemy.[11] This is a perfect example of how much more can be accomplished by decentralizing control. Western commanders would never dream of operating this way. The Russian infiltrators must have been equally skilled in the Russo-Japanese War of 1904-05, because the subsequent Japanese manual warned of Russian interference during the relief of the sentinel.[12]

If the mission of a Japanese patrol were the night reconnaissance of an enemy position, the patrol would make its move during a sentinel's relief or patrol's passage.[13] A *ninpo* master might have fallen in behind that patrol.[14]

Of course, some of the Japanese outposts were stationary. On Guadalcanal, the Nipponese hid in treetrunks and treetops. On Iwo Jima, they hid in tiny, hollowed-out hummocks.[15] Some of those hummocks may have been prefabricated.

Russians

During WWII, the daytime Russian sentry post was manned by two pickets. With one in charge, these pickets would often occupy separate locations. One would climb up in a house or tree while the other stayed on the ground.[16] They would challenge anyone who

Figure 13.1: Russian Observation Posts

(Source: *Podgotovka Razvegchika: Sistema Spetsnaza GRU,* © 1998 by A.E.Taras and F.D. Zaruz, pp. 374, 375)

came near. They communicated in undertones during the day and signals or whispers at night. The nighttime sentry post was manned by three people and operated quite differently.

> The listening post does not stop anyone and questions no one; individual persons and small enemy reconnaissance patrols (2 to 3 men) are let pass. . . .
>
> The listening post comes back at the end of their tour of duty, for this they look for a previously arranged signal and return independently.[17]
>
> — *Soviet Combat Regulations of November 1942*

Sometimes Russian pickets were dispatched from a squad outpost. During the day, they were placed 100 meters out and at night, 50.[18] Their long-term outposts disappeared into the microterrain. (See Figure 13.1.)

In Afghanistan, the Russians operated somewhat differently. According to a DoD-published study, "The security outposts functioned around the clock. During the day, one man per squad or tank was on watch while a two-man patrol worked the area."[19]

An Outpost Need Not Remain Passive to Hide

The U.S. listening post normally comes in after spotting enemy. Its mission is to alert the parent unit without disclosing its own location. However, it could call in preregistered fire and/or trigger command-detonated mines without violating either part of that mission. (See Figure 13.2.)

There is a good chance the Japanese tank killer teams on Iwo Jima were double hatted as sentry posts.

How the Eastern Infiltrator Bypasses a Sentry

Professional night fighters are keenly aware of what any potential opponent can or cannot see. Of particular interest are the blind spot in the middle of his retina and the limits to his peripheral vision. These limits can be quickly ascertained by forming the figure "T" with one's arms. With index fingers pointed up, the arms are then moved forward until both fingers become visible. By watching

Figure 13.2: An Outpost Can Fight without Revealing Itself

(Source: *FM 20-32* (1976), p. 151)

the base of the sentry's skull, the *ninja* can predict which way his head and eyes will move,[20] and thereby stay out of view. When threatened, the sentry can defeat the technique by spinning his head and body around simultaneously.

A skilled infiltrator will exploit any defensive weakness. Such weaknesses fall into five categories: (1) interruptions to watchfulness, (2) diversions, (3) changes in lighting conditions, (4) microterrain avenues of approach, and (5) various combinations of the other four.

Natural Interruptions to the Sentry's Watchfulness

Every time a sentry's attention is diverted from his assigned sector, an infiltrator can capitalize on the opportunity. One of the best times to crawl up on a U.S. sentry is when his leader checks the lines or his relief takes over.

Almost anything the sentry does—from eating to scratching, to stretching, to urinating—constitutes a temporary gap in his defenses. When a sentry dreams of home, it's a gap.

Sentry Diversions

Everything from a passing helicopter to distant flashes will temporarily divert a sentry's attention. In a quiet sector, many can be orchestrated. The most common impromptu diversion is a pebble thrown into a different location. Below are listed a few others.

To work, a diversion need not capture the sentry's conscious attention. Only required is that his focus be drawn to the wrong thing.

Sudden Changes to Lighting Conditions

The human eye takes a while to adjust to a sudden change in the brightness of ambient light. That change creates a temporary gap for a short-range infiltrator. A stationary sentry is temporarily blinded every time he lights a cigarette, sees lightning, or looks at a cooking fire. His normal night vision is also interrupted by NVGs and thermal-imaging devices. There are additional chinks in his armor when a full moon is obscured or an aerial flare dies out.

Even in the daytime, or under artificial illumination, or during full moonlight, any deep shadow will constitute a gap. Any movement within that shadow will be invisible to the sentry. His eyes can apparently not process both light and darkness at the same time.

Microterrain Avenues of Approach

The most obvious avenues of infiltration are wet places, watershed ditches, ripples in the ground, rotting tree trunks, uncleared vegetation, sewers, etc. Just as a deer can avoid being seen in the daytime, so too can a skilled infiltrator disappear into his surroundings.

Combinations of Gaps

When a defender positions himself behind the military crest of a hill, he allows his adversary to crawl up behind that crest unobserved. All that adversary then needs to pass the sentry is a sudden change in lighting conditions, interruption to watchfulness, or diversion.

Someone trained in sentry removal, could easily manipulate his quarry. He must remember, however, that humans recognize shape whereas many animals do not.

A Disturbing Thought

Regularly used by every *ninpo* practitioner is *saiminjutsu.*[21] While he normally reserves it for self-hypnosis (to better focus on the job at hand), he may also apply it to his quarry.[22] Hypnosis is not a form of sleep; it simply causes its recipient to concentrate on one area at the expense of others.[23] As hazing is one of techniques of *saiminjutsu,*[24] the Eastern short-range infiltrator or his accomplice may try to divert a sentry's attention with repetitive sights or sounds. The sentry who is overly susceptible to suggestion might put himself and his unit at great risk.

With Contact Patrolling

- *How does the Eastern unit handle a chance encounter?*
- *Why can't its quarry return accurate fire?*

SOVIET MOTOR RIFLES PRIVATE, SUMMER 1945-70

(Source: Courtesy of Cassell PLC, from *World Army Uniforms since 1939*, © 1975, 1980, 1981, 1983 by Blandford Press Ltd., Part II, Plate 1; *FM 21-76* (1957), p. 89)

The "Low-Tech" Security Patrol

To dissuade a more powerful attacker, one would first have to disrupt his momentum. The underdog in football alters momentum through goal-line stand or unexpected gain. The weaker opponent in war changes it through rearward attrition or spoiling attack.

That's why the "low-tech" army needs highly proficient security patrols. Each must be able to seek out and successfully attack a much larger maneuver element. To get within attacking distance, it must also be able to annihilate any unit of equal or smaller size. All of this takes continual practice of well-tested techniques.

Why the Eastern Unit Gets Surprised Less Often

To be forewarned, the "low-tech" army must watch the adversary's patrol bases and shadow his forays. While patrolling during the 1920's "Banana Wars," Marine legend Chesty Puller would periodically drop off a fire team to dissuade any followers. In Vietnam, enemy self-defense forces probably monitored the whereabouts of every American unit—however small—to enter their area. That would explain why so few main-force VC or NVA units got ambushed. More recently, the technique has reappeared in Afghanistan.

> Taliban and al-Qaeda fugitives have paid teenage Afghans to act as spies. The agents position themselves outside known U.S. special-operations bases near Kandahar and near Khost in eastern Afghanistan and notify their handlers when special-operations patrols leave the compounds.
>
> In at least two incidents, Green Beret A teams have confronted armed Afghan men who appeared to be following the soldiers. In one case, an officer shot an Afghan who raised his weapon as if to fire.[1]
>
> — *Washington Times,* 15 January 2003

Where the occupier uses its special operators for combat instead of intelligence gathering, as both the Russians and Americans have done in Afghanistan,[2] an Eastern irregular is seldom surprised. Even with good intelligence, his maneuver elements are tough to spot. Under constant threat of bombardment and ambush, they develop sophisticated movement techniques. Their point security elements can soon see, hear, smell, or otherwise sense any irregularity in their surroundings. Just as an undersized animal will do in the woods, their smallest patrol will seize the initiative, albeit briefly, in a meeting engagement.

The Meeting Engagement with a Larger Unit

When an Eastern security patrol spots a larger enemy force on the move, it makes just enough contact to delay it. If the two are moving along the same trail, the Eastern unit will set up a hasty ambush as the Japanese did on Guadalcanal.[3] If the two are not moving along the same trail, the Eastern unit will often sneak closer. The WWII German squad would work as far forward as possible without shooting.[4] As soon as it took fire or ran out of cover, it would open up with its light machinegun.[5] Whether this attack by fire was followed by a ground assault would depend on the specific circumstances.[6] Either way, the gun had to displace after it had divulged its location.[7] While the easiest direction would have been rearward, the security patrol generally required the permission of its headquarters' to withdraw.

On sighting a larger unit, the squad-sized WWII Soviet patrol was only slightly less aggressive than the German. If undetected, it reported its sighting to higher headquarters and kept the enemy force under surveillance. This gave its parent unit an artillery attack option. The squad knew how to withdraw under fire but could only do so with their commander's permission. It did so by splitting into two groups (with the machinegun and squad leader in one and the rest of the squad in the other). Those groups alternately covered each other and moved rearward.[8] If a Soviet security patrol actually collided with the larger unit, its instructions were to assault to break through.[9] If the enemy force were close, the assault would probably take the form of a stand-up rush. If the enemy force were farther away, the assault would probably be by fire and movement. Each squad would place its organic machinegun on its flank and then alternate moving forward with the gun (as depicted in a painting in Moscow's Armed Forces Museum).[10] Archival film footage in the same museum shows individual Soviet soldiers getting down every time small-arms fire comes anywhere near them. They then jump back up as soon as the fire is directed elsewhere. Of note is the relatively greater degree of decision-making leeway granted to the Eastern rifleman in this circumstance.

To break the momentum of an enemy maneuver force, the Asian patrol will snipe—with a single shot—that unit's leader or loft a string of mortar rounds. Even when not preregistered, these rounds can be amazingly accurate. When a VC mortar platoon commander switched sides in January of 1971, he revealed how he could hit certain types of targets so accurately and quickly. He operated his 82mm mortar in much the same way the Japanese had used their smaller-caliber "knee" mortar.

[The VC] told how he had mortared Dat Do earlier in the

> year. His rounds had landed in a string that stretched across both sides of the [ARVN] District Headquarters to the hospital . . . He carried the tube and his few soldiers carried the baseplate and the mortar rounds. No tripod was needed. At a given point he stood, legs apart, behind the baseplate, held the tube in his hands and raised his thumbs to be used as sights.[11]

By "direct laying" his mortar from somewhere on the same axis with a linear target, he was quickly able to place an accurate sheaf. To cover the entire target, he had only to tilt the mortar slightly between rounds. One Marine lieutenant experienced mortar rounds that walked up and down roads and front lines as if they had eyes.[12] The first launch tube had to be up the road from him, and the second to one side of his lines. The North Vietnamese may not have invented the technique. Pictured in Moscow's Armed Forces Museum is an 82mm mortar without a sight.[13]

The Meeting Engagement with a Comparably Sized Unit

The WWII German squad would rush an opponent of equal size. While it tried to do so with bayonets and grenades only,[14] it still brought its machinegun.[15] The WWII Russian squad did likewise, except that it left the machinegun behind to cover the maneuver.[16]

During an unexpected engagement, the Asian also prefers a "boom" to a "bang." Whereas one connotes mortar or mine, the other means ground contact and reinforcement. The North Vietnamese patrol often initiated contact with an RPG. U.S. Marine Sergeant Les "the Rock" Ford had an RPG fired at him at the start of a chance encounter.[17] He was not the only Allied soldier to see this.

> The enemy's quick reaction contact drill [in Vietnam] was excellent, firing an RPG round immediately on contact. I remember it fly past my right shoulder and thinking how beautiful it looked at night, just like a large firecracker.[18]
>
> — 2d Lt. Karl Metcalf, Royal Australian Regiment

Russian Spetznaz troops now have a hand-held automatic grenade launcher.[19] Its four grenades would be lethal to Western scouts and could be made to sound like a short string of mortar rounds.

Figure 14.1: The Eastern Soldier Is Elusive

(Source: *A Concise History of the Unites States Marine Corps 1775-1969*, by Capt. William D. Parker, sketch by Capt. Donald L. Dickson, Hist. Div., HQMC,1970, p. 62)

The Meeting Engagement with a Smaller Unit

When an Eastern patrol spots a smaller adversary first, it will try silently to annihilate it. In a meeting engagement with a few enemy soldiers, the WWII Russian squad would fall on them with bayonet only and take as many prisoners as possible.[20] When the sighting was simultaneous, the larger Eastern force demonstrated at the center and sent forces around both sides to encircle its prey. This is just another example of the "encirclement and annihilation" thread that runs throughout Asian, Russian, and German tactics. (See Figure 14.1.) The idea was to lessen the quarry's direct fire by attacking him from every direction at once. To avoid his indirect fire, one had only to close the noose.

By U.S. admission, the Japanese so revered encirclement as to differentiate it from double envelopment and disregard the opposing unit's size.[21] On his famous patrol behind Japanese lines on Guadalcanal, Chinese-trained Evans Carlson routinely sent people around both flanks on a loop wider than his opponent was using.[22] For the WWII Germans, a double envelopment was doctrinally correct. They also practiced encirclement. At the platoon level, it was accomplishing by opening up with the light machinegun in the middle and sending maneuver squads around both sides.[23] To complete the encirclement as quickly as possible, the flanking squads would advance in column with their machineguns up front.[24] Then by tightening the noose in the following way, they could deny the enemy accurate return fire.

> [T]he Gruppen [infantry squads], without firing their weapons, work forward as close as possible to the enemy. These Gruppen carefully exploit all available cover and concealment which the terrain offers: ditches, wooded areas, bushes, etc. They should advance by marching, running, or crawling as the situation demands. . . .
>
> If possible, the *Zugfuhrer* [platoon commander] employs several Gruppen advancing from various directions . . . In this way the defensive fires of the enemy will be scattered.[25]
>
> — WWII German small-unit infantry manual

Some 40 years later in Afghanistan, the Russians would routinely open up with a machinegun and then move it slightly during their double envelopment. This was done so that the machinegun would not receive accurate return fire.[26]

The *Coup de Grace*

Once the smaller prey is encircled, the Eastern force has well-tested ways to finish it off. If the Eastern force is large enough, it first employs a second circle.

> Their [Soviet] normal methodology called for an inner encirclement force to hold the trapped force in place while an outer encirclement force pushed out from the encircled area to put distance between the trapped forces and an enemy rescuing force. Only after the two forces were in place, would the Soviets fragment and meticulously destroy the trapped force. . . . In Afghanistan, the enemy . . . could usually slip through the Soviet encirclement.[27]
>
> — DoD-published Soviet military academy study

Then, there is the *coup de grace*. It is as safe as it is lethal. To escape the prey's supporting arms, the "low-tech" force tightens the noose. To escape the prey's small arms, the "low-tech" force tightens the noose in a unique fashion. It allows its riflemen to individually crawl and randomly shoot from every direction. As the shooters are behind cover and their fire well aimed, they don't shoot each other by accident. Under intermittent fire from all sides, the quarry can never acquire a target and soon becomes disoriented. Then the prey is segmented (as the Marines were on Operation Buffalo),[28] and the process is repeated for each segment.

Where Only One Squad Is Involved

Where only one squad made contact, most Eastern armies preferred a single envelopment. The WWII Russians would alternate moving the light machinegun and the rest of the squad.[29] On flat ground, the troops would advance to one side of the gun only.[30] In the Afghan hills, they advanced under the watchful gaze of a BMP (Soviet armored personnel carrier) gunner.

Where only one WWII German squad was involved, it would either bound forward with its machinegun or occasionally use a single envelopment with its machinegun as a base of fire.

> The squad column *(Schutzenreihe)* formation is used for approaching the enemy during the fire fight when only the light machinegun is firing and the *Schutzen* are held back.[31]
> — WWII German small-unit infantry manual

Of note, the members of an Eastern squad will not always move forward *en masse* or standing up.

When It Comes Down to Assault by "Fire and Movement"

The WWII German squad preferred to move forward in column so that their machinegun could more easily shoot around them. Only when they came under intense fire, would they advance individually by bounds or crawling.[32]

While one might think the Russian disinterested in fire and movement, quite the opposite is true. To keep from getting shot, individual soldiers were allowed to rush randomly forward. Notice how they escaped detection on the next rush.

> In open terrain and under heavy enemy fire the squad leader must let the soldiers run forward in short bounds and support them with fire. Each soldier *individually* goes to earth promptly after arriving at his position by exploiting hollows, shell holes, and other cover and fires on direction of the squad leader or *individually* [italics added].
>
> The guide man always moves first. . . .
>
> The length of the distance covered in the momentary runs and therefore the overall speed depends on the terrain and the enemy fire. The more open the terrain and the stronger the enemy's fire, the faster and shorter the dashes must be.
>
> The dash should occur suddenly. . . . At the end of the dash the soldier falls to the ground like a stone, crawls unnoticed to the side, and takes up a firing position.[33]
>
> — *Soviet Combat Regulations of November 1942*

Asians also hastily assault by fire and movement. As each element stops, it disappears.

> On Operation Buffalo in Vietnam, the North Vietnamese used fire team rushes. Their three-man fire teams were so well camouflaged as to become almost invisible when they stopped in the waist-high elephant grass.[34]

Double-Column Pursuit

The WWII Germans would pursue an enemy patrol with several units across a wide frontage.[35] Russians from the same era pursued or advanced with two parallel columns.[36] The WWII Japanese routinely moved forward in two or more parallel columns.[37] The whole idea was to be in a position to pass, cut off, and encircle any opposition.

This differs from the U.S. Army's "bounding overwatch" procedure. Eastern forces try to get behind their opponent, rather than just outflank him. That way, their attack can take the form of an ambush.

15 On Point

- *How would an Asian forward-security element operate?*
- *For what does the point man look?*

NORTH VIETNAMESE ENLISTED MAN, 1975

(Source: Courtesy of Osprey Publishing Ltd., from *Armies of the Vietnam War 1962-75,* Men-at-Arms Series, © 1980 by Osprey, Plate H, No. 1; *FM 21-76* (1957), p. 79)

The Function of a "World-Class" March Security Element

Only in theory is the Western point man the eyes and ears of his unit. How can a recent recruit detect boobytrap, ambush, or strongpoint when he has no idea what to look for? Only rarely is he even allowed to choose the safest route to the next treeline.[1] Sadly, his Eastern counterpart enjoys more responsibility and therefore develops more ability. The Eastern private's natural instincts are more highly tuned. He is expert at recognizing human sign. He can track that sign to its source. He can make a tactical decision for his squad. He can also virtually disappear.

Armies from the East depend more on their scouts than do those from the West. They can't afford an unexpected encounter with their more powerful adversary. This scouting expertise extends all the way from Berlin to Indonesia.

> The *mujahideen* . . . were natural scouts.[2]
> — DoD-published Soviet military academy study

With Additional Responsibility Comes More Stress

To walk point without getting killed, one needs considerable patience. Asians are known for their patience, but Americans aren't—they must be rotated hourly.

> [On point,] stress keeps you sharp, and that helps keep you alive. The sense of danger throws all of the body's systems into high gear. Adrenaline pours into the bloodstream, the heart rate goes up, muscles tense for fast action. The animal that you live in screams *do something!*
>
> But the brain says no! Instead of running or fighting, the things your body wants to do, you concentrate on the job at hand. . . .
>
> Slowly, deliberately, you go forward. Every movement is a trained response, thought out well beforehand. . . . More water is consumed inside of you as the body's functions stay in high gear. Damn but you can get tired and thirsty on point![3]
> — U.S. SEAL point man in Vietnam

How the Easterner Employs His Keener Senses

To detect trouble before it happens, the point man must know the lessons of Chapter 5 and more. All of his senses must be fully engaged, and his brain free to assess their impulses. His equipment can in no way impede this process.

To survive for long on point, he must be able to spot the tiniest inconsistency in his ever-changing surroundings. To do so, he has to understand the behavior of native animals, birds, and insects.[4]

He must also be able to pick up the slightest movement at the very edge of his peripheral vision. Tiger hunter Corbett hints at the extraordinary level of skill required.

> On one occasion the darting in and out of the forked tongue of a cobra in a hollow tree, and on another occasion the moving of the tip of the tail of a wounded leopard lying behind a bush, warned me just in time that the cobra was on the point of striking and the leopard on the point of springing. On both these occasions I had been looking straight in front, and the movements had taken place at the extreme edge of my field of vision.[5]
>
> — James Corbett

The point man must do more than distinguish a barely audible aberration from the cacophony of background noises. He must be able to pinpoint that aberration.[6] Upon this skill, his very life will depend.

> Animals who live day and night with fear can pinpoint sound with exactitude, and fear can teach human beings to do the same. Sounds that are repeated, for instance a langur [monkey] barking at a leopard, . . . are not difficult to locate nor do they indicate immediate danger or call for instant action. It is the sound heard only once, like the snapping of a twig, or low growl, or the single warning call of a bird or animal, that is difficult to locate, that indicates immediate danger, and calls for instant action.[7]
>
> — James Corbett, *Jungle Lore*

The point man must be able to assess his chances of moving noiselessly through all types of vegetation.[8] He needs to do what he can to keep resident animals from breaking cover or otherwise sounding the alarm. He should be familiar with their language, if not able to speak it. With a thorough understanding of regional bird calls, he can secretly talk to his partner like the American Indians used to do. At his disposal are calls that are rarely heard or from nonindigenous species.

> There is no universal language in the jungles; each species has its own language, and though the vocabulary of some

> is limited, as in the case of porcupines and vultures, the language of each species is understood by the jungle folk [other animals]. The vocal chords of human beings are more adaptable than the vocal chords of [most] jungle folk, and for this reason it is possible for human beings to hold commune with quite a big range of birds and animals. The ability to speak the language of the jungle folk . . . can . . . be put to great use.[9]
>
> — James Corbett

To "feel" for hidden dangers in the trail, U.S. special operators sometimes traveled barefooted in Vietnam.[10] They even made a valiant attempt to harness their sense of smell. Unfortunately, to smell adversity is much easier said than done. It requires delineating occasional odors from routine ones.

> Lying there in the mud and humus of the jungle floor, you pay attention to the area around you in a different way. Any small warnings of an enemy presence have to be filtered out from the sensory assault of the jungle. There's the earthy rotting smell of the ground less than a foot below your nose. Or is that the body odor of a Vietnamese, a man who habitually flavors his food with *nuoc mam,* that fermented fish sauce that's so common in this part of the world? If it is a Vietnamese, is he VC or a friendly?[11]
>
> — U.S. SEAL point man in Vietnam

As a *ninja* would do, the point man must take care never to silhouette himself, throw an unnatural shadow, or cast a reflection. Constantly aware of the wind's effect on scent and sound, he must continually reassess his route. In Asia, his opponent will have keener senses than his own. Jim Corbett had an interesting way to lessen the threat.

> [I]t [a man eater] approaches its intended victims upwind, or lies up in wait for them downwind. . . .
>
> In all cases where killing is done by stalking or stealth, the victim is approached from behind. This being so, it would be suicidal for the sportsman to enter [a] dense jungle . . . without making full use of the currents of air. . . . [O]wing

> to the nature of the ground, in the direction from which the wind is blowing, the danger would lie behind him, . . . but by frequently tacking [moving back and forth] across the wind he could keep the danger alternately to the right and left of him. . . . I do not know a better or safer method of going upwind through dense cover.[12]
>
> — James Corbett

Finally, the point man must actively cultivate his sixth sense. He need not know from where it comes, only that it can save his life.

> On this occasion I had neither heard nor seen the tigress, nor had I received any indication from bird or beast of her presence, and yet I knew, without any shadow of a doubt, that she was lying up for me among the rocks. I had been out for many hours that day and had covered many miles of jungle with unflagging caution, but without one moment's unease, and then, on cresting the ridge, and coming in sight of the rocks, I knew they held danger for me, and this knowledge was confirmed a few minutes later by the karker's warning call to the jungle folk, and by my finding the man-eater's pug marks superimposed on my footprints.[13]
>
> — James Corbett

How the Eastern Point Team Operates

The WWII Germans used two-man point and flank security elements at maximum visible range from the column.[14] The Japanese also used a two-man point (Figure 15.1). As they were highly skilled at traveling through the jungle, a few Westerners tried to learn from them.

> When I am scouting and come to an opening in the [Guadalcanal] jungle, and have to cross it, I generally run across quickly and quietly. Going slow[ly] here may cost a scout his life. Different types of terrain call for different methods
>
> Here is the way Japs patrol. I was out on the bank of the river with another man. We were observing and were

carefully camouflaged. We heard a little sound and then saw two Japs crawl by about 7 feet away from us. These Japs were unarmed. We started to shoot them, but did not do so as we remembered our mission. Then, 15 yards later came 8 armed Japs. They were walking slowing and carefully. We did not shoot as our mission was to gain information. When I got back, we had a lot of discussion as to why the two Japs in front were not armed. . . . I believe they were the point of the patrol.[15]

— Plt.Sgt. C.C. Arnt, Marine scout on Guadalcanal
FMFRP 12-110

Figure 15.1: Point Security Is a Two-Man Job

(Sources: *MCOP1550.14D* (1983), p. 10-8; *FM 22-100* (1983), p. 164)

How the Easterner Reconnoiters Different Ground

While on the move, the WWII Russians sent two-man point and flank security elements out to the limits of sight and sound, not to exceed 300 meters during the day and 100 meters at night. Better to support each other, the point and squad moved by bounds.[16] In other words, the scouts did not move to the next bend in the trail until the main body had caught up. This leapfrog motion made possible routine listening. The WWII Russians had ways to scout all types of terrain.

> Small woods are scouted as they are traversed and while moving along the edge of the woods; a larger, but not thick woods, is searched by a line of scouts. Thick woods are searched along parallel trails and fire breaks ("aisles"); special attention must be paid to tree trunks, tree tops, windfalls and underbrush. . . .
>
> In scouting a height, the scout should not move onto the ridgeline/crest. . . .
>
> In the examination of a deep stream, the scouts investigate it thoroughly and along both banks.[17]
>
> — *Soviet Combat Regulations of November 1942*

Japanese riflemen had been shown how to cross different types of terrain since 1913—so that they could function as messengers. Of particular note, they were also taught how to use linear terrain features like the streets of a road map. In effect, every Nipponese private was instructed in the advanced land navigational technique of "terrain association."

> [The messenger traverses the countryside by] moving along a prominent extended physical object (as river, mountain, forest, etc.) . . .
>
> After entering the physical object (woods, etc.), do not mistake the direction on exit. . . . If possible to pass around the flank of the object, it is preferable to going through it. The interior of villages and woods are important, but it is best not to enter them . . . ; roads in the interior of a village are complicated, and it is often easy to lose direction. In important cases, the messenger will go down to the stream to verify the road. . . .

... For example, in going from A to B, when the road is indistinct and cannot be used, follow along the stream which flows in the direction A—B. . . .

By this method, or by the direction of mountain ranges, rice-fields, ravines, etc., the general direction can be kept.[18]

— Japanese night-fighting manual, 1913

How the Eastern Point Team Disappears from View

The point man must conceal or disguise the following eight aspects of his signature: (1) shape, (2) shine, (3) shadow, (4) surface, (5) silhouette, (6) movement, (7) sensory signature, and (8) ultraviolet or infrared signature.[19]

Outlines that are obviously human must be distorted. Most recognizable are one's upper torso and weapon, but the inverted "V" of his legs can also give him away. Crawling is the least-visible means of lateral movement.

Routinely carried items can—through their reflection—give away a point man. Most dangerous are his shaving mirror, glasses, compass face, watch crystal, binocular lens, and knife blade. However, under certain lighting conditions, his canteen cup, entrenching tool, and cooking pot can be a liability.

So as not to draw attention to contrasting surfaces, the point man must stay in the shadows. While U.S. forces seldom worry about the shadows they cast, Eastern forces have been trained to monitor their shadows. As early as 1913, Japanese privates were asked to assess naturally occurring shadows for abnormalities.[20]

To blend with natural surroundings, the point man employs disruptive patterns and light-absorbing materials. All the while, he must remember that, in full moonlight, black clothes can look almost white.

The point man must not silhouette himself. This takes considerable practice. He must move at a crouch through the lowest ground available. He must walk just above the military crests of hills. He must crawl across elevated places that cannot be avoided. He must avoid "false horizons"—areas that are more open or lighter in color than those around them. When nearing a danger zone, he should keep a shaded treeline or other dark backdrop to his rear.

The point man conceals his movement by staying behind things, on low ground, and in the shadows. At night, the lighting conditions will dictate how much of his movement can be detected with the naked eye. In full moonlight, he must move very slowly to remain unnoticed.

For he who serves on point, any noise and odor attributable to the human condition must be minimized. Wherever possible, the point man stays downwind of potential trouble.

Now, the nighttime point must additionally worry about being picked up by night-vision and heat-sensing devices. He can minimize his signature by wearing garments that are specially treated to minimize the reflection of heat and ultraviolet (UV) radiation. Or he can wash untreated camouflage clothing with UV-brightener-free soap.[21] Even without special clothing, he can still operate with relative impunity in the rain or fog and under a cloudy sky or foliage canopy.

When It Comes Down to Gun Play

The NVA point man often carried an RPG to handle trouble without the telltale sound of small arms. If he didn't have an RPG, he slung his rifle over his shoulder and held it waste high. As he was quick to shoot, any opponent had to be quicker. A professional tracker of Communist guerrillas in Africa preferred the "cover shoot" technique. After the first exchange of gunfire, the guerrilla would often hit the ground and roll behind cover. The tracker would then place an additional "double tap" (two rounds) into every potential area of cover.[22]

To generate surprise, Eastern forces have never needed rifles with silencers. Still, they would be of great help to Americans. The SEALs asked for them in Vietnam, but their request was denied.[23]

There Are More Subtle Risks to the U.S. "Want-to-Be"

As was shown in Chapter 2, there is a whole portfolio of *ninja* technique for killing point men. One involves hiding behind a trail-side tree and moving around its opposite side as the point man passes. The ease with which the quarry's neck can be broken is chilling.

Only the most stout of heart and sound of mind should attempt to walk point in the East without months of peacetime practice. To this day, the U.S. Marine Corps has no formal school on walking point. Within the U.S. Army, there are only a few unit courses on the subject.

About Tracking an Intruder

- *Will an accomplished tracker look for every footprint?*
- *Where should he keep the tracks in relation to the sun?*

CAMBODIAN "KHMER ROUGE," 1960's-70's

(Source: Courtesy of Osprey Publishing Ltd., from *Armies of the Vietnam War 1962-75*, Men-at-Arms Series, © 1980 by Osprey, Plate C, No. 3; *FM 21-76* (1957), p. 77)

An Eastern Force Will Follow Its Opposition Around

Combat tracking has been defined as following another human being from his marks or scent. Thanks to technology and firepower, it is an art that has been largely forgotten in the West. In the "low-tech" East, it is alive and well. In the early 1950's, Viet Minh deserter Ghia Xuey showed the German contingent of the French Foreign Legion what their underfed and mild-mannered foes could accomplish.

He moved in the jungle like a panther. Nothing escaped his

searching eyes; no crushed grass, no awkwardly lying bough or bent twig. He noticed marks on the ground, however faint, and he could read a trail as ordinary people read books.[1]

There Is Only One Way to Beat a Guerrilla in Close Terrain

From 1948 to 1957, the British Army faced an insurgency on the Malay Peninsula. As the Malayan People's Liberation Army (MPLA) was predominantly Chinese, one suspects Chinese Communist instigation.[2] From 1962 to 1966, the British had to protect the Malaysian Federation states on the island of Borneo from Indonesian expansion.[3] In both cases, they owe much of their success to following enemy sign to its source. In actuality, indigenous attachments did most of the tracking.[4]

In the United States, most of the man-tracking expertise belongs to Native Americans (Figure 16.1), the U.S. Border Patrol, and volunteer "search and rescue" units. However, there are also

Figure 16.1: America's Best Trackers Are of Eastern Origin

(Source: *Corel Gallery Clipart,* Totem Graphics, Man Miscellaneous, 28V018)

immigrants who have mastered the art. Among the most experienced is David Scott-Donelan who spent the better part of 28 years tracking Communist guerrillas for the Rhodesian and South African armies. From his book *Tactical Tracking Operations,* much of the meat of this chapter has been extracted.

Tracking-Team Composition and Training

Tracking is so mentally exhausting as to be best accomplished by a four-man team—one tracker, two flankers, and one controller. That way, the tracker and flankers can spell each other at regular intervals. This interval can be as little as 30 minutes.[5] Most useful is a "Y" formation with the flankers slightly to the front and side of the tracker, and the controller to his rear. If either flanker encounters difficult terrain, he can fall in behind the controller. In open country, where there is no possibility of ambush, everyone can walk on line. In this formation, all four can look for a hard-to-follow trail at the same time. Team members must be able to communicate by hand-and-arm signal. The following four signals are essential: (1) tracks lost, (2) tracks found, (3) require flanker participation in lost-track search, and (4) quarry seen.[6] The controller monitors the team's location, manages its members by exception, and only upon request decides what to do next.

To perform the requisite number of stoops and bends, a tracker must be extremely fit. To extract enough clues from the jumble of misinformation, he must have full command of his senses. Those whose vision cannot be corrected to 20/20 should not attempt to track someone. Even then, the glare from their glasses may get them killed. Color blindness would also be a significant handicap except at night. Then, it can provide better delineation between shades of grey. Finally, a lack of depth perception would also be a disqualifying factor. Trackers have to be able to distinguish between like-colored surfaces at different distances. They must sometimes force their eyes to look behind the forward-most shrubbery.[7] Without good peripheral vision, team members could not perform their functions and simultaneously monitor other members. There would be too little communication.

Tracking is a skill that must be constantly exercised or lost. Minimally recommended is following a one-mile track two or three times a week.[8]

How to Enhance One's Sight

Depth perception can be improved by shifting one's gaze from the closest tree to those progressively more distant. Alternately, one can swing his gaze from left to right until its maximum range has been reached.[9] As the "S" is widened, one's vision will also improve. There are other visual acuity exercises for formally training trackers.[10] One is to send each trainee up a trail laced with clues and then ask him to recount their sequence. Another is to see how many camouflaged trip wires and pressure-release devices he can find without triggering them. Still another is to have him focus on gradually more distant stars. The last is to ask him to identify ground abnormalities 20 yards to his front while both walking and running.

To efficiently operate in close terrain, the tracker must constantly practice looking past the first barrier of foliage.

Where to Locate Sign

Sign will be most evident in the following locations: (1) thick undergrowth [lanes/damage in direction of movement], (2) steep slopes, (3) newly tilled land, (4) stream/pond banks [water splashed on rocks/vegetation], (5) logs lying across streams/obstacles, (6) pond banks [water discolored], (7) shoulders of roads and trails, and (8) lines of drift [draws, powerlines, firebreaks, ridgelines].

Where Tracking Is Difficult

Tracking will be the most difficult in the following locations: (1) along hardened or well-used trails, (2) along large rivers, (3) near inhabited or friendly force areas, (4) in cultivated areas, (5) where sunlight is directly overhead, (6) under poor visibility, (7) during heavy rain or wind, and (8) at night.

The Rules of Tracking

In the United States, one of the most popular techniques of tracking is the Step-by-Step Method. It has three rules: (1) Don't advance beyond the last footprint found; (2) Don't destroy a clearly identifiable footprint; and (3) Use a sign-cutting stick—one the length of the subject's stride.[11]

Taught at the Scott-Donelan School is an eight-rule method: (1) Correctly identify the quarry's trail; (2) Mark the start point; (3) Don't step on a footprint; (4) Don't overshoot the last known imprint; (5) Obtain ground-level confirmation for any above-ground sign; (6) Monitor one's own precise location; (7) Keep visual contact with other team members; and (8) Try to anticipate what the quarry will do.[12]

Sidelighting

The best time to track is when the sun is either rising or setting. It is then that any footprints between the tracker and the sun will be sidelighted, and the soil indentations outlined by shadow. (See Figure 16.2.) To keep the track in the best light, the tracker can also use the Native-American trick of moving from side to side.[13] During midday, prints can be shaded from the direct rays of the sun and sidelighted with a mirror.[14] While effective for peacetime tracking, this method can quickly compromise the element of surprise in war.

On all trails, there are four "signs of passage:" (1) that caused by the weight of the quarry at ground level, (2) evidence of a human being above ground level, (3) that which may have been caused by man or beast, and (4) litter. The final three can be used to locate the first, but they are only conclusive in conjunction with the first.[15] Most often attributed to the first group (conclusive sign) are the flattening/disturbance of surface material by the quarry's foot and the regularity of its impression.[16] They are ground changes that smaller and more sure-footed creatures are unlikely to cause. While older tracks may not have sharp edges to cast a shadow, they can often still be seen. Because of the way light is reflected, compressed soil looks lighter in color. The tracker must move so that the sun will always give him a good view of the tracks. Even if he has to look over his shoulder to do so, he will make better progress.[17]

At night, tracking is sometimes possible under bright moonlight. Trackers can also sidelight with a flashlight on a stick.[18] In enemy territory, the flashlight should have an extended cowl, red beam, and always be directed at the ground. In the deep jungle,

Ground-Level Sign of a Particular Quarry's Passage

Ground
- Footprint impressions (or portions thereof)
- Ground indentation from stepped-upon branch or rock
- The soil made more reflective from compression
- An imprint in the dew or frost
- Wet marks on the far side of a stream
- Different colored dust from the quarry's feet
- Mud transferred from boots to road surface
- Shiny ground

Foliage
- Stepped upon tiny ferns/grasses/plants
- Crushed or broken dead leaves
- Bent blades of grass
- Crushed twig
- Out-of-place twig
- Dirt stuck to the bottom of a trod-upon plant
- The bruise on a low growing leaf

Insect life
- A disturbed ant hill or trail

Roots and rocks
- Scuff marks from rubber or leather soles
- Scratches on rocks, fallen trees, and roots
- Moss or bark dislodged from a root
- Mud smeared on stones, logs, and roots
- The wet underside of an overturned rock

Pebbles
- A pebble pressed into the ground

Streams and ponds
- Cloudiness caused by recent disturbance of bed
- Algae removed from rocks or stream bed

Above-Ground-Level Sign of Human Passage

Foliage
- Broken twig
- Changes to vegetation color
- Cut/broken/bent vegetation
- Crushed or bent grass
- Buds or berries knocked from a small bush
- Stripped leaves
- Dust or dew brushed from plants

Table 16.1: The Various Types of Trail Detail

Foliage (continued)
- Intertwined vegetation
- Dragged vegetation displaying the paler underside of leaves
- Snagged vines pulled in the direction of travel
- Bits or clothing snagged by thorns

Insect life
- A broken spiderweb
- Disturbed insect nest
- Altered ant trail

Trees and rocks
- Mud smeared onto tree bark or low lying leaves
- Scratches on bark of trees/branches or moss of rocks
- Displaced bark revealing lighter color of unprotected wood

Sign of Not-Necessarily-Human Origin

Ground
- The collapse of a trench or slope
- Drag and scuff marks on hard ground
- Sand and mud particles dropped from sole of boot
- Flattened animal droppings or other debris

Foliage
- Leaf overturned to show its wet, dark underside
- Displaced twigs

Trees and rocks
- Scuff marks on trees
- Sand deposited on rocks

Pebbles
- A stone dislodged from its socket

Discarded Material

Ground
- Paper, packets, or used bandages
- Match sticks or cigarette butts
- Mark, hole, or crusty deposit from urine
- Feces
- Blood stains
- Ration containers
- Sputum

Insect life
- Drawn to urine or feces

Table 16.1: The Various Types of Trail Detail (Continued)

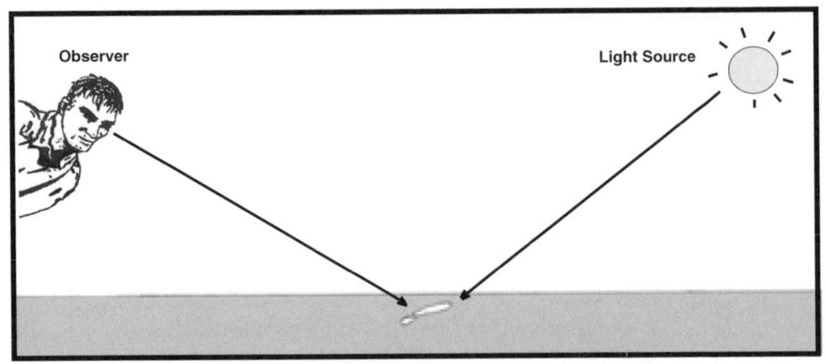

Figure 16.2: Optimal Viewing Conditions

(Source: *FM 21-76* (1957), p. 38)

phosphorescent growth will sometime throw off enough light to permit some tracking. (For most of the signs that can make up a trail, see Table 16.1.)

Assessing the Age of Tracks

Without practice in a particular locale, a tracker can't accurately estimate a footprint's age. The sharpness of sole pattern is the best indicator. (See Figure 16.3.) In moist earth, a man's foot will initially push up a tiny ridge of soil around its imprint. Soon, the sun and wind start to erode this ridge. If its crumbling can be observed, the quarry is not far ahead.[19] Then, the ground-level edge of the print will start to dry out. It will change to a lighter color. Finally, it will start to erode, and its composition and other dirt will sift into the print, softening its edges. (See Figure 16.4.) Unfortunately, this rate of erosion depends on the quality of soil, degree of sunlight, and force of the wind. Additionally, how well the track is protected from the elements comes into play.

It often takes testing to determine the age of a track—e.g., watching how long bent grass takes to right itself. When cracks or bruises in that grass are still green, the tracks are relatively recent. When water has not yet reentered a footprint in the swamp, that print is fresh. If the water in a footprint is still clear, the trail is fresh.[20] Knowing the time of the last shower also helps to determine a track's age. If a track in a dry locale has collected water, it was made after the last rain. Other indicators of a track's age are rain pockmarks and superimposed leaves or animal tracks. All time estimates are based on formerly corroborated evidence. That evidence will vary by locale and season. A legendary tracker of man-eaters describes the assessment process as follows:

> [In a fresh track] the pile or nap of the dust where it took the weight of the leopard is laid flat and smooth, and . . . the walls of the dust surrounding the pads and toes are clear cut and more or less perpendicular. Presently, under the action of the wind and the rays of the hot sun, the nap will stand up again and the walls will begin to crumble. Ants and other insects will cross the track; dust will drift into it; bits of grass and dead leaves will be blown or will fail onto it, and in time the pug marks will be obliterated. There is no hard and fast rule by which you can judge the age of a track. . . . But by close observation and by taking into consideration the position of the track, whether it is exposed or in a sheltered spot, the time of day and of night when certain insects are on the move, the time at which winds normally blow, and the time at which dew begins to fall or to drip from the trees, you can make a more or less accurate guess when the track was made.[21]
>
> — James Corbett, *Jungle Lore*

Even an experienced tracker can sometimes be fooled as trails mature. While certain climatic conditions can help to preserve tracks, others can accelerate their erosion. For example, a light rain can "round out" a footprint, making it look older than it really is.[22] Above-ground sign is affected differently by time. Crushed or bruised vegetation will eventually wilt and dry out. However, the process differs between areas of shade and sun. Over time, sunlight will bleach most litter.

What Else Signs Can Tell

Footprints can tell the tracker many things about his quarry.

Figure 16.3: A Recent Print in Soft Soil
(Source: Courtesy of Paladin Press, from *Tactical Tracking Operations*, © 1998 by David Scott-Donelan, p. 42)

They will often indicate whether the subject is young or old, big or small, female or male, fresh or tired, in good health or bad, nervous or confident, moving quickly or slowly, and laden heavily or lightly.[23] Much of this information can be gleaned from the initial and final contact point of the foot. For example, if the initial points are deeper and farther apart than usual, the quarry is moving more quickly. As the heel print decreases and the toe print increases, the subject's speed has increased. If only the ball of the foot is visible, the subject is running at full speed.[24] Women are more often pigeon toed.[25] By comparing one's own print depth to that of the quarry, one can also closely estimate the subject's weight. Normally spaced prints with an exceptionally deep toe print indicate the subject is carrying a heavy load.[26] Unusual sole wear would tell of some infirmity. Grass will be bent in the direction of movement.[27] Stepped-on pebbles will push up dirt away from the direction of travel. Also indicative of that direction will be the moisture left behind when the quarry exits a stream. If the trail leads around ground obstacles and contains little sign, it is the result of a daylight movement. If the trail leads through shrubs and deadfall and shows signs of tripping or staggering, then it was probably made at night.

Enemy soldiers often wear distinctive footgear. Persons traveling in a group will leave at least one distinct set of footprints—that of the last person in column. The average pace will vary with con-

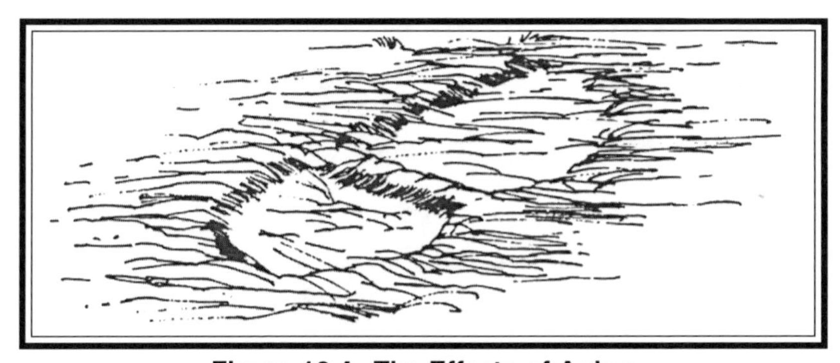

Figure 16.4: The Effects of Aging

(Source: Courtesy of Paladin Press, from *Tactical Tracking Operations*, © 1998 by David Scott-Donelan, p. 42)

ditions and quarry speed. When in doubt, 36 inches will work as a rough estimate.[28] By counting the footprints in two average paces and dividing by two, one can determine how many enemy soldiers have passed by. James Corbett only counted the number of heel prints.[29] (See Figure 16.5.)

The quarry may stop to rest. On a hot day, he will do so in the shade of a tree and leave the imprint of his behind or full body. How long he rested can sometimes be estimated by shadow movement. Careful analysis of the area may reveal clues to the loads being carried.

Sounds and Odors May Also Come into Play

Under the right wind and atmospheric conditions, the tiniest sound can carry a long way. That presents both opportunity and difficulty for the tracker. He must be able to identify which of the many sounds may have come from his quarry. Among the most obvious might be the rustling of vegetation, staccato of voices, or rattling of equipment. Once during the Rhodesian War, a tracker heard his quarry unloading a Soviet "PPSh" submachinegun at almost 1000 yards.[30] To complicate the issue, nighttime sounds can seem four times closer than they really are.[31]

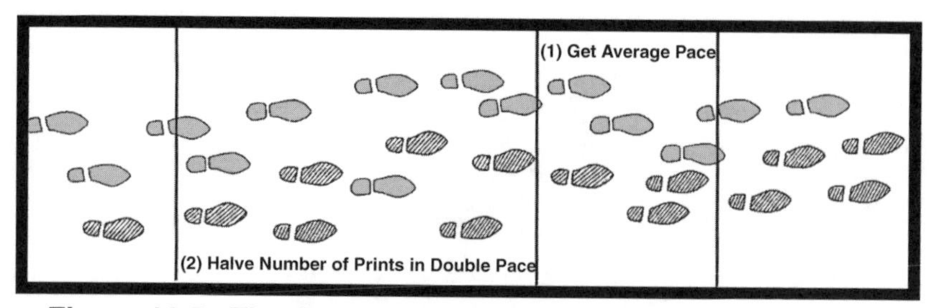

Figure 16.5: The Average-Pace Method for Counting Heads

(Source: Courtesy of Paladin Press, from *Tactical Tracking Operations*, © 1998 by David Scott-Donelan, p. 57)

Nighttime Marine security patrols in Korea could often locate their opposition through the odor of garlic. While acting platoon leader with 3d Battalion, 7th Marines in Vietnam in 1968, Sgt. Les Ford came to a familiar fork in the trail. He put his nose to the ground to assess both routes. Instead of following the branch that smelled of sweaty feet, he tossed a grenade up it. Then all hell broke loose. The enemy had been waiting in ambush.[32] While chasing Communist guerrillas in Rhodesia, a small patrol caught a whiff of human sweat mixed with rifle oil. By establishing from which direction the wind was coming, it was able to locate seven sleeping guerrillas.[33] One could say then that any unusual odor could be of tremendous benefit to a tracker. Among the most important might be the smell of fire, food, soap, sweat, urine, defecation, military equipment, insect repellent, and weapon lubricant.

An Ongoing Investigation

The tracking team will constantly seek answers to the questions—who, what, where, how, and why.[34] Without so doing, it may waste valuable time following the wrong trail or stumble into an ambush. Here are the five things they must correctly ascertain: (1) quarry's footprint, (2) number of people being pursued, (3) specifics of each person's footprints, (4) initial direction of travel, and (5) approximate age of the tracks.[35]

To close the gap, the tracker will also need the quarry's speed. Both track age and quarry speed can often be approximated from earlier quarry sightings. All the tracker needs then is a way to cover the required distance in the allotted time.

Lost-Print Procedures

Sooner or later the trail will go cold. When it does, the tracker must clearly mark the last footprint found. Then, in descending order of utility, here are the various ways he find can it again.[36] He can move forward about 30-40 yards along, but slightly to the side of, the subject's most likely route of advance. Most probably, that route will be either in the same direction or along the clearest path. If the tracker doesn't reacquire the tracks, he moves back to the point of departure and repeats the process along other possible routes. In the next method, the tracker tries to "cut" the quarry's trail by making gradually larger concentric circles around the last known track. He starts from a point 30 yards back. Normally, he will reestablish the trail within 500 yards of the known track. As a last resort, the controller gets the flankers involved in the "cutting" process. He has each do a circular search pattern that will intersect that of the tracker.

If all of these methods fail, the team can move in the enemy's previous direction or toward his possible destination. Depending on the circumstances, it will choose either a straight line or "line of drift." If the quarry were following linear terrain features (ridgeline, streambed, draw, or finger), one would search a zigzag route. Or the team could search the paths, streams, powerlines, and firebreaks that intersect the quarry's direction of travel. Lack of sign would indicate that he had changed course. This is sometimes called a "cross-grain" search. Then there is the "box" search.[37] By searching the boundaries of progressively smaller boxes, the team can systematically shrink the area from which the quarry must have moved laterally. Finally, the team can backtrack to where the subject may have left the trail after retracing his steps. Or it can go all the way back to his starting point to look for additional clues to his final destination. Backtracking along enemy trails in Vietnam would have produced some very interesting results.

Closing the Gap

To close the time-distance gap with a quarry, the tracker must leapfrog ahead sometimes as much as 30-40 feet at a time. However, he must first mark the last footprint found. Even when time permits, looking for every successive footprint would overly tire

Figure 16.6: Russian Tracker

(Source: *Podgotovka Razvegchika: Sistema Spetsnaza GRU,* © 1998 by A.E.Taras and F.D. Zaruz, p. 173)

the tracker' eyes. He must study the ground at his feet as little as possible.[38] Instead, he looks ahead for sign. To appreciate tracking's finer points, the novice must first break the habit of always looking down. Even the British Army's jungle tracking course advises advancing quickly:

> - Search ground to determine general direction of trail
> - Look through undergrowth to determine furthest sign
> - Move toward furthermost sign and repeat process
> - Observe left and right of trail for possible deceptions
> - Confirm and maintain direction by occasionally looking back.[39]
>
> — British Combat Training Center lesson plan
> Paluda, Malaysia

While British Commonwealth troops search ahead instead of following tracks step by step, they try not to imagine sign. They stay initially within five paces of last definitive clue and frequently rotate trackers.[40] As their technique probably comes from the Gurkhas or native troops, it may be practiced throughout Asia. What happens in Asia also happens in Russia. (See Figure 16.6.)

How the Foe May Counter

When the Truong Military Corps No. 559 took over responsibility for the Ho Chi Minh Trail in 1959, its construction crews had some interesting ways to keep their mission secret. When in single file, the lead man would gently separate the leaves and grasses with a stick, and the last man would walk backwards to restore them to their original state.[41] There are many ways to counter a tracker. They fall into four general categories: (1) increasing the time/distance gap, (2) disguising/concealing sign, (3) reducing sign, and (4) counterattacking.[42]

The least efficient is the first. Moving faster leaves more obvious clues that the experienced tracker can use to leapfrog ahead. To insure the widest possible gap during the French and Indian Wars, Rogers' Rangers would keep moving until dark.[43]

Tracks can be concealed or disguised in many ways. An area can be brushed out. Of course, it will then be devoid of all animal sign and weathering. When a leafy branch is used too long for this purpose, it can turn into a bundle of twigs that will leave an unmistakable trail. By following in each other's footsteps, those being pursued can disguise their number. Their column can follow (or offset along) hard ground and streambeds. In the process, it may create long-lasting underwater footprints, crushed/broken water plants, unnaturally wet rocks, and stream bank skid marks (Figure 16.7). The muddy water alone will alert anyone downstream to the ruse. The last man in column can restore displaced vegetation to its natural state with a stick. He can cover his and his comrades' tracks with leaves. Or he can drive large numbers of people or animals over the tracks. Initially choosing a heavily traveled trail would accomplish the same thing. Various foot coverings can be used. They will disguise the sole pattern, but also add width to the imprint. Among the more ingenious coverings are leaves. Shoes can be removed or interchanged with wide pads and animal feet.

Another trick is to walk backwards. Unfortunately, as the foot's initial strike shifts to the toe, soil particles will reveal the actual direction of movement. On soft ground, the heel print will also be deeper and the pace shorter.[44]

The lone quarry can step from rock to rock. He can also follow roots, fallen trees, rocky ground, or paved road. Or he can step on the harder ground at the base of each bush. If he tiptoes, there will be no heel marks. He can suddenly change direction. A partial reversal in direction is the most productive. Even an experienced tracker will lose time looking for the next print to his side and rear. Then, there are composite techniques. The quarry might move past a tree, retrace his exact footsteps, and then jump to the blind side of

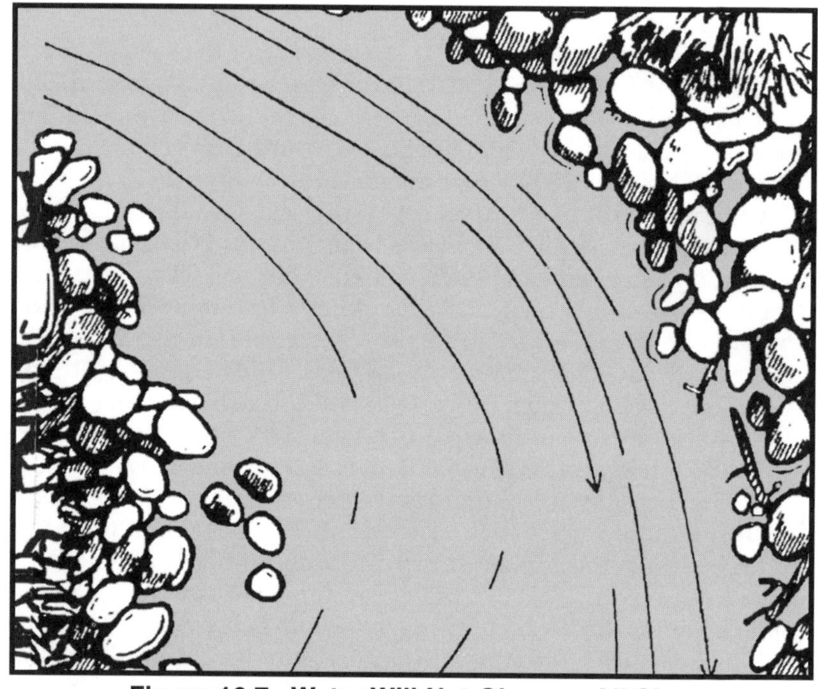

Figure 16.7: Water Will Not Obscure All Sign

(Source: *FM 21-76* (1992), p. 8-27)

that tree (a method used by some animals).[45] Or he might move 20-30 meters back along the stream he just came up and then carefully exit through a feeder stream or thick brush.

Next, there are ways in which sign can be reduced. While traveling across swamps or soft ground, Rogers' Rangers would simply spread out abreast.[46] The group can break up into subgroups. Or every member of the group can take off in a different direction to rendezvous later. Individuals can drop off at intervals, using rocks or sun-baked ground to exit without a trace. Every other man can carry his companion piggyback. Of course, this will cause the tracks to get suddenly deeper. A lone quarry can crawl off on his hands and knees or roll along the ground. However, he will leave behind hand prints, boot toe indentations, and weapons/equipment drag marks. He can swing off the trail on a vine or low-hanging limb. This will leave telltale clues—a sudden disappearance of tracks, broken branches, and missing or scuffed bark.

Finally, the tracker can be discouraged or eliminated altogether. This is most easily done through sniper, boobytrap, or ambush. While raiding the Viet Minh's rear areas, the German contingent of the French Foreign Legion would simply snipe from several hundred yards anyone following them.[47] The commander of Rogers' Rangers advised circling back on one's tracks to ambush those doing the trailing.[48] In Haiti and Nicaragua, Marine legend Chesty Puller would simply drop off a small welcoming party. Another way to discourage unwanted company is to start a brush fire in one's wake.

In Which Ways Can Modern Technology Help?

On occasion, the Global Positioning System (GPS) device may save a tracker a few minutes that he might otherwise have spent on land navigation. As it works best in the open, it will also make him easier to snipe. As the GPS user no longer needs terrain features as navigational aids, he gradually looses his appreciation for the tactical value of terrain.

While the tracker should not let his equipment dictate his actions, he may find some of the newer gadgets helpful. Commercially sold in America are infrared sensors that can supposedly detect a partially hidden human being at 150 yards on a 65° day.[49] Also available are sound enhancement devices. Some can magnify—

up to six times—the sound of voice exchange, magazine loading, safety flipping, or footsteps.[50] With high-quality binoculars, the tracker can improve his vision in dim light.

To detect an ambush, one's natural senses must all be functioning equally. For this reason, trackers should use sensory enhancement devices sparingly.

17 While Stalking a Quarry

- *How can one secretly approach a sentry from the front?*
- *What difference does the wind direction make?*

NORTH KOREAN SERGEANT, SUMMER 1950

(Source: Courtesy of Osprey Publishing Ltd., from *The Korean War 1950-53*, Men-at-Arms Series, © 1986 by Osprey, Plate A, No. 1; *FM 21-76* (1957), p. 150)

The Art of Stalking Is As Old As Hunting Itself

Over the centuries, the hunter has routinely had to stalk his prey. To get close enough to kill it, he was forced to obscure any visible, audible, or odorous hint of his presence. To accomplish the first, he used camouflage and stayed hidden. To do the second, he dressed carefully and move silently. To manage the third, he stayed downwind of his quarry.

Just spotting a wild animal in its natural habitat can be difficult. Often, only part of its head, legs, or back will even show. The hunter must search for unusual colors and shapes. With bear, he may look for a splotch of dead black—a shade that occurs seldom in nature. For deer, he may try to locate a sun-dappled horizontal line—an equally rare occurrence in the forest.[1] If he is puzzled by something he sees, he can try the American Indian trick of moving his head from side to side. "This allows the eyes to 'feel' the object, and [with] the binocular properties of human sight . . . to reach out and touch the surface."[2] Perhaps the viewing of an object from a slightly different angle emphasizes its outline. It would certainly help to establish its range. After eliminating telltale odors, the Native American stayed downwind of his quarry. He also kept his feet pointed forward and most of his weight on his toes.[3] Once he had spotted his victim, he froze.[4] He noticed that his quarry seemed unaware of subsequent slow motion.[5] James Corbett once shot—by slowly reversing the direction of his rifle—a tiger that was about to pounce on him from behind. As several of Corbett's other stalks will attest, the gap between hunter and beast can most successfully be narrowed by crawling.[6] Corbett may have learned many of his stalking methods from the very predators he was chasing.

Routinely outgunned in war, the distant cousins of the American Indian have had to retain their stalking ability. To close with an Eastern infantryman, the U.S. soldier will need every trick he would use on a wild animal.

Stalking Is Vital to the Eastern Way of War

Having originated on the Asian mainland, *ninjutsu* is as old as Japan itself. The *ninja's* role was largely to sneak past sentries. He could more easily approach that sentry by keeping some object between them. Among the ancient *ninja* techniques is assuming the shape of a stump, bush, or tree.[7] As early as 1913, Japanese messengers were trained to advance by successive rushes between objects. In so doing, they had to leave unaltered the shadows of those objects.[8] Seldom would their interim destinations be in a straight row. While sneaking up on someone, Japanese scouts may have also zigzagged between objects. So doing would have minimized their lateral movement. It would have also made it harder for their quarry to confirm a sighting. When someone detects movement at the corner of his eye at night, he looks for additional movement in the same direction. The constant reversal of direction by the scout would have confounded the sentry's intuition. At any rate, most modern stalking technique seems to have evolved from this zigzag maneuver.

Just as the Kodiak island sportsman must carefully stalk his 1600-pound quarry, so too must an Eastern assault force sneak up on a Western-style defense. To prevail over so much firepower, its lead elements must get—unnoticed—to within grenade range of the Western watchstanders. This was what the Japanese did on the night of Chesty Puller's famous defense of Henderson Field on Guadalcanal,[9] and this is what the Chinese did 18 years later at the Chosin Reservoir.[10] For each member of the lead echelon, this effort is tantamount to stalking the watchstander in his lane. Of course, he must also escape the notice of those defending other lanes.

Eastern Sentries Provide More of a Challenge

Until the U.S. infantryman can stalk an equipmentless sentry, he should not worry about how to approach one with thermal imaging or night vision goggles. Neither device is all-powerful. The first cannot penetrate heavy rain, fog, or vegetation. The second is of no use against someone with aerial illumination at his back. First mastered must be the mechanics of the old-fashioned stalk.

A well-placed and fully camouflaged Eastern sentry can be harder to sneak up on than a wild animal. Just pinpointing his location will be a chore. Only occasionally will any part of his upper torso be visible. More often, he will be peering out from the shaded area beneath a rock, inside the knothole of a tree, or behind the bushiest part of an elevated branch. Or he may be a roving sentry so completely covered in leaves as to be indistinguishable from the other foliage while stationary. There will be only the slightest indicator of his presence, and the Western stalker must be prepared to wait for it. It may be the slosh of his canteen water, rumble of his stomach, or snap of a twig.

Once spotted, the Eastern sentry must be secretly approached. Depending on the number of trees, rocks, and bushes available, the stalker should keep one between himself and his quarry at all times. If he must cross open ground during the day, he can sometimes do so quickly when his quarry turns around or looks down. It cannot occur within the sentry's peripheral vision. At night, openings can best be crossed slowly. In low light, the human eye cannot detect slow movement. Whether day or night, a well-camouflaged crawler should enjoy the most success. He will have, for additional cover, the folds in the ground.

The Western Version of Night Stalking

The U.S. military manuals liken night stalking to very slow "night walking" at a crouch.[11] (See Figure 17.1.) Night walking is what a deer does anytime. He raises each knee before advancing his hoof. Unfortunately, the deer is not a predator.

> Begin by looking around, then slowing lift the right foot approximately knee high and, balancing on the left foot, ease the right foot forward to feel for twigs and trip wires. Keep the toes pointed downward. The lead foot should touch the ground about six inches to the front. As the toes come to rest, the soldier feels for the ground with the outside of the toe of the boot. Then he settles the foot on the ground. As this step is taken, the boot is used to feel for twigs and loose rocks. Confident of solid, quiet footing, the soldier slowly moves his weight forward, hesitates, then begins lifting his left foot. . . . This method of balanced, smooth walking at night reduces chances of tripping over roots and rocks and reduces noise. . . .
>
> The soldier is usually watching the enemy. . . . When close to the enemy, squinting helps conceal light reflected by the eyes. Breathe slowly and through the nose. If the enemy looks in the direction of the stalker, he [the stalker] freezes, balanced or not. Movement should take advantage of the background to blend with shadows and prevent glare or contrast. Movement is best conducted during distractions such as gusts of wind, vehicles moving, loud talking, or nearby weapons fire. . . . The goal should be to get within 36 inches of the enemy in order to execute a rapid, silent kill with a knife or garrot *[sic]*. Often the final approach is made by crawling because it uses low concealment, provides a better view (looking from below up), and takes advantage of silhouetting and ambient light.[12]
>
> — U.S. Army, *FM 7-70*

Figure 17.1: The Western Approach to Stalking

(Source: *FM 7-70* (1986), p. D-13)

While this excerpt briefly touches on many issues, it may be a bit overly optimistic. Approaching—from the front—to within 36 inches of an alert German, Japanese, Chinese, or North Korean soldier would be beyond the capability of a *ninja* master. With the help of night vision devices, Western special operators or scouts may be tempted to dress up in "gilly" (bush) suits and move upright toward a suspected enemy position. During the Panama incursion, one Marine saw Cuban troops doing just that.[13] While this might occasionally work against a Western unit, it won't work anywhere in the East. There, the "low-tech" enemy will smell, hear, or see the 'high-tech" American coming well in advance. Part of his warning will be the American's continual readjusting of his uncomfortable head gear.

The Eastern Difference

One could do much worse than copying a technique from nature, albeit one of a prey animal. The U.S. method contains many of the basic elements of stalking and a smattering of detail, but it pays only lip service to a number of very complex issues—e.g., light/

shadow, shape, background, and noise abatement. It further ignores other crucial factors—like odor and wind direction. Finally, it makes the cardinal error of which most U.S. technique is guilty. It assumes that the quarry is of equal ability to the average American infantryman. Once and for all, he isn't. His senses are more finely tuned. He is better informed of his opponent's tactics. His location will be more difficult to ascertain. From the below-summarized Japanese night movement manual excerpts, one can see that every Nipponese private in WWII had *ninja*-quality surveillance training.

On a moonlit night, the sentry's eyes are dilated. If the moon is to his front, he can see less well into the shadows than if it is at his back.[14] Either way, he will have trouble seeing dark objects or movement in the deep shadows. In full moonlight, a sentry can also see well from darkness into light. Right after a cloud has obscured the moon, he cannot see anything well.

In the open on such nights, black clothing can look almost white. This is just one more reason for the stalker to stay in the shadows.[15] Of course, the same rules of vision apply to the stalker. He tries to keep the moon at his back. That way, his vision will be enhanced while that of his quarry will be impaired. The stalker will have better luck from late spring to early autumn, because of the greater abundance of foliage and shadow during that period. If he does not keep dark objects at his back, he will become silhouetted against a lighter background.

In the fall, the stalker could more silently move through dead leaves when the wind was blowing. He could synchronize his movements with the puffs of breeze. If downwind from his quarry, he would project less sound. When wind blew in the sentry's ears, it would temporarily impair his hearing.[16] Normally, there is more background noise at twilight or dawn.

Man's movement can only rarely be obscured by naturally occurring motion. In Vietnam, Marines saw North Vietnamese soldiers who, while stationary, could not be distinguished from the grass around them.[17] If, on a windy night, a similarly camouflaged stalker were to advance a few feet during every gust, he could theoretically cross a well-watched field over the course of several hours. However, the only shadow moving rapidly enough to do him much good would be that caused by the headlights of a passing vehicle.

As the initial stages of short-range infiltration are tantamount to stalking, Chapter 19 will provide adequate evidence of the

Easterner's skill. One of his more advanced techniques was observed on Operation Maui Peak during the Vietnam War. In full moonlight on a level slope sparsely covered by bushes, NVA skirmishers got to within grenade range of fully alert Marines in late 1968. They did so by dressing up like bushes and spending seven hours crawling 100 yards. Their movement was so slow as to be imperceptible to the human eye. Only by counting the clumps of vegetation could the Marines killed that night have avoided their fate.[18] (See Figure 17.2.)

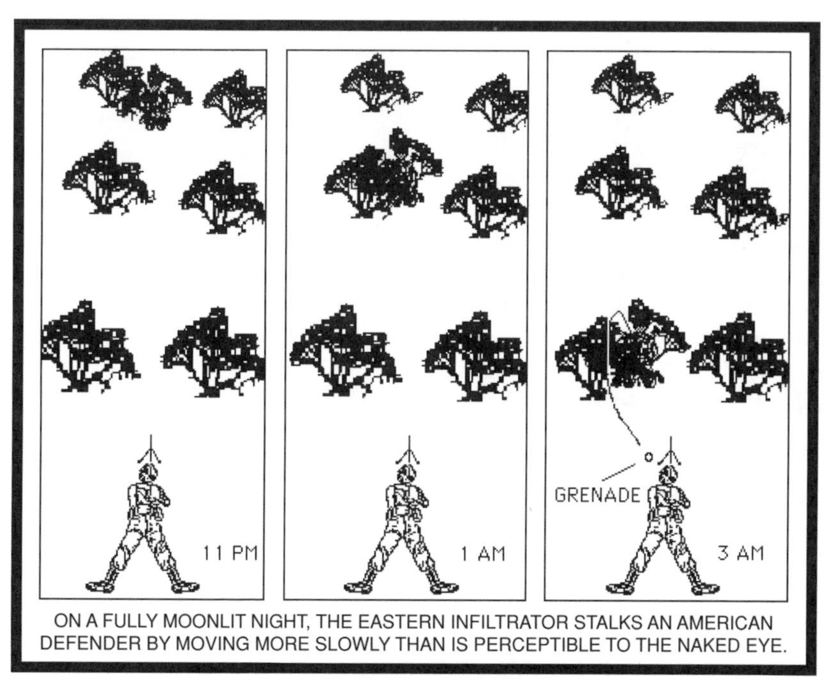

ON A FULLY MOONLIT NIGHT, THE EASTERN INFILTRATOR STALKS AN AMERICAN DEFENDER BY MOVING MORE SLOWLY THAN IS PERCEPTIBLE TO THE NAKED EYE.

Figure 17.2: Eastern Refinements

(Source: *FM 7-11B1/2* (1978), p. 2-iii-e-8.2; *FM 7-8* (1984), p. 3-28)

After stalking a defender, the infiltrator will attempt to bypass him. This is an art in itself and well beyond the scope of this book. Suffice it to say that it is complicated. For example, the *ninja* crouches while passing behind an upright sentry to stay below his line of sight.[19] The sentry's shoulders, in effect, block his peripheral vision. There will be less-than-optimal "homeland security" until U.S. forces admit to their short-range deficiencies.

Stalking a Moving Target

This chapter has been largely devoted to stalking a stationary sentry, but one can also stalk a moving quarry. It involves more than just tracking that quarry. James Corbett offers a brief glimpse into what might be involved. At a distance, he could monitor a tiger's movements through the chatter of birds. Up close, he could do so through the tiger's own noise.

> As there was no fear of the leopard leaving its cover at that time of day for the jungle two miles away, I tried for several hours to stalk it, helped by bulbuls, drongos, thrushes and scimitar babblers, all of who kept me informed of the leopard's every movement.[20]
>
> — James Corbett

> Having acquired the ability—through fear—of being able to pinpoint sound, that is, to assess the exact direction and distance of all sounds heard, I was able to follow the movement of unseen leopards and tigers, whether I was in the jungle by daylight or in bed at night.[21]
>
> — James Corbett, *Jungle Lore*

As was shown in Chapter 13, Eastern armies will sometimes use roving sentries. Penetrating their screen would take monitoring their whereabouts.

18 To Reconnoiter an Enemy Position

- *Could an assault force study its objective from the inside?*
- *What opportunities would it then have?*

CHINESE INFANTRYMAN, SUMMER 1951

(Source: Courtesy of Osprey Publishing Ltd., from *The Korean War 1950-53,* Men-at-Arms Series, © 1986 by Osprey, Plate H, No. 2; *FM 90-6* (1980), p. 1-9)

First Things First

Only in the Western World does "reconnoitering an attack objective" mean looking at it from high overhead or across a wide expanse. On the other three-fourths of the planet, prepared enemy positions are routinely reconnoitered from the inside.

To determine how best to approach an enemy enclave along the ground, one must have some idea of its strategic significance. Is the position's location of value, or its contents? Next, one must know his parent unit's probable method of attack. Without this insight, he could compromise the attack route. Finally, the scout must be clear on what the enemy's "front lines" might look like. Eastern armies will normally only pretend to use the outmoded linear defense. What kind of outpost screen will there be? Will it incorporate roving listening posts? Does the enemy have NVGs? Not getting these questions answered before leaving could result in a botched patrol.

The Eastern Defense

Eastern-style defenses look much different from those in the West. Many are so well camouflaged as to be hard to detect from 10 feet away. (See Figure 18.1.) They are not linear—not made up of a clearly defined row of fighting holes. Instead, they look like a belt of mutually supporting machinegun bunkers, each with its own tiny perimeter of fighting holes. Three of these squad-sized "forts" constitute a platoon-sized strongpoint, and three of the platoons a company—each embedded in the other. Where the lines of communication run below ground, the fort resembles a mound surrounded by five or six two-man holes. Between the forts are extraneous obstacles and dummy positions. The scout who sneaks through these dummy positions has not penetrated enemy "lines," he has occupied a carefully watched firesack. Where the enemy has miles of communication trench, the squad-sized fort is nothing more than a tiny, manned segment.

To avoid Western firepower, the "low-tech" Eastern foe has had to leave the surface unaltered and dig. The reverse-slope forts on Iwo Jima were so well hidden that U.S. Marines routinely walked right past them.[1] Their only evidence must have been a few brush-covered apertures. They may have also lacked the telltale rings of fighting holes. Only when the Marines tried to take the next ridgeline would the Japanese reverse-slope machineguns announce their presence.

To locate these tiny forts, the Western scout will have to be more closely attuned to nature than he is right now. No amount of "high-tech" equipment will replace his God-given senses. The only evidence of such a fort may be a murmur of voices, whiff of food, or feel of freshly turned earth. There may be no boobytraps, trip-flares, footprints, altered vegetation, and lighter-colored tailings for the scout to find.

Figure 18.1: The Well-Hidden "Prepared Enemy Position"
(Source: *FM 7-8* (1984), p. B-1)

The Eastern Outpost Zone

Even without technology, defenders can field a sophisticated early warning apparatus. Before WWI, just to the front of each Japanese stronghold was a string of one- or two-man outposts connected by roving sentinels. Every outpost occupant became quickly aware of whatever security patrols discovered to his front. This was a dynamic, protective screen not plagued by overcontrol from headquarters.

> When a patrol leaves the line of sentinels . . . , neighboring sentinels will be notified by moving sentinels or other means.[2]
>
> — Japanese night-fighting manual, 1913

Those various connecting files were in effect tiny security patrols. They apparently had authority to deviate from their appointed route. On one occasion during the Nomonhan Incident of 1939, Japanese defensive scouts were able to shadow Russian infiltrators.[3]

Locating and Penetrating This Sentinel Screen

It will take time to pinpoint these tiny outposts. Often, their only signature will be the muted movement and communication of a sentry relief or inspection visit.[4] Then there is the problem of moving between them without running into one of the roving connecting files.

> The patrol being hidden . . . , it should strive to discover the position of one sentinel; this being used as a base will assist in the discovery of the other posts and noncommissioned officer post. Having reconnoitered the intervening open ground, the enemy's method of security can be verified, and it can be judged whether or not it would be a good idea to enter the line of sentinels. To accomplish this, it is a good thing to follow directly after a passing moving sentinel or a visiting patrol.[5]
>
> — Japanese night-fighting manual, 1913

Making the Final Approach to the Objective

Not every defender sits around waiting to be attacked. The soldiers of many nations are more proactive on defense. The Japanese were not above launching a tiny sortie to check out any suspicious activity to their front.[6]

The reconnaissance team must leave no residual evidence of its passage near the intended point of attack. Eastern defenders are skilled at looking for sign. They have little trouble ascertaining its origin. Further, they can and do easily follow any tracks to their source.[7] The reconnaissance team's parent unit will conduct a much better attack if not ambushed on the way down to its objective.

The Chinese Method of Reconnaissance

Chinese patrols were especially good at locating the main line of resistance, unit boundaries, and weak spots. If their infiltrators could not easily acquire these elements of information, they were not above drawing fire to do so.

> In scouting a particular objective, [enemy] patrols normally took the shortest route to the site, usually moving in single file with no flank or rear guards. Once near the objective, the patrol separated, each member accomplishing his specific task. . . .
>
> Patrols usually moved along low ground or below the crests of ridges and mountain sides. . . . Patrol leaders often took the lead, but when they moved through dangerous ground, they used one- to three-man points.[8]
>
> — "The Chinese Communist Forces in Korea"
> U.S. Army, *Leavenworth Research Survey No. 6*

Just before the Chinese launched their "human-wave" attacks at the Chosin Reservoir in November of 1950, they dispatched squads that specialized in probing lines.[9]

The VC Way of Reconnoitering

The "outgunned" Eastern commander has little choice but to study his quarry. To have much chance, his troops must practice every assault through a life-size mock-up of the objective. For such a mock-up, the commander will need every detail of the defender's barrier and fire plans. That's why he has *ninja*-quality sappers (Figure 18.2). "When adequate information could not be gathered from reconnaissance outside the enemy camp, teams were prepared to go inside to gather the needed information."[10] To avoid drawing attention to the porosity of a Western defense, the Easterner keeps these capabilities secret. Below is one of the few firsthand accounts in Western literature.

> Once [in the Mekong Delta in 1965] I got into the middle of Cai Be [South Vietnamese Army] post where the district

Figure 18.2: Eastern Sappers Can Crawl through Barbed Wire
(Source: *FM 21-75* (1967), p. 32)

chief's office was. It was fifteen days prior to the attack and takeover of the post by our battalion. I was accompanied by two comrades armed with submachineguns to protect me in case my presence was discovered while I was nearing the post entrance. I was then wearing pants only and had in my belt a pair of pincers, a knife, and a grenade. At one hundred meters from the post I started crawling and quietly approached the post entrance with the two comrades following me. At twenty meters from the post, my comrades halted while I crawled on. At the post entrance there was a barbed-wire barricade on which hung

> two grenades. Behind the barricade stood a guard. I made my way between the barricade and the stakes holding up the barbed wire fence. I waited in the dark for the moment when the guard lit his cigarette. I passed two meters away from him and sneaked through the entrance. On that occasion I was unable to find out where the munitions depot was but I did discover the positions of two machineguns and the radio room. I got out at the back of the post by cutting my way through the barbed wire.[11]
>
> — Reconnaissance sapper
> 261st VC Battalion

Every VC conscript received periodic instruction on short-range reconnaissance.[12] The average level of expertise was painfully evident at Fire Base Mary Ann in March 1971.

> [The] local NVA [special-operations detachment] probably served as guides for the [VC] sappers during pre-assault recons and their attack itself. . . . It was the sappers alone, however, who negotiated the wire obstacles and infiltrated the base. . . . With weapons slung tightly across their backs, grenades attached to their belts, and faces and bodies blackened, they slid snake-like through the brush, silently, patiently, an inch at a time, listening, watching, gently feeling the ground ahead of them. They neutralized trip-flares encountered along the way by tying down the strikers with strings or strips of bamboo they carried in their mouths. They snipped the detonation cords connected to claymores, and used wire cutters on the concertina, careful to cut only two-thirds of the way through each strand, then breaking each noiselessly with their hands, holding it firmly so the large coil wouldn't shake.[13]

Who Else in the World May Have this Much Ability?

During WWII, every Russian squad had to know how to reconnoiter an enemy position.[14] Almost all Russian reconnaissance was by stealth. If the squad ran into trouble, it was expected to attack without shooting.[15] The Russians would sometimes station a single squad near their objective to watch it. Its mission was not only to look for a covered route and assault position, but also to warn of enemy movement or deception.[16]

In the Rural Assault

- *How do Eastern armies usually attack in the country?*
- *What have been their most recent trends on offense?*

MALAYAN RACES LIBERATION ARMY MAN, 1953

(Source: Courtesy of Cassell PLC, from *World Army Uniforms since 1939*, © 1975, 1980, 1981, 1983 by Blandford Press Ltd., Part II, Plate 104; *FM 21-76* (1957), p. 112)

How the Future Foe Will Attack

America's next major war will be fought at the southeastern flank of Europe, periphery of Asia, or somewhere between. Its foe will be a former Soviet client state, Chinese ally, or rogue movement wishing—without much technology—to project its will. For how to defeat the U.S., that foe will study its mentor or another Eastern army with abundant light-infantry experience. To predict the foe's methods, one must learn how those armies have fought at a firepower disadvantage. The first look should be at the least tactically astute of the group.

World renown for thinking big are the Russians. Because they like to attack with brute force in open terrain, it is widely believed that they can only maneuver at the operational level. Yet, this brute force is normally directed at a narrow frontage. Perhaps the point of the spear is more sophisticated than commonly thought. Soviet offensives are launched for three reasons: (1) to exploit faults, (2) to attack weakness, or (3) to dissipate strength.[1] For all three, some measure of surprise would be required. Total surprise is almost impossible with armor, even during a deafening barrage. It's more likely that tiny, dismounted elements precede Russians tanks (Figure 19.1)—much as they did Germans tanks in the Battle of the Bulge. Thoroughly to disrupt a defense, the Russians unleash a series of these narrow thrusts across a wide area.[2] At the tip of each thrust would be some semblance of tactical proficiency.

Russian maneuver forces use three types of lead elements—reconnaissance, forward detachment, and enveloping detachment. While the first has traditionally tried to avoid combat, the others are fully prepared to fight. However, they will only do so selectively. In a meeting engagement, they may wait for the main force to fix the adversary in place before becoming one arm of a double envelopment.[3] Normally, they will only assault a prepared enemy position to facilitate their parent unit's advance.

Operating independently at the front of a Russian column, the forward detachment contributes to its command and control. It also performs "reconnaissance-pull" through the enemy's forward defense zone (outpost belt). To do so, its commander relies on an extensive portfolio of deceptive measures.[4] He is chosen for his knowledge, flexibility, and initiative.[5]

When the forward detachment encounters the enemy's main line of resistance, it looks for gaps and again waits for its parent unit. At the head of the main body is an advance battalion with reinforced platoons that have been specially trained to function as assault groups.[6] While the main force attacks a strongpoint from the front, at least part of the forward detachment or follow-on "enveloping detachment" may try to sneak behind it. The main-force assault element will assault on a narrow frontage and then sometimes do a double buttonhook to widen the breakthrough.[7] It does so to exploit the enemy's newly exposed flanks and rear.

For a breakthrough in open terrain, the Russians will normally turn to armor with motorized infantry in trace.[8] Yet, for the modern Russian Army, the principle forms of maneuver are turning

Figure 19.1: Russian Scouts

(Source: *Podgotovka Razvegchika: Sistema Spetsnaza GRU,* © 1998 by A.E.Taras and F.D. Zaruz, p. 142)

movements and envelopments, not breakthroughs.[9] In close terrain, it is more likely to employ the surprise that only dismounted infantry can provide.[10] The Russian definition of "initiative" is much the same as Mao's—keeping pressure on the foe by whatever means.[11] While preferring the "scientific approach" to war, the Russians still require their commanders to instill flexibility into their planning and conduct of the offense.[12] For example, they might have "on-call" deceptions should surprise be compromised. Or they might rehearse ways to exploit chance opportunities.

Once behind the enemy's main line of resistance, the forward detachment may try to remove any impediment to the main force,[13] or simply sidestep enemy strongpoints to go after opposition headquarters.[14] It has been configured and trained to fight through hasty defenses.[15] Repeatedly to assault while maintaining pressure on withdrawing units,[16] it would need considerable short-range ability. Several tactically proficient lead elements operating in tandem could all but destroy the continuity of a defense in depth.[17]

Throughout the exploitation phase, each maneuver force would depend heavily on the success of its lead element. Many of the

WWII Soviet victories would not have been possible without tactical expertise at the tips of the spears.[18] While WWII Soviets were fairly skilled at infiltration, their modern-day counterparts still rely on small units in mountain, desert, arctic, and urban terrain.[19] In the cities, motorized rifle companies regularly form two platoon-sized assault groups.[20] Unfortunately in Afghanistan, the Soviets had so much trouble trapping *mujahideen* that they began to use their reconnaissance force for combat instead of gathering intelligence.[21] The resulting deficit in up-to-date information cost them the war.

The Multiple-Pronged Attack

Asians have been launching simultaneous, semi-independent attacks for a very long time.

> With large [defensive] bodies, especially, several points of attack must be selected, and independent attacking detachments be used for each point.[22]
>
> — Japanese night-fighting manual, 1913

Making concurrent thrusts at different spots along the same line has been an essential element of Soviet tactics since early in WWII. Those multiple prongs would often converge on the same location.

> The attack seeks to penetrate the enemy defense in two or more sectors and converge on a limited objective; its mission is to encircle or envelop enemy groupings and destroy them by simultaneous attacks from all directions. This scheme of maneuver is fundamental for the offensive operations for units of every size from the platoon to the army group.[23]
>
> — "Handbook on U.S.S.R. Military Forces,"
> U.S. War Dept., *TM 30-340*

For years, specially trained "forward detachments" functioned as the tips of these Soviet spears. Now, the multi-element approach has required—of the average Russian line outfit—more tactical expertise.

> [Maj.Gen.] Reznichenko and others now suggest that tactical missions call for securing objective along multiple axes through the depth of the enemy's defense, whose seizure fragments the defense and renders it untenable. At the tactical level, specifically tailored and tailored maneuver forces (usually forward detachments) had earlier performed this function . . . Today, and in the future, all tactical units and subunits are likely to operate in this fashion.[24]

Of course, Russia is not the only nation to discover the advantages of multiple thrusts. During WWI, the German column making the most progress would sometimes swing behind the unit opposing a sister column.[25] Furthermore, the lead squads in a WWII German attack were often roughly aligned but still operating semi-independently.[26] To tie down defenders on the flanks of the breakthrough, the Germans would frequently choose an attack frontage wider than the area needed for the main effort.[27] That way some of their narrow thrusts could probe for weak spots, while others functioned as feints.[28] It only makes sense to conceal one's true intentions among other possibilities.

Each Thrust Initiated by a Squad or Platoon

Near the end of WWI, NCOs controlled the German ground attack spearheads.[29] In fact, every German squad to participate in the highly successful Spring Offensives of 1918 knew stormtrooper technique.[30] Their tactical successes were not ignored by the Russians.

The Soviets did not use massed-infantry assaults to reduce enemy strongpoints during their highly successful foray into Manchuria in 1945. Instead they relied on fully supported, but tiny, assault elements to make the initial penetration.[31] Each division had a number of specially trained assault groups "of up to platoon size."[32] They also had plenty of sappers.[33] (Of great interest would be who trained them.) After sappers had breached each strongpoint's barriers and an assault element penetrated its lines, the rest of the assault force had little trouble entering. (See Figures 19.2 and 19.3.)

The Soviet night assault totally surprised the Japanese.

Figure 19.2: How Russians Cross Protective Wire

(Source: *Podgotovka Razvegchika: Sistema Spetsnaza GRU,* © 1998 by A.E.Taras and F.D. Zaruz, pp. 278, 279)

Figure 19.3: Barbed Wire Can Be Silently Cut
(Source: *FM 21-75* (1967), p. 32)

Because the time of the attack (at night), the climatic conditions (rain), and the infiltration style of the assault caused a general paralysis among the Japanese.[34]
— U.S. Army, *Leavenworth Papers No. 8*

Many former Soviet client states will still preempt larger attacks with tiny, thoroughly reconnoitered breakthroughs. In the Vietnam War, most opposition units had attached "special operations" personnel who could instruct assault squads in the modern equivalent of German stormtrooper technique. According to VC sapper Nguyen Van Mo, these special operators would often personally cut or bangalore the wire.[35] More recently, the North Koreans have followed the same trend.

> All NKA divisions and brigades will have a light infantry element which will be forward deployed to conduct conventional infantry tactics in the offense.[36]
>
> — "North Korea Handbook"
> U.S. Dept. of Defense, *PC-2600-6421-94*

If the Lead Element Could Get In Undetected

To break through enemy "lines," the U.S. uses a fully supported battalion or company (one with air, artillery, and armor backup). Remotely-detonated munitions have made this approach far too dangerous. (Pictured in Moscow's Armed Forces Museum are claymores rigged for command detonation in Afghanistan). To break through a heavily manned defensive sector, an Eastern nation relies on surprise. Both may end up committing a large force to the offensive, but the Eastern nation will make the initial penetration (and bypass the immediate danger) with a platoon or squad. This dichotomy of attack methodology creates differences in preliminaries as well. While the U.S. commander assesses his objective with binoculars, 1:50,000 map, and aerial reconnaissance, his Eastern counterpart does much more. He dispatches infiltrators to learn its every detail and watch its occupants. The difference is in the detail of the intelligence. In knowing where every defensive weapon is located lies additional tactical opportunity.

> During the fighting on the Kerch peninsula in the winter of 1942 the Germans captured Russian soldiers who had spent two nights and one day in the immediate vicinity of the German positions and who had been able to obtain a wealth of information during that time.[37]
>
> — U.S. Army, *DA Pamphlet 20-236*

It would make perfect sense for a tiny lead element intentionally to sidestep enemy strength. By so doing, it could not only avoid harm, but also reach more lucrative targets.

> The watershed of modern military tactics is WWI. . . . The other side . . . adopted specialized formations for attack by infiltration (so-called Hutier tactics) and follow-up by regular formations. Its tactical success . . . led to "blitzkrieg."[38]
>
> — Study contracted by the
> U.S. Defense Advanced Research Project Agency

The goal of these infiltration tactics was to bypass enemy resistance and push on as far as possible.[39] It's a point of official record that the British machineguns hardly fired a shot at the lead German elements on the first Spring Offensive of 1918.[40]

> The first wave was an infantry probe (from the accompanying division) whose purpose was to identify enemy positions for the next wave, about 250 meters behind. The second wave consisted of the elite storm companies and the flamethrower section, with additional infantry support from the division. This second wave attempted to penetrate the enemy zones by pushing through weak areas to envelop enemy positions. Supporting these efforts was the third wave, about 150 meters behind, which contained the storm battalion's heavy weapons and similar additional support from the division.[41]
>
> — U.S. Army, *Leavenworth Papers No. 4*

At the head of these storm companies were stormtrooper squads. When more than one company attacked concurrently, the semi-independent actions of their respective lead squads created tactical fluidity. On the subsequent Allied trenchlines, the German squads conducted what might best be described as "quick deliberate attacks." While these attacks happened rapidly, they met both deliberate-attack prerequisites—rehearsal and reconnaissance. Each squad had recently rehearsed its techniques so completely that it could interchange parts. Then, through "reconnaissance pull" (being responsive to what was encountered), it was able to transcend technique. Fortunately for the Allies, even highly skilled lead elements will eventually get tired.[42]

> Although the German infantry at Camrai . . . were organized and equipped for . . . attack "with limited objectives,"

> they did not have the advantage of being able to prepare for weeks . . . for an advance of a few hundred meters. Once the first [Allied] trench system was reached, there was not time for a detailed reconnaissance. . . . The decision to attack and the approach used was decided instantly, on the basis of a quick look at a large-scale map and a hasty analysis of the situation.[43]

The Noisy Way to Get In Unnoticed

To penetrate Allied lines in the Spring Offensives of 1918, the German squad used an indirect-fire deception. Its members had only to follow a well-tested formula: (1) crawl up to enemy barbed wire, (2) insert a bangalore torpedo, (3) detonate that torpedo during a precision artillery barrage, and then (4) assault with grenades and bayonets right after the barrage was lifted. To keep most defenders from realizing the ground attack, the Germans refrained from shooting their small arms. Very possibly, some went directly for the Allied machinegun positions.

> The essential elements of the tactics that [Capt.] Rohr developed in the course of these experiments were (1) the replacement of the advance in skirmish lines with the surprise assault of squad-sized "stormtroops," (2) the use of supporting arms . . . coordinated at the lowest possible level . . . to suppress the enemy during the attack, and (3) the clearing of trenches by rolling them up with troops armed with hand grenades.[44]

The Japanese assaults on Henderson Field at Guadalcanal bear a striking resemblance to Rohr's method. They were comprised of the following steps: (1) sneaking up to the protective wire, (2) bangaloring it during an artillery barrage,[45] and then (3) transiting the breach with bayonets and grenades alone.[46] However, there was now a refinement—an enhancement of the bombardment deception. If prematurely discovered, the front rank would assault with weapons blazing while the rear rank tossed concussion grenades.[47] This same cooperation between ranks would occur nine years later in a much colder place.

The first indication of a Chinese attack in Korea was often incoming hand grenades.[48] The so-called "human-wave" attacks at the Chosin Reservoir were really columns of squads from separate companies.[49] This should have come as no surprise to the Americans, because the Japanese regulations of 1913 had required attacking at night with "a line of company columns."[50] At the Chosin Reservoir, as on Guadalcanal, many the grenades were of the concussion variety and coming from the second row of attackers.[51] The Chinese liked to attack with alternating rows of burpgunners and grenadiers.[52]

An NVA command that specialized in assault technique must have provided the "special-assignment squads" that were attached to VC regional forces during the Vietnam War.[53] That these "special units" cooperated with artillery certainly suggests stormtrooper technique. During Operation Buffalo, NVA soldiers assaulted throwing TNT charges after a mortar and artillery barrage.[54] In 1975, the "special task force and artillery" neutralized Hoa Binh Airbase, the civilian airfield, and Mai Hac De storage area to facilitate the main force's attack on Buon Me Thuot.[55] In other words, skilled infantry elements paved the way for Russian-style armored spearheads.[56] Alternately called "special forces,"[57] these infantry elements employed attacks of the decidedly deliberate variety.

> The special task forces who were to strike at Hoa Binh base had also completed their scouting activities and rehearsed their mock battle.[58]
>
> — *The Victorious Tay Nguyen Campaign*
> Foreign Languages Publishing, Hanoi

Even the ragtag Eastern militias may now know how to do the stormtrooper assault. In Afghanistan, the *mujahideen* would spring from nowhere, creep to within grenade range during an attack by fire,[59] and then assault throwing grenades.[60] It would be interesting to know what kind of grenades.

The Quiet Way to Make a Penetration

The WWII Russians and Germans could both slip through a lightly held sector (long-range infiltration), but the Asians could sneak through one that was heavily defended (shot-range infiltration). As WWII progressed, the Japanese resorted less often to the stand-up "banzai" charge. On Okinawa, the first major counterattack was one of short-range infiltration. On 12 April 1945, tiny elements from four battalions successfully snuck between U.S. holes, hid for a while in caves and tombs, and then attacked those lines from the rear. The least successful of these four battalions still managed to penetrate 500 yards.[61] A large-scale, last-ditch infiltration was also planned for the thousands of naval troops stationed at the southern tip of the island.[62]

To capture a Japanese strongpoint in Manchuria in 1945, the Soviets would first try to reduce the defenders' bunkers with point-blank fire from tanks or towed artillery.[63] Then, they would sneak a small force close enough to assault.[64] If that assault failed, they would block the defenders' avenue of retreat/reinforcement and send in a nighttime "infiltration" attack.[65] However, this infiltration attack would not always take the form of a crawling skirmish line. At Hutou, some of the "infiltrators" became forward observers for the supporting arms,[66] while others poured burning gasoline down the exhaust vents of the Japanese underground positions. This became standard procedure for driving the Japanese to the surface.[67] At Hailar, Soviet "infantry sapper assault units . . . reduced the [defense] zone pillbox by pillbox."[68]

One well-respected historian describes the Chinese conduct of the Korean War as an "endless succession of platoon infiltrations."[69] The official U.S. Marine Corps history credits the Chinese soldiers with infiltrating at night better than any other soldiers on earth.[70]

> The Chinese Communists . . . made frequent and effective use of infiltration. . . .
>
> Small units often infiltrate their members individually under cover of darkness and regroup them at a previously designated point. The size of the infiltrating force may vary from a few individuals to a regiment, depending on the mission.[71]
>
> — "Handbook on the Chinese Communist Army"
> U.S. Army, *DA Pamphlet 30-51*

Fifteen years later, their southern neighbors displayed just as much affinity for this silent form of assault.

> A favorite method [in Vietnam] was attacks by a company (of about a hundred men) led by a platoon of sappers (twenty to thirty troops). Often the sappers went in alone, for the purpose of blowing things up and then getting out.[72]
>
> Sappers conducted thousands of assaults during the war. Most attacks by regular infantry units were led by sappers, who then performed the traditional role of leading the infantry through enemy barbed wire, minefield, and fortifications.
>
> Sapper assaults were fast, being over in less than an hour in most cases. . . .
>
> . . . There was usually a fire-support force that fired mortars, machineguns, and sometime recoilless rifles at the camp just when the assault force was through and ready to do their worst inside the camp. While the fire-support weapons would wake everyone up in the camp, it would also make them think it was just a mortar attack. Thus distracted by their effort to get into a slit trench or bunker (often not taking their rifles with them), [the defenders could give] the sappers . . . less interference.[73]

As the American military cannot attack this way, it has never officially acknowledged that its competition can. A tiny contingent of fully trained infantrymen can accomplish amazing things. Although fully alerted on the night of 26 August 1966 at Cam Lo, the U.S. Marine defenders could do nothing about the first wave of enemy sappers.

> They [the NVA] snuck on through before we ever illuminated the area. . . . [A]s you know, they're real[ly] proficient at moving at night . . . very silently, very slowly and very patiently. . . . [The NVA] did get through even though our people were waiting for them. They crawled in between the holes, and our people never even realized that they passed through their positions.[74]

— Lt.Col. Westerman (CO, 1st Battalion, 4th Marines)
U.S. Marines in Vietnam, Hist. & Museums, HQMC

The Marines were not the only ones to have trouble with this type of attack. Below are listed the most glaring examples of what a few people—with the ability to covertly enter a U.S. perimeter—can accomplish.

> [E]nemy sappers enjoyed a record of chilling successes during the Vietnam War. Air bases were a favorite target, and in February 1965, the VC employed mortars and demolition teams to reduce the army's Camp Holloway near Pleiku to a shambles. The shelling hit barracks buildings, and eight Americans were killed, another 126 wounded. One sapper body was left behind near the flight line where twenty-five planes and helicopters had been damaged or destroyed. Records do not indicate ARVN casualties at the base, but they must have been heavy.
>
> In October 1965, a ninety-man VC raiding force penetrated the Marine air facility at Marble Mountain near Da Nang. Almost a quarter of the sappers were killed or captured, but not before they and their comrades had killed three Marines wounded ninety-one, and practically destroyed a helicopter squadron. Nineteen choppers were blown up on the airstrip, and thirty-five damaged.[75]

> In March 1971, [NVA] sappers went through 101st Airborne Division units on the perimeter of the Khe Sanh Combat Base, which had been reopened to provide helicopter support to the ARVN in Laos during Operation Lam Son 719. . . . The base rocked for hours with burning fuel and exploding munitions from two ammo storage areas. . . .
>
> Less than a month later . . . , a small sapper team that got in and out without a fight blew up 1.8 million gallons of aviation fuel at the army's huge logistical complex at Cam Ranh Bay.[76]

> During the night of 27-28 March 1971, a Viet Cong sapper company infiltrated Fire Support Base Mary Ann, the forwardmost position in the 23d Division (Americal). Snipping through the defensive wire and entering the base without alerting a single guard in a single perimeter bunker, they killed thirty U.S. soldiers and wounded eighty-two in a humiliating defeat that sounded the death knell for the reputation of the once proud U.S. Army in Vietnam.[77]

The Less-Apparent Combination Technique

For Eastern assault technique, one normally thinks of a lone sapper relying on stealth or a stormtrooper squad on indirect-fire deception. However, there is evidence of a third technique. It has been underreported because it is a combination of the other two. For lack of a better name, it will henceforth be called the "inside-out" technique. In short, the attacking unit sneaks a few sappers into an objective to facilitate a forced entry. The sappers then attack outward as the parent unit assaults inward.

On a larger scale, this dual approach has been used throughout Eastern history. Didn't the Germans precede their armored columns with American look-alikes at the Battle of the Bulge? As covert-operations and "diversionary" troops,"[78] couldn't Russian Spetznaz facilitate the passage of armored spearheads?

> Spetznaz agents have no specific dress. Instead, they adapt their clothing to blend in with their surroundings (in most cases, urban dress).
>
> Specialists in stealthy night perimeter infiltration and sentry removal, all Spetznaz are experts in the Russian hand-to-hand combat art of sombo.[79]

Even on the small scale, there is ample evidence of the combination technique. The leading edge of the Chinese assault forces at the Chosin Reservoir on the night of 27 November 1950 behaved like infiltrators.

> The first files of Chinese skirmishers crept down from the heights opposite the slumbering Marine lines along the northwest, north, and northeast arcs of the Yudam-ni perimeter.[80]

Some of the Chinese who covertly penetrated Marine lines went after U.S. mortar pits and command bunkers.[81] However, others attacked outward. While beating back the Chinese human-wave assaults, frontline Marine positions received concussion grenades and "burpgun" fire from the rear.[82]

The "inside-out" attack was also evident in Vietnam. The one described below was a "live-fire" practice run for the attack on Fire Base Mary Ann.[83]

> During the first week in February 1971, a sapper force walked right into a hilltop perimeter manned jointly by the ARVN and a reconnaissance platoon from the . . . Americal Division. The VC entered through the ARVN side . . . and demolished the position from [the] inside out.[84]

The Main Attack Delayed

The Asian soldier practices the art of delay in conjunction with a "false face." He wants to mislead his opponent, wait for him to make the first move, and then capitalize on his almost certain mistake.

Just as guerrillas, "low-tech" Eastern soldiers will only fight when almost sure to win. They are more likely than their Western counterparts to discontinue a futile assault. As was documented in Chapter 12, most are allowed to pull back when continuing to fight holds no promise. By U.S. admission, WWII Germans sometimes had the option to turn around in mid-assault.[85] The North Vietnamese would also discontinue an attack for which too much surprise had been lost.

Possibilities from Encircling a Quarry

Eastern units like to surround an opponent completely. They do so to keep him guessing as to the direction of the main attack. After all, feints would be easier to launch from a circular formation. As the Golden Hordes did in their invasion of Russia, Easterners will send several prongs from different directions to converge on the same target. This can be done without fratricide as long as the assault troops stick to grenades, bayonets, and well-aimed (mostly downward) shots.

> In ascending order of effectiveness, the [Japanese] envelopment may be single, double, or an encirclement.[86]
>
> — "Handbook on Japanese Military Forces"
> U.S. War Dept., *TM-E 30-480*

> Encirclement had always been a German specialty.[87]

To preclude breakouts, several generations of Russians have used double encirclements.[88] This has permitted multisided attacks on the small scale as well as the large. At several battles in Manchuria in 1945, the final Soviet assault came from three sides at once.[89]

For annihilation, the NVA also preferred the double encirclement. In the initial battle of their 1975 push, they used this tactic against Saigon's 22d Division at Phuoc An.[90]

A Frontal Feint to Facilitate Rear Infiltration

With a ruse used against the White Army at the Isthmus of Perekop in 1920, the Red Army defeated fortified positions at Viborg (Finland) in 1940 and Sakhalin Island in 1945. Their ploy was to conduct a frontal demonstration while closing with their opponent from the rear across a water obstacle.[91] Throughout WWII, the Russians honed their individual and small-unit infiltration skills.

> Most German officers who fought the Soviets on the Eastern Front acknowledged their "natural superiority in fighting during night, fog, rain or snow," and especially their skill in night infiltration tactics.[92]
>
> — U.S. Army, *Leavenworth Papers No. 6*

By U.S. determination, the Japanese concurrently ran a main attack, secondary attack, and feint during WWII.[93] Most noted for their frontal "banzai" rushes, the Japanese have yet to be credited for their rear-area activities.

> To increase the effectiveness of the envelopment, the Japanese often sent a small force around to attack the enemy rear.[94]
>
> — "Handbook on Japanese Military Forces"
> U.S. War Dept., *TM-E 30-480*

According to a U.S. Army study of Korea, the Chinese routinely held a defender's attention with an overt frontal attack while sending covert elements around both flanks.[95] However, this frontal assault was no suicide mission. Led by a skirmish line of scouts, it could often get within 15 yards before being detected. After the initial mass effort, its parts would take turns attacking during lulls in the firing.[96] They were, in fact, the lead squads of company columns of squads.[97]

A thousand miles to the south, the Chinese-supported Viet Minh occasionally assaulted as a human wave. After surrounding a hill occupied by the French Foreign Legion, they took cover behind the hill's military crest and then rushed forward from every direction when the moon went behind a cloud.[98]

Fifteen years later, the NVA would launch just enough human waves to keep their opponents honest. While U.S. defenders prepared for all-out assault, sapper teams and stormtrooper squads nibbled away at their strategic assets. The Asian will show a Westerner only what he wants him to see. This little-known aspect of the Maoist "mobile offense" must have grown out of the ancient Chinese "cloud" maneuver.[99] By this means, a large unit could rapidly dissipate or reform. As Mao's "mobile war" philosophy was the product of his guerrilla experiences,[100] the two attack stages never completely diverged. That would explain why actual (vice feigned) Communist assault forces look, to the Westerner, like impromptu guerrillas.

> As defined by Giap, "[M]obile [offensive] warfare is the fighting way of concentrating troops of the regular army in which relatively big forces are regrouped and operating on a relatively vast battlefield, attacking the enemy where he is relatively exposed . . . , advancing very deeply and then withdrawing very swiftly" (Giap, *People's War—People's Army,* 1962, 106). In essence, mobile warfare involved large units conducting operations against the enemy, but it did not direct fighting set-piece battles. . . . Put simply, mobile warfare and guerrilla [warfare] were inseparable and mutually supporting (Giap, *People's War—People's Army,* 1962, 107-108).[101]

A Forerunner of Things to Come

Only in the Western World does warfare equate to "high noon in the middle of the street." Easterners have a different perspective on chivalry. They fight when and where they are least expected. They don't mind being surrounded in the process.

> In another instance [during WWII] . . . after the Russian attack had been beaten off, the German battalion commander found that a Russian rifle platoon had been left behind in the village after all other troops had withdrawn and that the men had concealed themselves in the dung hills near the farm buildings. Their mission was to observe the Germans after their entry into the village and communicate the information to the parent unit which was hiding in a nearby woods with the intention of launching a surprise [night] attack.[102]
>
> — U.S. Army, *DA Pamphlet 20-236*

The Emergence of an "Underground" Offense

Near the end of WWII in the Pacific, the beleaguered island defenders could most easily raid their tormentors through subterranean passageways. They could also best launch defensive spoiling attacks from bypassed caves. The Japanese had set the stage for the purely offensive "tunnel" attack of the Viet Cong. The tunnel was, after all, a foolproof way to bypass every barrier and defender.

A few months after Gen. Kuribayashi's classic defense of Iwo Jima, the vice-chief of staff for the Japanese Army outlined the tactics for Okinawa:

> First, we cannot hope to match the enemy's strength on the ground, at sea, or in the air. Therefore, we should attack the enemy from "underground."
>
> Second, in order to prevent the enemy from using his superior ground, sea, and air strength to the full, we should make use of the cover of night to pull up behind the enemy or penetrate his lines. Enemy troops would be thrown into a state of confusion, unable to distinguish friend from foe.
>
> Third, the greatest threat above ground is enemy tanks. We have only a few antitank guns, and these would be quickly destroyed by the enemy's bombardment. . . . Therefore, the Japanese army has formulated new "patented" antitank tactics. These involve hand-carried makeshift explosive devices containing ten kilograms of yellow [explo-

> sive] powder. . . . Delivery of these explosives would be in the nature of a suicide attack.[103]
>
> — Gen. Atomiya, Imperial Japanese Army

For a number of reasons, the Japanese could not prepare Okinawa for what was to come as well as they had Iwo Jima. While both islands had soft limestone in abundance,[104] Okinawa was much larger. Still, the defensive plans were similar. Stretching across Iwo at 1000-yard intervals were three bands of underground bastions.[105] Heavily defended on Okinawa were several east-west ridgelines spanning the narrowest part of the island (just north of Shuri Castle). To thwart U.S. direct-fire weaponry, the Japanese planned to defend many of these rocky escarpments from the reverse slope.[106]

> Japan's 32nd Army had spent the greater part of a year turning them [the escarpments] into formidable nests of interlocking pillboxes and firing positions. Connected by a network of caves and passageways inside the hills, their positioning enabled defenders to shift their strength constantly in response to attack.[107]
>
> — Col. Yahara, Japanese veteran of Okinawa

While most of the bastions in each of Iwo Jima's three defensive "belts" were probably linked by tunnel at the time of the U.S. landing,[108] those on Okinawa were not. The larger island's heavier vegetated and deeper gullies allowed the lateral shifting of some Japanese units above ground.[109] This was fortunate for the Americans.

> Our defense [on Maeda and Nakama hills] was not fully protected because the connecting underground network between the hills had not yet been finished.[110]
>
> — Col. Yahara, Japanese veteran of Okinawa

On Iwo, there had also been an underground thoroughfare linking the three defense belts with each other.[111] How else could a Kuribayashi-led force of 300 have mounted a surprise attack on Airfield Number One two weeks after the official end to hostilities.[112] His last stand had been 4000 meters to the north.[113] On Okinawa, the plan was to excavate a number of defensive positions that could be used as underground waystations during the north-south shifting of units.[114] Some 25,000 Okinawan laborers were put to work,[115] along with many soldiers. In the 10 months before the landing, the 44th Independent Brigade prepared seven separate positions.[116] Of note is how fast it could dig one.

> In defending ordinary field positions it takes only a couple of days to study the terrain, assemble required ammunition, and prepare proper fortifications. It takes more than a week, however, even for fresh troops to establish strong underground defenses.[117]
>
> — Col. Yahara, Japanese veteran of Okinawa

In these subterranean positions, the defenders of Okinawa had little trouble surviving the initial U.S. onslaught. While Japanese survivors were quick to admit their mistakes, most resolutely praised the protection from bombardment of their underground fortifications.

> The effectiveness of the Blue artillery was countered successfully to a great extent, by the elaborate system of underground fortifications. Heavy bombardments, such as came before attacks, caused relatively low casualties.[118]
>
> — Japanese POW on Okinawa

At least some of the forward bastions on the several-hundred-yard-deep Shuri "Line" were linked by tunnel to those behind them.[119] By U.S. admission, the Japanese used a daunting, triangular system of defenses,[120] in which the front of each position was crisscrossed by fire from the two behind it. Sugar Loaf, Half Moon, and Horseshoe formed an arrowhead pattern at the western end of the line. Once Sugar Loaf had finally been reduced by assault, secret passageways from Half Moon and Horseshoe may have facilitated underground forays behind U.S. lines. Veterans of the campaign remember all too well the unwanted Japanese activity.

Often on Okinawa, the Japanese would allow themselves to be overrun and then attack from below. This was standard operating procedure for Japanese antitank troops.[121] They would wait in a spider hole (or tunnel opening) for the forward echelon of U.S. ground pounders to pass, pop a smoke, and then run forward to affix a satchel charge to an American tank. To escape U.S. supporting arms, whole Nipponese units may have also waited below to attack American forces passing overhead.

> We received many reports of valorous fighting [for the Shuri Line], such as "Our soldiers jumped out of their caves as soon as the enemy tanks passed, crawled forward, and engaged in hand-to-hand combat with enemy soldiers."[122]
>
> — Col. Yahara, Japanese veteran of Okinawa

Further Refinements to the Subterranean Foray

To prevent further U.S. incursions beyond the 38th parallel in Korea, the Chinese dug 1250 kilometers of tunnels toward the end of the Korean War.[123] These tunnels provided ample chance for offensive action.

> Whether we can defend our positions is a question that was resolved last year. The solution was to hide in grottos. We dug out a two-level fortification. . . . Once the enemy entered the surface positions, we started counterattacks and inflicted heavy casualties on them [from below]. By this crude means we were able to seize and take away the equipment left by the enemy.[124]
>
> — *Mao's Generals Remember Korea*

> The Chinese liked to construct tunnels and caves at launching-attack points.[125]

> The Chinese would send out small teams to dominate no-man's land, take over exposed outposts, and mine daytime outposts abandoned at night, create disturbances in enemy rear. They would also keep up continuous sniping.[126]
>
> — *Mao's Generals Remember Korea*

To capture the main airstrip at Dien Bien Phu in May 1954 (Map 19.1), the Viet Minh shifted from overland night attacks to gradually encroaching trenchlines. To nullify their foe's supporting arms, they hugged him. To remove his mutual support, they segmented him.[127] Soon, the sheer profusion of ditches made those

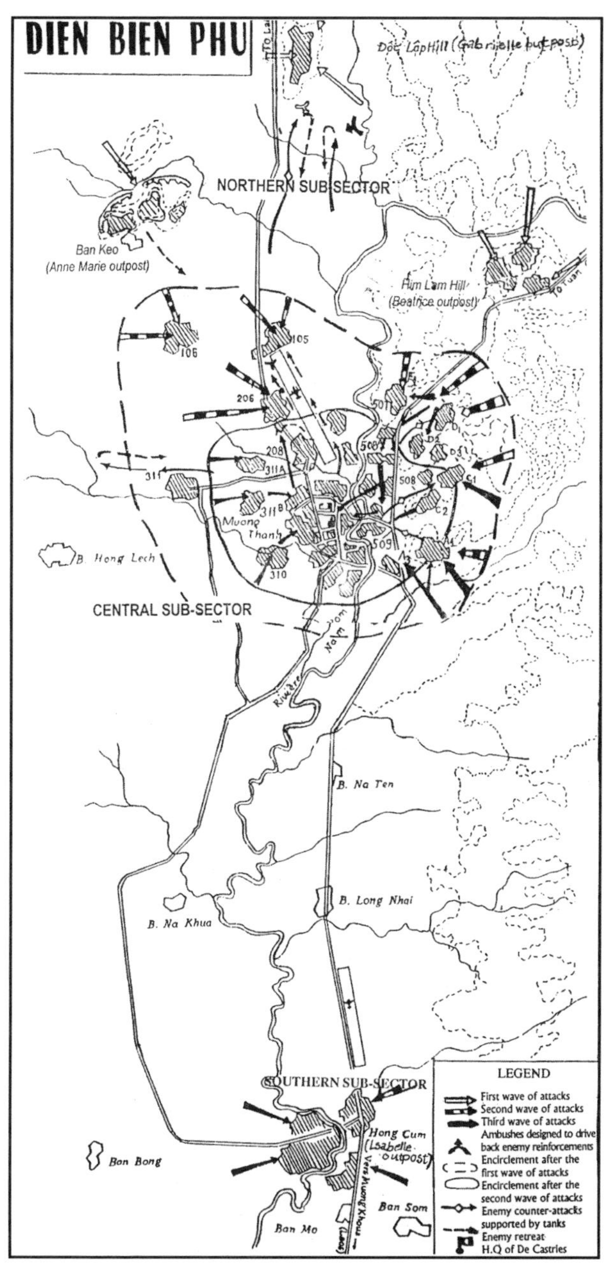

Map 19.1: Tightening Circle of Trenchlines at Dien Bien Phu

(Source: Courtesy of The Gioi Publishers, from *Dien Bien Phu,* by General Vo Nguyen Giap, © 1999 by The Gioi Publishers, Map Insert)

occupied difficult to ascertain. Hidden among the miles of "communications trench" were not only well-seasoned assault elements,[128] but also semipermanent support facilities.[129] Low berms blocked linear observation,[130] but only one thing could prevent aerial observation—indentures in the trenchline's sides.

> *The system of positions of attack and encirclement* included the communication axes running around the Muong Thanh [central] sub-sector . . . ; many lines of communication trenches . . . from the surrounding hills to the proximity of the enemy's front line; many cross communication trenches to increase liaison possibilities. . . .
>
> In about twelve days, our troops were able to dig a length of over 100 km of communications trenches and combat trenches and our positions, but despite the bombing and strafing by his aircraft and artillery, our troops advanced unchecked nearer and nearer the enemy . . . The success of this construction enabled us to close in upon the enemy on flat ground, solve the problem of food transport, keep firm the front, fight unremittingly day and night, and restrict the effect of the enemy's artillery and aircraft as much as possible.[131]
>
> — *Dien Bien Phu*
> The Gioi Publishers, Hanoi

In effect, the Viet Minh had ringed Dien Bien Phu with a vast web of ditches that extended to within 100 yards of every French position and, in some cases burrowed beneath them.[132] Many had been closely encircled to cut off resupply and reinforcement.[133] In addition to covered approach routes, the Viet Minh had prepared covered attack points.[134] Their assault positions were within 100 meters of French lines,[135] had preregistered artillery to their fronts,[136] "curbed the effect of enemy fire,"[137] and probably had expanded subsurface dimensions. For the really tough objectives, that subsurface staging area kept on going. "During the Siege of Dien Bien Phu, Vietnamese sappers dug their way under French perimeters to strike deep into the heart of the compound."[138] To capture the high ground at the east end of the core French defenses, the Viet Minh most definitely employed a tunnel.

> On May 6, at 8:30 P.M., our troops attacked Hill A-1. In the

> preparatory stage, our sappers had dug an underground trench leading to the centre of the hill and introduced there one tone *[sic]* of explosives. With the powerful coordination of this explosion, our troops attacked this position from various directions, put out of action the defending unit composed of paratroopers of the foreign legion, and *occupied this last height.*[139]
>
> — *Dien Bien Phu*
> The Gioi Publishers, Hanoi

The French had fallen victim to their own cultural orientation. To an industrialist flying high over Dien Bien Phu, their plan would have made perfect sense. They had lured their ill-equipped, undisciplined, and cowardly foe into a set-piece battle where the side with superior technology would prevail. As the French are a brave and intelligent people, other Western populations should take note of the French government's miscalculation.

The NVA/VC Improve on the Subterranean Attack

Hundreds of miles to the south and fifteen years later, the Viet Cong would continue the tradition of suddenly appearing from nowhere. Among the oldest tricks of the *ninja* is to predict where an opposing force will bivouac and then hide there below ground.[140] He soon has free access to anything he wants.

> At least once during the Vietnam War, an American army base was inadvertently built over a Vietcong tunnel complex. Whether this occurred because of coincidence or the Vietcong's anticipating the Americans' intent to build the camp on the site and then simply extending their vast system of tunnels to the site, the result was the same: enemy sappers appeared and disappeared at will within the perimeter of the base.[141]

A Sophisticated "Tunnel" Attack Emerges

Western soldiers have occasionally burrowed to emplace obstacle-breaching charges. Eastern soldiers keep going. In Vietnam, the Viet Cong and North Vietnamese often used tunnels to enter their objectives.

> During the American occupation [of Vietnam], tunnels were dug near, into, and completely around American bases. In places, these access tunnels allowed sappers to come and go from American installations at will.[142]

In September 1966, a Marine lieutenant saw an enemy sapper disappear into the ground 50 yards beyond the Dong Ha airstrip barbed wire.[143] Some years later, a former VC admitted to having tunneled up to the wire at the northern end of the An Hoa airstrip.[144] There was tunneling at Khe Sanh as well.[145] Map 19.2—the North Vietnamese version of what happened at Khe Sanh—shows no troop movements near the main airfield but extensive fortifications within 400 meters of it. *Asserchiamento* means encirclement in Italian, and *Ta Con* is the North Vietnamese name for Khe Sanh. As Marine security patrols didn't encounter any NVA trenching, there is a good chance that this encirclement was subterranean. Though not drawn to scale, another North Vietnamese schematic (Map 19.3) shows feeder tunnels to what must have been a concentric, underground fortification.

> According to a Reuter [news service] report dated March 18th, 1968, the siege of Khe Sanh was a three-level encirclement: in the air, on the ground, and under the ground.[146]
>
> — *The 30-Year War*
> The Gioi Publishers, Hanoi

A Khe Sanh defender, Marine Corporal Bob O'Bday, later reported the following: (1) no sightings of trenches on his many security patrols, (2) no rumors of trenchlines that close to the perimeter, (3) shellholes with false bottoms that may have been used as mortar positions, (4) rectangular patches of subsurface dirt that had been made to look like tilled fields, (5) many surrender pamphlets, and (6) digging sounds from the bottom of his perimeter hole.[147] As the battle for Khe Sanh was only a feint, the enemy may have been content with sabotaging U.S. materiel. Still, as with the "human wave,"[148] the Communists would often carry successful feints to fruition.

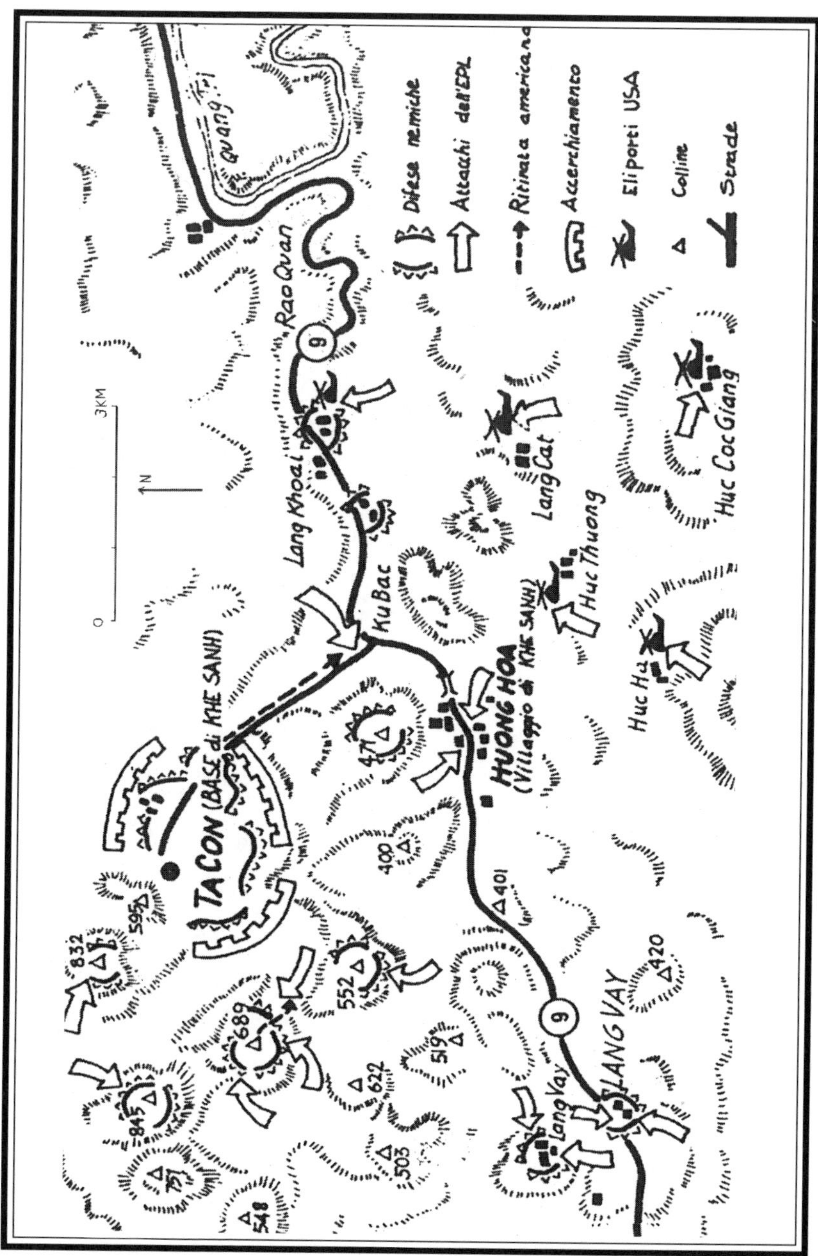

Map 19.2: Subsurface Fortification Encircles Khe Sanh

(Source: *The War 1858-1975 in Vietnam*, by Nguyen Khac Can, Phan Viet Thuc, and Nguyen Ngoc Diep, Nha Xuat Ban Van Hoa Dan Toc Publishers, Hanoi, Figure 540)

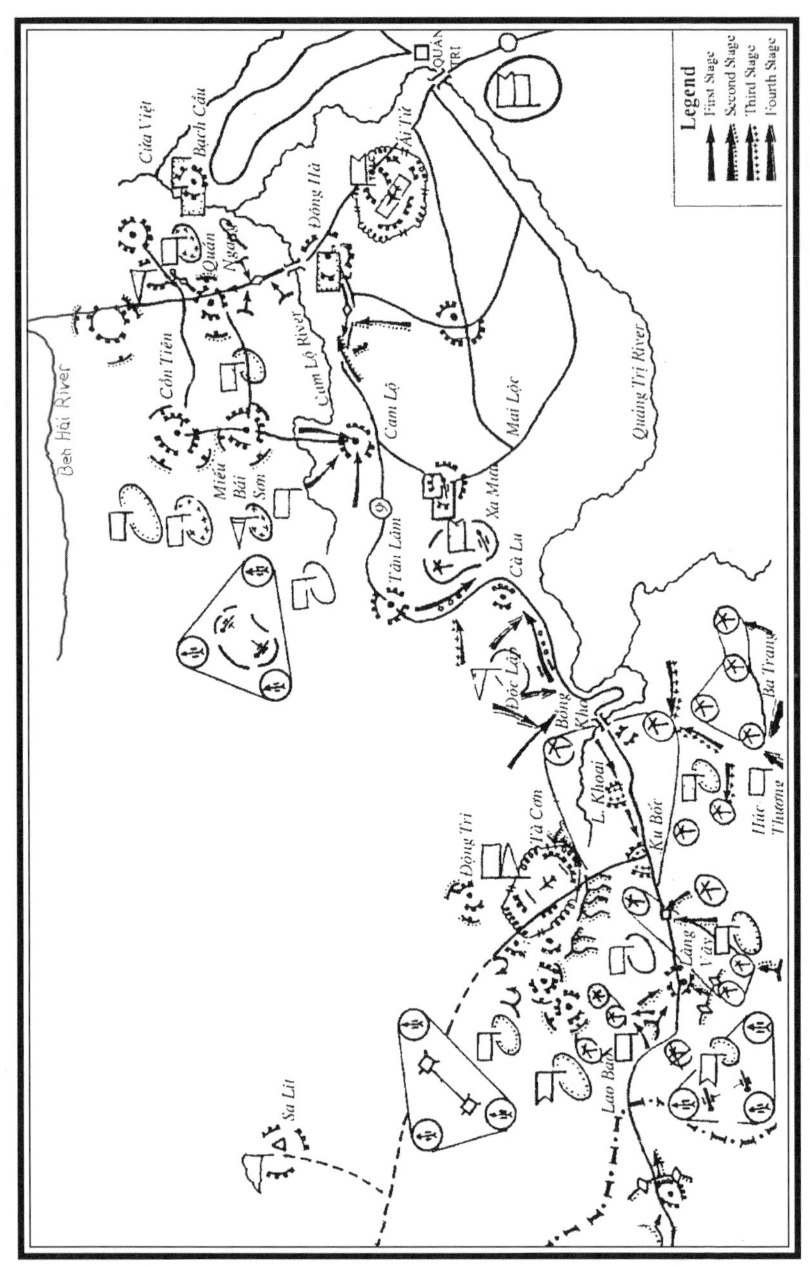

Map 19.3: Evidence of Feeder Tunnels at Khe Sanh

(Source: *The Tet Mau Than 1968 Event in South Vietnam,* by Ho Khang, © 2001 by The Gioi Publishers, Hanoi, p. 74)

Ominously, the North Vietnamese had accomplished with tunnels at Khe Sanh what their fathers had done with trenchlines at Dien Bien Phu.[149] They had closely ringed their objective. There's no telling what might have eventually occurred if U.S. forces had not abandoned the airfield. With hundreds of *ninja*-trained soldiers pouring out of the ground inside the wire, the U.S. firepower advantage would have been of little use. Something of a tactical precedent had been set.

More Recent History May Portend the Future

The infiltrators from North Korea still like to burrow beneath obstacles. "Ordinarily the primary barrier fence was climbed over or tunneled under."[150] To bypass minefields and artillery concentrations, North Korean units have only to attack through bigger tunnels.

> [I]n 1971, Kim Il-song *[sic]* specifically ordered the construction of infiltration tunnels along the DMZ: "one tunnel can be more powerful than ten atomic bombs put together and the tunnels are the most ideal means of penetrating the South's fortified line" (Republic of Korea, *Defense White Paper 2-1990,* 75).
>
> The engineer battalion of each infantry division deployed directly on the DMZ was assigned the task of digging two infiltration tunnels. . . . Aside from the four located and neutralized tunnels, ROK/U.S. intelligence currently [as of 1998] estimates that there are eighteen suspected active tunnels in various stages of completion along the DMZ.[151]

The Unmistakable Common Denominators

Eastern armies must attack from close range. Most lack the systems to do it any other way.[152] Prerequisite to this style of warfare are advanced individual skills.

> The common [Chinese] tactical style manifested itself in the use of surprise, deception and camouflage, movement, pa-

trolling and reconnaissance, and *individual skills* [italics added].[153]

— "The Chinese Communist Forces in Korea"
U.S. Army, *Leavenworth Research Survey No. 6*

As Western technology has continued to advance, the Eastern trend has been toward: (1) covert maneuver elements operating in conjunction with an actual or feigned holding attack, and (2) some form of "below-ground" offense. For the latter, the Easterner can either tunnel beneath his opponent's front lines or simply allow that opponent to "overrun" subterranean attack points. Either way, he can steal or destroy that adversary's strategic assets with impunity. The heavy guns with which the Viet Minh leveled Dien Bien Phu had been captured from U.S. forces in Korea.[154]

20 For Attacking Cities

- *How might a built-up area be most safely captured?*
- *What must the urban attacker be able to do?*

NORTH KOREAN PRIVATE, 1950

(Source: Courtesy of Osprey Publishing Ltd., from *The Korean War 1950-53,* Men-at-Arms Series, © 1986 by Osprey, Plate A, No. 2; *FM 90-10-1* (1982), pp. B-7, B-29)

There Is Greater Risk in Attacking Built-Up Areas

As the "terrorist" threat increases, Americans must remember the inherent dangers of offensive urban combat. Their attack experience with large cities has been limited to Manila in 1945, Hue City in 1968, and Baghdad in 2003. In June 2002, a U.S. battalion spent four days in simulated urban combat at George Air Force Base in California. It did not fully satisfy its expectations.

> In the assault on "al-George," however, it took 980 Marines to roust just 160 rebels from urban terrain. And despite

> wielding a 6-to-1 advantage, the Marine force still took about 100 casualties.[1]
>
> — *Wall Street Journal,* 22 August 2002

> The Marines put some of their ideas to the test in a recent exercise on a shuttered Air Force base in southern California. In it, a battalion of 1,100 troops, backed by tanks and helicopters, tried to capture base housing from a simulated enemy force, played by 200 eager reservists. . . .
>
> In taking the city, about 100 of the [regular Marine] attackers were killed. . . . Several helicopters also went down before the Marines captured the town, and more were killed as the defending forces began using truck bombs and other guerrilla activities. . . .
>
> One grisly lesson: The medics did not have enough equipment to treat all the head and shoulder wounds—injuries received when soldiers were shot trying to peek around a corner.[2]
>
> — Associated Press, 23 September 2002

Because small-arms fire can come from so many locations in the city, forward motion is extremely risky. No amount of covering fire can defeat the first bullet; only individual movement can do that. To complicate matters, the enemy can often shoot right through walls, floors, and ceilings.

> In [early] 1995 in Grozny [Chechnya], foes shot each other through floors and ceilings, every wall could conceal an ambush.[3]
>
> — *Armed Forces Journal International,* May 2000

Assault forces must be particularly wary of obviously contested structures. As was the case at Waco, occupied buildings are difficult to enter unnoticed. Any new aperture must be immediately exploited. Even upon successful entry, the assault team may be watched through makeshift periscopes, cracks in the woodwork, or pin pricks in the wallpaper. Or their progress may be monitored through sound alone. Then the assault team may face command-detonated mines or small-arms fire through ceilings, floors, and walls. While performing the "buttonhook," "high-low," "stack," or other "dynamic-entry" method, it may run afoul of other enemy initiatives. The room's occupant may have a bunker in one corner, concertina along the walls, boobytraps at the doors, and a covered exit route. Such an adversary will survive the attacker's grenade, toss one of his own, and make good his escape. The only way to beat him is with the overpressure of a shape charge. Structures defended like this must be bypassed or leveled with standoff weaponry. Unfortunately, there may be moral considerations. When houses are occupied by both criminals and innocent civilians, policemen can't kill everyone inside and neither can infantrymen. Every attempt must be made to safeguard those noncombatants. With or without technology, urban assault troops must assume this extra risk.

The Eastern Trend

Of the major Eastern armies, the Russians have arguably the most experience in attacking (or counterattacking) through built-up areas. During WWII, they fought for several cities between Stalingrad and Berlin. Unfortunately, much of what they learned in WWII never made it into their organizational memory. By 1995, they had little street-fighting expertise left. In the aftermath of the Chechen disaster, Gen. Mikhail Surkov admitted that "street fighting tactics are absent from the manuals of the Russian armed forces."[4] As would soon be evident in Afghanistan, the problem went much deeper than tactics. The Russians had lost Zhukov's confidence in NCOs.[5]

> In handing the Russian military a humiliating defeat [in 1995], the Chechens adapted to modern [urban] conditions their traditional forest guerrilla tactics. . . .
>
> For a force of guerrillas, the new urban "forest" provides many . . . opportunities for sniping, mines, boobytraps, and ambushes, while it negates the enemy's superiority in . . . armor, artillery and . . . air power. Urban fighting also cruelly exposes the shortcomings of an army accustomed to relying on major units acting together under a rigid hierarchy of command. Even more than in modern warfare in general, units operating in cities get separated and broken down to the section and even subsection level, throwing tre-

> mendous responsibility on junior officers, noncommissioned officers (NCOs), and individual soldiers.[6]
>
> —*Armed Forces Journal International,* August 1998

Several of Russia's former satellites have made better use of their firsthand experience with urban offense. Seoul changed hands several times during the Korean War. North Vietnam easily captured Hue City in 1968 and several other South Vietnamese cities (to include Saigon) in 1975.

Because of the extreme risk in frontally assaulting someone's neighborhood, Eastern armies have developed some alternatives. By U.S. admission, the WWII Russians not only reconnoitered each block before attacking it, but also used short-range infiltrators as their lead element.[7] As they disliked advancing along sidewalks,[8] they must have done some moving below ground.

During the second of the Warsaw Ghetto uprisings in WWII, one encircled Jewish enclave secretly joined the other through the sewers. To root out a few firmly entrenched defenders, their tormentors soon realized they would have to flood and burn the entire Ghetto.[9] This would do little to win the hearts and minds of its other residents.

The Asian Solution

To avoid the casualties of openly assaulting a city, the Asian will secretly attack it. He covertly enters through all available routes and then blends with the local populace. (See Figure 20.1.)

> [O]n September 29, 1768, Gorkha troops [predecessors of the Gurkhas] infiltrated Kathmandu [Nepal] while the population was celebrating a religious festival and took the town without a fight.[10]
>
> — U.S. Army, *DA Pamphlet 550-35*

The North Vietnamese operate in much the same way. They attack a city from the inside out.[11] Only after supplies have been staged, commandos positioned, and key buildings seized, will their spearhead battalions strike toward the city's center. Most built-up areas provide a veritable maze of underground rooms and passageways. Within those rooms and passageways can be hidden hundreds of commandos and tons of ammunition. The North Vietnamese call this the "blooming lotus" maneuver. It automatically envelops the Western-style defense.

> This methodology was developed in 1952 in an assault on Phat Diem. Its key characteristic was to avoid enemy positions on the perimeter of the town. The main striking columns move[d] directly against the center of the town seeking out command and control centers. Only then were forces directed outward to systematically destroy the now leaderless units around the town. . . .

Figure 20.1: Every Town Has Underground Drainage

(Source: *MCRP 3-02H* (1999), p. IX-1)

> This approach contrasts sharply with Western doctrine, which traditionally would isolate the town, gain a foothold, and systematically drive inward to clear the town. This sets up a series of attrition-based battles that historically make combat in built-up areas such a costly undertaking.[12]
>
> — "'Urban Warrior'—A View from North Vietnam"
> *Marine Corps Gazette,* April 1999

At the outset of the 1968 Tet offensive, NVA spearhead battalions had little trouble night-marching into Hue City. All the bridges, intersections, and gates along their avenues of approach had been captured in advance. Those battalions then consolidated all but a few of the key buildings and compounds. Map 20.1 shows their actual routes.

In 1975, the NVA launched their final offensive of the Vietnam War at Buon Me Thout. Again they attacked from both outside and inside the town. After capturing the airport, they seemed content to encircle the town and dug in—possibly to further the illusion that Buon Me Thuot was just a feint.[13] Only then did their "special forces" assault from the city's center and their main force supported them with spearheads.[14] Still believing the primary target to be Kontum or Pleiku, the South Vietnamese never bothered to reinforce Buon Me Thuot.

> Of the five principal targets inside Buon Me Thuot—the airfield, the artillery and armor bases, . . . the base camp of the 23rd Division . . . , and the Mai Hac De [supply] depot, we gave priority to the 23rd Division's base camp and rated the artillery and armour bases as important. . . .
>
> Our forces would move in four prongs, backed by strong reserves. Later, another prong was added by the advanced command. . . .
>
> Attacks were to be made in the following order: special forces, assisted by infantry, would close in upon the airfield inside the town, take it by surprise and from there occupy the main intersection. . . .
>
> Infantry units would . . . launch lightning attacks at Chu E Bua, Chu Duc, and Peaks 491 and 596 [as a] springboard from which to encircle and finally attack the enemy.
>
> After the infantry and special forces had occupied the

assigned targets, the artillery would pound the headquarters of the 23rd Division . . . throwing their command into confusion and forcing them into passive defence.

While the enemy were coping with these attacks . . . , our sapper units would relatively easily clear a passage to the outskirts of town, while the spearhead units . . . would speedily [enter].[15]

— *The Victorious Tay Nguyen Campaign*
Foreign Languages Publishing, Hanoi

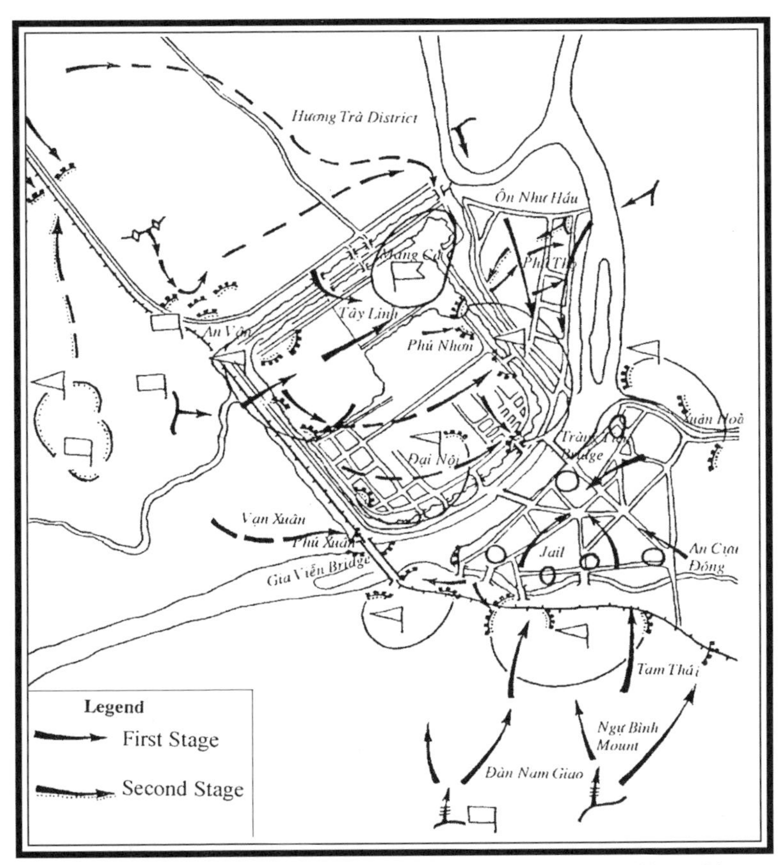

Map 20.1: The North Vietnamese Tet Attack on Hue City

(Source: *The Tet Mau Than 1968 Event in South Vietnam*, by Ho Khang, © 2001 by The Gioi Publishers, Hanoi, p. 82)

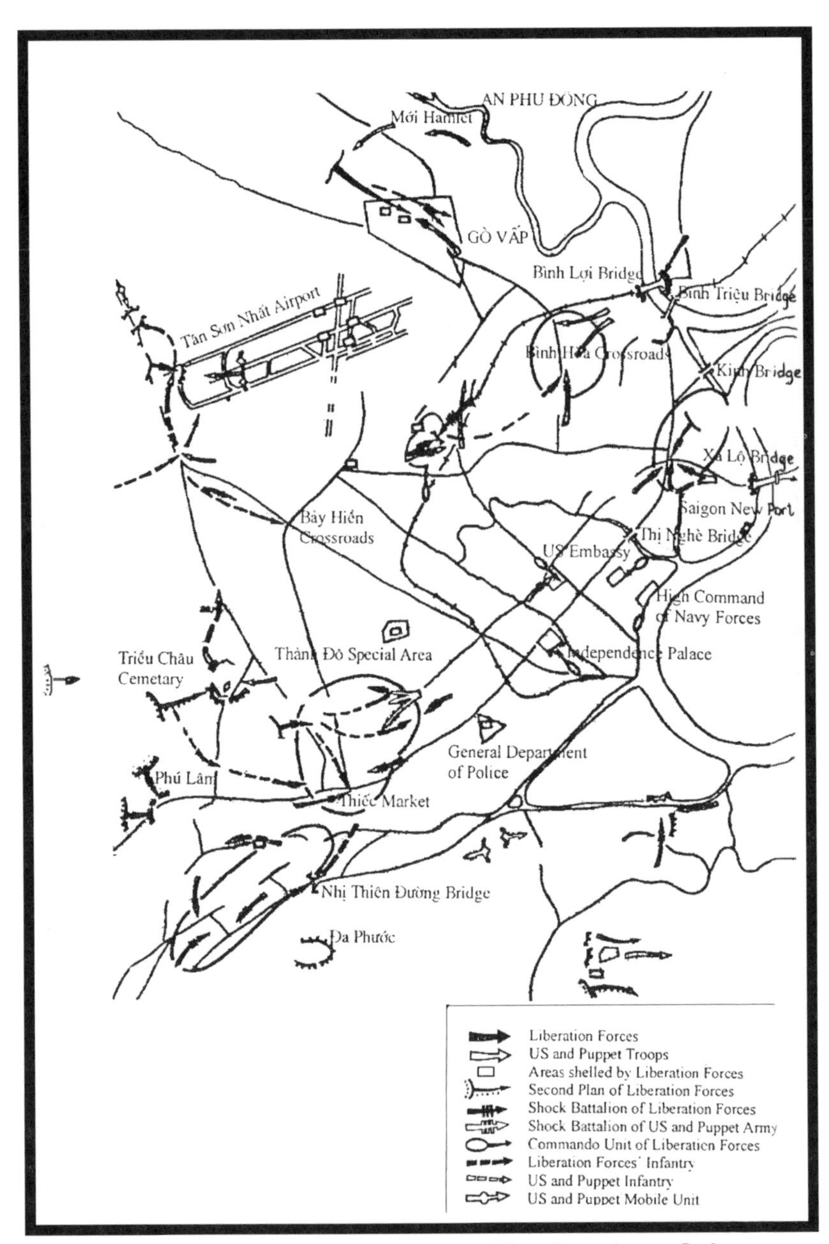

Map 20.2: The North Vietnamese Tet Attack on Saigon

(Source: *The Tet Mau Than 1968 Event in South Vietnam*, by Ho Khang, © 2001 by The Gioi Publishers, Hanoi, p. 88)

To sneak up on the South Vietnamese capital, the NVA had the Cu Chi infiltration route. It must have provided no shortage of underground rest stops and storage areas, as the battle was won before it commenced. For the 1975 assault on Saigon, the Communists used a maneuver scheme much like that of seven years before. In 1968, commandos had tried to seize nine key objectives and hold out until their spearhead battalions could arrive (see Map 20.2). Among those objectives were Independence Palace, Army General Staff, Navy Command, Tan Son Nhut Airfield, U.S. Embassy, Police Headquarters, Capital Special Zone, and Chi Hoa Jail.[16] All the while, other main-force units had tried to keep the majority of U.S. and ARVN forces occupied at the outskirts of the city. However in 1968, the Communists had lacked both the timing and wherewithal to pull off their attack plan. By 1975, they had done their homework, and their Western-minded counterparts had learned nothing from their previous experience. This time, the VC/NVA were to dispatch commandos and spearheads against only five key objectives and then fan out to capture other important facilities. For the inside effort, they had a larger and more sophisticated "special task force."[17] Its elements launched surprise assaults against the following: Tan Son Nhut airfield, Army General Staff, Presidential Palace, Saigon Special Military Sector, and General Police Headquarters. Others seized bridges and intersections along the spearhead column approach routes. Still others reconnoitered and then helped spearhead elements subsequently to capture secondary targets. Of note is how easily thousands of personnel and tons of supplies were prestaged within the city.[18] Without extensive space below ground, this would not have been possible. Through it all, the South Vietnamese never suspected the buildup. If they had, they would have better defended Saigon's center.

> In the meantime [in 1975], the special task force, commandos and armed units from Saigon itself had moved to their appointed places around key targets. These included six special task force groups (equivalent to six regiments in size) which had for several years operated behind enemy lines, four battalions and many special task force groups which had been operating in suburban villages, and 60 groups from inside the city. In addition, there were about 300 armed individuals and large segments of the popula-

> tion who had been organized and put under the command of the special task force. Our forces also included many companies and battalions from the underground Saigon army, together with two regiments from Gia Dinh . . . [and] Nam Bo provinces.
>
> Hundreds of Party cadres, including many armed ones, had infiltrated into the city to join with local officials in organizing the masses for an eventual uprising as soon as the army began its large-scale assault on Saigon. In general, the balance of forces before the Saigon battle was one to three in our [Communist] favor.[19]
>
> — Gen. Hoang Van Thai

For the 1975 spearheads, the North Vietnamese used armored columns with embedded infantry and artillery support. On 30 April, those spearheads attacked Saigon from all four of the cardinal directions. Of particular interest is how commandos and special task force elements paved the way for the highly successful eastern spearhead.

> In the east, from 0500 hours a detachment from the 2d Army Corps, supported and guided by a commando group, crossed the Dong Nai bridge in the direction of Saigon. After wiping out enemy resistance at the Thu Duc Office[r] Training Camp, the advance detachment made contact with Battalion 81 of the special task force guarding Rach Chiec bridge, then made for the Saigon bridge (Tan Cang bridge had been captured by our special task force on April 29). After crushing enemy resistance at Thi Nghe bridge the advance detachment made straight for the Independence Palace under the guidance of a commando team.[20]
>
> — Gen. Hoang Van Thai

Among the VC/NVA ruses in the final assault on Saigon were regular soldiers wearing civilian clothes, special task force personnel dressed like ARVN soldiers,[21] and Communist pilots bombing with South Vietnamese planes.[22] It was no coincidence that these ruses furthered the impression that a popular uprising was in progress. The "blooming lotus" had become the state of the art for urban offense.

The Risks of Only Partially Implementing the Solution

On 31 December 1994, the Russians sent one armored spearhead into the center of the Chechen capital without enough preliminary infiltration. When their pocket collapsed, they lost most of their armored vehicles (along with their human cargo).[23] The *Wall Street Journal* reported 102 out of 120 "tanks" destroyed before the fighting ended,[24] but the total may have included personnel carriers.

While armored vehicles help to penetrate a city's peripheral defenses, they soon become useless without infantry protection. For the "inside-out" maneuver to work, there must be thousands of infiltrators prepositioned within the city. There must also be highly proficient infantrymen in the spearhead columns. To remove an impediment or expand a beachhead, they will need advanced small-unit offensive technique—e.g., a way to cross a heavily contested street. Just to protect their flanks, they will need advanced small-unit defensive technique—i.e., ways to defend a city block.

The plan for the 2003 assault on Baghdad contained all of elements of a "blooming lotus," but not to the same degree as the Saigon example. While it made provision for both infiltrators and urban thrusts, there was little apparent connection.

> Relying on spies, electronic sensors and other intelligence to pinpoint Saddam and other top leaders, coalition special operations forces could infiltrate the Iraqi capital from all directions.
>
> Armor-tipped infantry columns would blast into the heart of Baghdad along several corridors and swiftly isolate key areas of the sprawling city.[25]
>
> — Knight Ridder News Agency, 26 March 2003

As U.S. armored columns penetrated the city's center from both west and east in early April 2003, a few Hispanic U.S. Special Forces personnel disguised as Arabs were roaming the streets to call in air strikes on pockets of resistance.[26] While the first armored thrust was probably just a show of force, the second may have intended to topple the statue next to the press-occupied Palestine hotel. Either way, both were daring. Still, an Asian army would have sneaked thousands of commandos into the heart of Baghdad and then assaulted every strategic target simultaneously. As those commandos breached their objectives, they would have been reinforced by the elements of any number of urban thrusts. If U.S. forces had failed to demoralize their opposition, they would have had to capture 200 square miles of close terrain.

> Such an assault will rely heavily on hundreds of teams comprising a dozen infantrymen and a tank. A tank's heavy machineguns can protect soldiers against ambushes. Infantry can guard tanks from rocket-propelled grenades fire from rooftoops and basements. . . .
>
> Those troops who do secure individual buildings won't exactly knock on doors that can be booby-trapped. . . .
>
> Good intelligence will be vital, says Randolph Gangle, a retired Marine Corps colonel who heads its Center for Emerging Threats and Opportunities in Quantico, VA. Sensors that help locate tanks can't differentiate between a soldier with a rifle and a mother cooking dinner.[27]
>
> — *Christian Science Monitor,* 7 April 2003

That the city fell so easily may have had more to do with worldwide prayer than tactical expertise. To have achieved so quick a victory—without a miracle—against a more dedicated opponent, the Coalition would have needed the following: (1) more commandos at Baghdad's center with state-of-the-art building assault techniques, and (2) more urban-thrust reinforcements with state-of-the-art urban consolidation techniques. This is the lesson of Grozny.

> A belief in the all-powerful nature of high-tech weapons and long-range bombardment is wonderfully appealing to contemporary Americans. It both flatters their justified pride in American technology and suggest the possibility of repeated victories without the risk of serious casualties. However, the U.S. will not always have the ability to pick and choose its wars, and the key lesson of Chechnya is that there will always be military actions in which a determined [self-reliant] infantryman will remain the greatest asset.[28]
>
> —*Armed Forces Journal International,* August 1998

21 During an Urban Defense

- *Where does the Easterner hide strongpoints in a city?*
- *How do strongpoint occupants escape encirclement?*

GERMAN "PEOPLE'S" GRENADIER, 1945

(Source: Courtesy of Cassell PLC, from *World Army Uniforms since 1939,* © 1975, 1980, 1981, 1983 by Blandford Press Ltd., Part II, Plate 193; *FM 90-10-1* (1982), p. B-9)

The Urban Defender Has the Edge

More than once over the last 65 years, built-up areas have been successfully contested by loosely controlled and underequipped small units. In a type of "urban swarm," reinforced Russian squads stopped whole German divisions at Stalingrad. Twice during WWII, Jews armed only with squirrel rifles, grenades, and fire bombs stymied German armies at Warsaw. Only by flooding and razing that ghetto did the Germans finally capture it.

In effect, walls and openings canalize and expose whoever moves through them. Against a larger foe, the urban defender has only to establish a line that will bend but not break. After sufficiently bleeding that foe, he can straighten the line. A belt of platoon-sized strongpoints would work if they could cover the vacant areas with supporting-arms, machinegun, and antitank fire. Those vacant areas would become "firesacks" into which the attacking force would be inexorably drawn. The platoons would emerge to interrupt the attacker's momentum, fight to expend his strength, withdraw to escape his encirclement, and counterattack to blunt his penetration.

If properly prepared, those platoon-sized strongpoints might not have to back up at all. A structure can be so rigged with mirrors, obstacles, and mines, as to preclude capture. Automatic-weapons fire could knock down hundreds of people crossing a street. However, one need not wait until the enemy is that close. Enemy momentum can be interrupted with ambushes forward of the main line of resistance. Antitank men can hide in the ventilation systems of abandoned buildings, and automatic-weapons gunners in the manholes beneath junked cars. After killing a tank or spraying a staging area, those "fighting observation posts (OPs)" could easily return to their parent unit's position through the sewers. Within each strongpoint, certain structures could be rigged for demolition. If their rubble were covered by plunging fire, the overall position would remain intact.

America's foes do not share its aversion to rearward motion. Tactical withdrawal is, after all, an important aspect of maneuver warfare. Urban defenders should be able to shift forwards or backwards. When surrounded, they should be able to hold out until dark and exfiltrate along preplanned routes.

Stalingrad

The classic defense of a built-up area occurred at Stalingrad in 1942. After backing the Russians into several enclaves along the Volga River, the Germans could not finish them off. Within the exploits of those enclave occupants lie the seeds of advanced defensive technique. Of note, that technique was predicated on decentralized control. As each Russian replacement boarded the ferry for the dangerous crossing of the Volga, he was handed a pamphlet entitled "What a Soldier Needs to Know and How to Act in City Fighting." In the world of "big-picture" thinkers, its contents have become somewhat of a tactical treasure. The leaflet listed the individual methods with which Lt.Gen. Vasily Chuikov's battered 62d Army had successfully contained the vastly superior (in numbers, conventional training, and materiel) German 6th Army. Chuikov had allowed his men to function on their own initiative in cleverly improvised small-unit actions that took advantage of the rubbled terrain. The pamphlet's instructions were concise: (1) to neutralize German artillery, "Get close to the enemy"; and (2) to escape German observation and small-arms fire, "Move on all fours," "Make use of craters . . . ," and "Dig trenches. . . ."[1]

As with any good defense, this one had offensive characteristics. They were not the exaggerated counterattacks that one might expect from a Soviet army, but widely dispersed, loosely controlled, and initiative dependent. Their first target was the German outpost line. Every night, small packs of Soviet snipers would creep forward to do their damage. Every morning, they would pull back to secret vantage points. Soviet scouts formed killer teams with a more personal approach.

> Parties of three to five scouts used the . . . ravines . . . to infiltrate enemy lines . . . [and] pounce on outposts.[2]

When the Germans did blindly advance, their maneuver elements were swarmed by reinforced Russian squads. Partially by necessity, those squads had mastered the art of annihilation.

> These [squad-sized] "storm groups" would creep close to their target, wait . . . and then charge.[3]

One wonders why Chuikov focused more on the techniques of the individual than the squad. Perhaps, he wanted to make sure his squad leaders followed suit in delegating authority. He went so far as telling each Russian soldier how to penetrate a contested space: (1) throw a grenade, (2) enter, (3) throw other grenades into every corner, and (4) direct submachinegun bursts at anything left while moving forward.[4]

Within each Soviet enclave were mazes of "strongpoint" buildings connected by trench. Relatively flat areas between the buildings were covered by antitank and submachinegun fire from basement or first-floor apertures. Uneven areas between the buildings were covered by mines, mortars, and plunging machinegun fire from upper-story windows. Until the heavy fire was needed, Soviet snipers kept the Germans at bay (presumably from elsewhere) while strongpoint defenders hid. Each little maze of strongpoints created an impenetrable island. Protected by interlocking fields of fire from rearward positions, small bands of resolute Russians could and did hold off much larger German forces "almost indefinitely."[5]

Chuikov expected his troops to turn every occupied house into a fortress, complete with barrier plan and covered routes of reinforcement or egress. His men seeded the surrounding areas with mines and dug trenches between the buildings. Then they took up the position most advantageous to their weapon. Artillerymen covered the open approaches that were distant, while heavy machinegun and mortar crews covered the open approaches that were close. Submachinegunners defended each structure's immediate access to lower floors. All withheld their fire until a German unit entered their kill zone. In one such ambush, two machinegunners at a crossroads house took an entire German battalion under fire (presumably along its axis). A perfectly logical combination of technique and trust had made it possible for two Russian privates to kill scores of enemy troops and live to tell the tale.[6]

Published on 1 November 1945, the U.S. War Department's *Handbook on U.S.S.R. Military Forces* may show how Stalingrad's buildings were defended. (Notice the underground escape route in

Figure 21.1: The Russian Appreciation for Urban Microterrain

(Source: *Podgotovka Razvegchika: Sistema Spetsnaza GRU*, © 1998 by A.E.Taras and F.D. Zaruz, p. 147)

Figure 21.2.) To this day, the Russians retain a healthy respect for urban microterrain (Figure 21.1). In the late 1960's, Russia's Southeast-Asian protege may have utilized some of the same techniques.

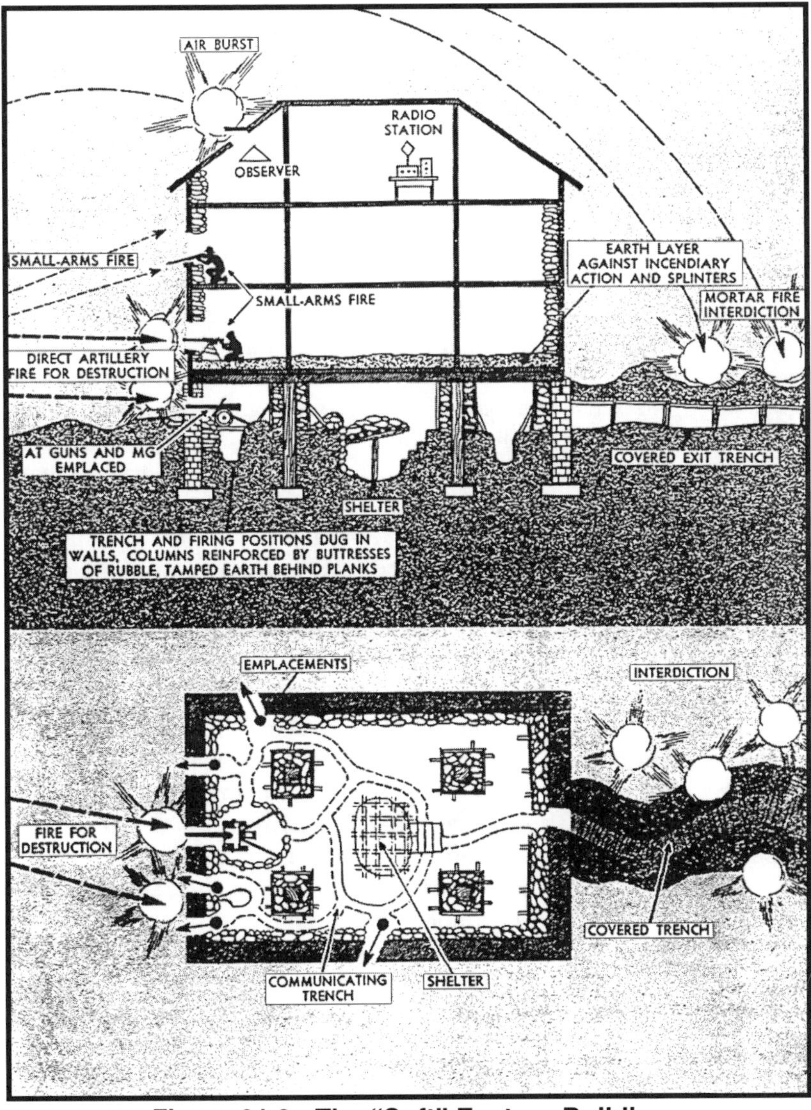

Figure 21.2: The "Soft" Eastern Building

(Source: *Handbook on U.S.S.R. Military Forces*, TM 30-340 (1945), U.S. War Department, p. V-124)

Manila

Much of what follows has been extracted from U.S. Army war chronicles and studies of the 1945 battle. Luckily for U.S. GIs, Japanese urban-defense know-how was only partially applied at Manila. The commander of Japanese ground forces in the Philippines had ordered the city abandoned, and only at the last minute did the Japanese naval commander decide to contest it.[7] Once the unwanted battle had been joined on 3 February 1945, the 9500-man Japanese Naval Defense Force resisted the U.S. juggernaut for four weeks.[8] Of note, it had no armor or aircraft, and little artillery, communication, or overall control.[9] While it has been speculated that the naval troops were also deficient in doctrine and training,[10] their actions speak otherwise. Within the 15-square-mile city, they created semi-independent matrices of mutually supporting strongpoints. To hide them, they created the illusion that the whole city was defended. They assigned one combatant per building and may have also patrolled with gun trucks.[11] As well as a diversion, each building's occupant may have functioned as a forward observer for supporting arms.[12] At the center of many of these matrices were important buildings[13]—e.g., the New Police Station, the Philippine General Hospital, City Hall, the General Post Office, Philippines University, Rizal Stadium, the Manila Hotel, and the Legislative/Finance/Agriculture Complex.

During the subsequent defense of Manila, these Japanese marines exercised considerable initiative. They cut firing slits through the foundations of buildings at ground level.[14] To hide and protect crew-served weapons, they placed their bunkers deep inside rooms.[15] Such a bunker could also participate in the defense of a building's interior.[16] To keep the Americans from assaulting their structures from the top down, they placed machinegun nests on flat roofs,[17] tossed aerial bombs or molotov cocktails from upper floors,[18] destroyed stairs,[19] dropped grenades from holes in the ceilings,[20] and fired their rifles up through holes in the floors.[21] Just to move forward, U.S. troops would often have to turn into rubble a structure at their front.[22] In effect, the Japanese naval security forces had turned all open areas into kill zones and unoccupied areas into "firesacks."

> [In the Paco district] Japanese observers were present in almost every building. At street intersections, machinegun

> pillboxes were dug [recessed] into buildings and sandbagged so as to cover their approaches. Artillery and anti-aircraft weapons were placed in doorways or in upper-story windows. Most streets and borders of streets were mined. . . . The streets were a fireswept zone forcing Americans to move between streets and within buildings.[23]
>
> — U.S. Army Combat Studies Institute research

Like the Russians had done with the Stalingrad industrial sites, the Japanese heavily defended the Provisor Island electricity plant and Paco Railroad Station. All around the train station were machinegun posts. As at Iwo Jima, the defenders had surrounded each machinegun with "foxholes with riflemen."[24] Like the Germans and Russians, the Japanese had composed their strongpoints of embedded subordinate units. On open ground, the final participant was the semi-independent, squad-sized fort. Within each building, the strongpoint hierarchy may have extended all the way down to the two-man team. The occupiers of each room may have had their own fire, barrier, and withdrawal plans.

> Sandbags and barricades blocked all ground-level doors and windows. Interiors were also fortified as in other strongpoints.[25]
>
> — U.S. Army Combat Studies Institute research

To canalize armor and blunt momentum, the Nipponese employed road blocks. One consisted of a mine field, roadbed-driven steel rails, and wire-attached trucks. The whole arrangement was often covered by the interlocking fire of four machineguns.[26]

What has just been described comes close to advanced maneuver warfare methodology. All that is lacking is a rearward, "leap frog" motion to previously prepared fallback positions and a way for strongpoint occupants to escape encirclement. The only evidence that a tactical withdrawal took place was the intentional demolition or incineration of defended buildings.[27] That the Japanese occasionally counterattacked does not disprove the theory. By 12 February 1945, the U.S. 37th Infantry Division and 1st Cavalry Division had encircled Manila.[28] Upon reaching the 16th-Century Intramuros Fort (the most probable last stand for a diehard defender), they encountered little resistance. Either every Japanese marine had already given his life for his Emperor, or some had complied with their initial instructions to abandon the city.[29] That the Japanese Naval Defense Force commander and his staff subsequently committed suicide in no way proves that his frontline fighters did likewise. The Japanese Army was still poised north of the city to defend Luzon. Shortly after the Intramuros' 40-foot-thick walls were breached, U.S. assault troops were deluged by 2000 refugees. The Americans were too busy to check identification papers. The very same thing had occurred when the Pasig River (the defender's principal barrier) was first crossed.[30] Old forts had secret passageways that led under their walls from royal chambers. One can easily imagine how the Japanese rear party made good its escape.

> The hardest fighting in Intramuros was the 129th's effort to capture Ft. Santiago in the northwest corner of the old wall. They fought room to room, and then through subterranean dungeons and tunnels . . . The regiment did not secure the last of the fort's [most obvious] tunnels until 1200 on 25 February (U.S. Army, *The 129th Infantry in World War II,* 1947, 108; Smith, *Triumph in the Philippines,* 1963, 298; U.S. Army, *37th Infantry Div. Report after Action,* 1945, 83).
>
> During the fighting in Intramuros, some Japanese troops attempted to exfiltrate wearing U.S. uniforms and carrying M1 rifles.[31]
>
> — U.S. Army Combat Studies Institute research

Gen. Kuribayashi's had his headquarters in the "Gorge" on the north end of Iwo Jima,[32] and not in the tier of huge caverns on its northeast coast.[33] Rear Admiral Iwabuchi had his headquarters in the Legislative Building complex in Manila,[34] and not at the most promising location for a mass exfiltration—the Intramuros' Ft. Santiago. To escape a closing noose on *terra firma,* one has only to let it pass overhead.

Berlin

Having already fought for Stalingrad and Warsaw,[35] the Soviet armies that closed in on Berlin in the Spring of 1945 were no strangers to urban attack. Before crossing the Oder, Zhukov further insisted on special instruction in street fighting for all of his assault troops.[36] Yet, to take the city, those armies suffered 305,000 casualties from 16 April to 8 May 1945.[37] While some of those losses may have occurred along the city's approaches, the total is still mind-boggling. By this late in the war, the German capital was being protected by hastily trained *Volksgrenadier* (people's infantry) battalions,[38] a hodgepodge of other understrength units, *Volksturm* (home guard) volunteers, and Hitler Youth. To the *Volksturm,* the Fuhrer had assigned the job of defending street by street.[39] It was comprised of men over 60, boys between 16 and 18, the "medically unfit," and possibly even women.[40] By some accounts, only half had small arms with which to confront the Russian tanks.[41] For the final battle, one army would be well organized, well trained, and well supported, while the other would not. From the Soviet casualty total comes an indication of the power of innovative technique and decentralized control. Against a veritable cascade of Russian bombs, shells, and tanks, the German citizen soldiers would have little else.

> As dawn broke on the morning of 25 April, 464,000 Soviet troops, with 12,700 guns and mortars, 21,000 *Katyusha* multiple rocket launchers, and 1,500 tanks and self-propelled guns, stood ready and waiting to begin the final assault on the heart of Berlin. To the north, south, east and west, the Soviet spearheads were within four miles of Hitler's chancellery and the Reichstag. Supporting them [were] . . . thousands of planes.[42]

Of the 1500 tracked vehicles (from the First and Second Guards Tank Armies) to enter the built-up area initially,[43] two-thirds may have been lost. Just before the final assault on the Reichstag, another entire tank army (the Third Guards) had to be committed.[44] At first, the tanks operated in columns, and then in pairs, but seldom with infantrymen in front.[45] This was to cost them dearly. The Russians were adhering to their favorite tactic—tank assaults preceded by bombardment.[46] Although the hastily assembled neighborhood defenders had few heavy weapons of their own,[47] they did have an ample supply of durable buildings, underground thoroughfares, and *Panzerfaust* antitank rockets. While mounting the rubble unprotected, many a Russian tank must have succumbed to a belly shot. The defenders were also good at ambushing Russian armor. The *coup de grace* was a random barrage of antitank munitions from every direction. Those munitions would so confuse the quarry as to deny him accurate return fire. In this way, a Soviet infantry unit had destroyed a German armored column on the grassy Russian Steppes.[48]

> The tanks, in particular, started out disastrously, moving along city streets in columns. The defenders had only to knock out the leading tank with a well-aimed *Panzerfaust,* and the others, strung out in line behind it, were trapped, easy prey with their sides exposed to fire from the buildings on either side.[49]

With three huge rings of defensive works around the city,[50] and Berlin's waterways forming concentric circles around the Reichstag, the Russians became convinced that the inner city had a sophisticated defense plan. It may have, but the plan's execution depended more on initiative than orders.

> Never before in the experience of warfare had we been called upon to capture a city as large and as heavily fortified as Berlin. Its total area was almost 350 square miles. Its subway and other widespread underground engineering networks provide ample possibilities for troop movements. The city itself and its suburbs had been carefully prepared for defense. Every street, every square, every alley, building, canal and bridge represented an element in the city's defense system.[51]
>
> — Marshal Zhukov

Probably to canalize the Russian armor, the Germans did build tank traps and barricades in abundance.[52] German snipers had been assigned to some buildings,[53] and whole units to others. In effect, strongpoint matrices were sprinkled throughout a sea of partially defended buildings. Most likely to be heavily fortified were the structures around flak towers, canal bridges, and S-Banh stations.[54] Yet, many were only temporarily occupied. Was it by design or necessity? The Russians did discover "cellar escape routes."[55] On the walls were the German inscription, "We withdraw but we are winning!"[56]

Under fire from Soviet advance units, Skorning's combat group was soon in danger of being surrounded. The regimental commander ordered them to pull back to the northern side of the canal, where Skorning set up his strongpoint at a narrow bridge. . . .

Soviet troops soon crossed the canal, and Skorning and his men were forced to retreat under fire, back through the graves of St. Luke's and Emmaus churchyards toward the S-Banh [tunnels] skirting the airport, part of the city's inner defense ring.[57]

With no more *Panzerfausts* left, Skorning's men had no chance of holding them [the Russian tanks] off. Under fire from Soviet gun positions, they escaped along the S-Bahn [tunnels] to Tempelhof Station, where his regimental command post was situated in a public air-raid shelter.[58]

The Germans were withdrawing to the center of Berlin, blowing the canal bridges as they went.[59] One might assume that they had no choice. Yet, there is evidence that they were tactically withdrawing to soften up their opponent for counterattack. In some neighborhoods, the Germans had already taken offensive action.[60]

From his vantage point [Gen.] Chuikov had a clear view of the rings of defense works built along the canals and railway lines which curved around the city centre, where every building seemed to have been transformed into a fortress. The most powerful defenses appeared to be in the area of the old city walls, built in the eighteenth century. . . . The Landwehr canal and the sharp bend in the Spree [River] with its steep, concrete-lined banks, formed a protective screen around the government buildings, including the chancellery and the Reichstag—the bull's eye in Chuikov's target.[61]

Some of Berlin's defenders may have even escaped the Russian encirclement—through the city's vast subterranean network. While many German units had their command posts in S-Bahn and U-Bahn tunnels,[62] Hitler's last-minute decision to flood the ones near the Landwehr Canal didn't improve their chances of escape.[63] However, when all seemed lost, Hitler forbade surrender but suggested that his frontline troops "break out in small groups" to join the forces still fighting outside the city. He issued a formal order to that effect before committing suicide.[64] Soon many of the *Volksturm* were throwing away their rifles and melting into the civilian population.[65] On 2 May 1945, all that remained of the Berlin garrison—some 70,000 men—surrendered to the Soviets at the center of city.[66] Still alive below the city were some two million other Berliners.[67]

Seoul

During the Korean War, Seoul changed hands several times. To retake it in September 1950, U.S. Marines had to cross a contested river (the Han).[68] While the North Koreans made the Marines "fight for every inch of ground" and launched several spoiling attacks,[69] their overall motion was rearward. As the Japanese had done in Manila, Seoul's defenders succeeded with roadblocks in blunting American momentum. However this time, a specially trained rear guard was more obviously at work.

> Even as air reconnaissance revealed the main body of Kim Il Sung's [North Korean] army fleeing northwards, Communist rear guards fought on to delay the advance of [Marine Gen. O.P.] Smith's regiments, extracting their price, yard by yard at each rice-bag barricade.[70]

Just as "retreat combat" had become Japan's most formidable defense near the end of WWII,[71] so too had tactical withdrawal come to signify the Maoist mobile defense. This was not a retreat in the Western sense of the word, but rather a carefully choreographed attempt to overextend and exhaust a more powerful pursuer. A slightly different account of the North Korean pullout reveals a more orderly event.

> [A]lthough the bulk of the KPA [Korean People's Army] forces in the city were withdrawing, units were left behind to fight a desperate rearguard action. Fighting behind a succession of street barricades, they exacted a heavy toll among the ranks of the U.N. forces for every meter of territory they yielded. Against such bitter resistance, it took the U.N. forces three days to gain complete control of the town.[72]

As at Manila, the U.S. troops discovered many buildings occupied by a single enemy soldier. While this soldier may have occasionally fired his rifle, his role in the overall scheme of things had to be more than that of a sniper.

> It seemed like every building in Seoul housed an enemy sniper.[73]
>
> — S.Sgt. Lee Bergee, Seoul veteran

> George Company was on one of the main streets which led through the outskirts of the city. I was about halfway back in the column and on the right. It seemed as though all of the fighting was on the point. . . . As I'd come up [upon] a side street or alley, I'd take a careful look around the corner before I crossed the open area. At one such crossing I thought I saw a head and rifle barrel in the window of a small, one story building. . . . Taking a deep breath I jumped through an opening [in its wall]. As I landed inside, I saw there was indeed an armed North Korean in the room. . . . He wore a pair of dark brown trousers and a light brown shirt. He did not have a cap. A rope attached the belt. His tennis shoes were low cuts. His rifle had a five-round clip in it. There was no other ammunition on him. Five rounds of ammo. No extra![74]
>
> — PFC Fred Davidson, Seoul veteran

There are other similarities between the two battles. Again in Seoul, U.S. troops encountered hordes of civilians at inopportune times.

> Suddenly, in trying to get away from a firefight in their neighborhood, hundreds of women and children would mob into our area, blocking us off.[75]
>
> — PFC Win Scott, Seoul veteran

As at Manila, U.S. troops had trouble locating the source of enemy fire. They soon discovered the reason. They were facing extremely narrow fields of fire from recessed positions.

> I can remember going down this street and coming to an intersection. Later, we called it Blood and Bones Cor-

> ner. . . . One of our fire teams made it across without problems. It was when the next team tried to cross that all hell broke loose. The intersection was situated in such a way that the only rifleman who could get into the fight was the Marine at the corner. As soon as he went down, another man moved up and took his place. That's the way it continued. . . .
>
> We were called back and told to get out of there. . . . The fire team that tried to get back across the street was wiped out except for the squad leader . . .
>
> When we got back to that corner after the North Korean machinegun had been outflanked, South Korean civilians . . . helped us carry our wounded.[76]
>
> — PFC Jack Wright, Seoul veteran

On the rocky barrens of Iwo Jima, the Japanese had routinely taken Marine assault units under fire from all sides.[77] In the built-up terrain of Seoul, the North Koreans did likewise. Often, the only answer was to level a building.

> There was no front and no rear. Thank God we had tanks with us. Without them we'd still be fighting there.[78]
>
> — PFC Win Scott, Seoul veteran

The North Koreans had been smart to back up. The U.S. troops were not the least bit hesitant to unsheathe their supporting arms. "The slightest resistance brought down a deluge of destruction blotting out the area," wrote an eyewitness, R.W. Thompson, of the *Daily Telegraph*.[79] Shortly after the Chinese entered the war in November of 1950, Seoul fell again to the Communists. When the Americans finally retook it during Operations Killer and Ripper in February and March of 1951, the rubbling was even worse than it had been before. As before, the Communists conducted an orderly, fighting retreat.

Hue City

The urban equivalent of a Maoist mobile defense became more evident in the Vietnam War. After the North Vietnamese had used the "blooming lotus" (or inside-out) method to capture the center of

Hue City in February 1968, they were able to hold it for almost a month. Within the Citadel walls, they established—on alternating streets—strings of strongpoints.[80] So that U.S. forces could not tell which blocks were heavily defended, those between were manned by roving shooters.[81] The occupants of one such street—the narrow Mai Thuc Loan (or "Phase Line Green")—were able to keep a reinforced U.S. battalion at bay for four days.[82] To do so, it used preregistered, long-range machinegun and mortar fire from both ends of the street. It also employed narrow, overlapping bands of random small-arms and RPG fire from bunkers between the houses, within the rooms, and sometimes even behind the Americans.

The erstwhile defenders of Hue City made good use of the Japanese-dug passageways within the dirt-filled Citadel walls. Mai Thuc Loan exited the Citadel at the Dong Ba Gate. That's where the NVA machineguns were. Every time the Marines would bomb or capture it, more NVA would emerge from its interior.[83] The Communists may have also moved between strongpoints on Mai Thuc Loan through the covered sewage trenches. Look at Map 21.1 again more closely. It may reveal the precise locations of the tunnel entrances through which the NVA rear guard escaped from the Imperial Palace and southern corner of the Citadel on the nights of 22-23 February 1968.[84] The book *Phantom Soldier* provides a more detailed description of the battle.

Grozny

On 31 December 1994, a few thousand lightly armed and loosely controlled Muslim rebels defeated a Russian assault on Grozny. The Russians had the Chechens outnumbered, outgunned, and surrounded,[85] but they had forgotten how Stalingrad had been defended. To avoid casualties, they had become too dependent on technology and firepower. Against a quick-witted opponent in built-up terrain, such weapons are largely useless.[86]

> As for artillery, using it in urban conditions is useless. . . . Tanks and infantry fighting vehicles are also helpless on the streets. . . . Vehicles are fired on at point-blank range from grenade launchers; view ports are covered with tarpaulins; vehicles are set on fire—all of these methods have

already been honed by the rebels.[87]
— Gen. Mikhail Surkov
Armed Forces Journal International, August 1998

The Russians virtually flattened the center of the Grozny after losing almost every man and vehicle in their armored thrust.[88] The bombardment lasted until February 1995.[89]

> [T]he 6,000 Russians troops ran headlong into approximately 15,000 urban guerrillas. . . . The Chechens waited until the armored columns were deep into the confines of the urban sprawl before initiating their ambush with a hail of hit-and-run rocket-propelled-grenade (RPG) attacks. Within 72 hours, nearly 80 percent of the Maikop Brigade were casualties, while 20 of their 26 tanks and 102 of their 120 armored vehicles were destroyed. The 81st Motorized Rifle Regiment also came under ambush as it entered the town from the direction of the northern airport. . . . For the next 20 days and nights, the Russians fired up to 4,000 artillery rounds an hour into the city while they struggled to extract their remaining troops.[90]
> — *Marine Corps Gazette,* October 2001

"The Chechen [rebel] campaign plan revolved around fighting the technologically superior Russian Army asymmetrically by employing urban guerrilla tactics."[91]

> These [Chechen] militia forces would "march to the sound of the guns" from their [suburban] villages at the onset of the Russian invasion. The defense was a shifting, mobile defense laid out in three concentric circles with the Presidential Palace at . . . [their] center. . . . The Chechens would position small teams behind the Russian forces to harass their progress through the city's 100 square miles of urban high-rise sprawl. The Russian Army planned to fight in a traditional, linear style, but the Chechen defense would force them into a noncontiguous and nonlinear fight.[92]
> — *Marine Corps Gazette,* October 2001

Grozny would be defended in much the same way that

Stalingrad had. It would be subdivided into tiny sectors. Each sector would be defended by several, loosely controlled maneuver elements.

> The Chechen standard hunter-killer team consisted of an RPG gunner, machinegunner, and sniper. Three to five hunter-killer teams would work together in a sector.[93]
>
> — *Marine Corps Gazette,* April 2000

The standard method of engagement would be ambush. The Chechens would surprise their quarry from every direction at once. Under random fire from sewer opening, ground level, and rooftop, the victim would have trouble acquiring a target and become quickly confused. Armor was handled in the same way it had been at Hue City—with a barrage of RPG rounds from every side. The ambushers always had a preplanned escape route.[94]

> RPGs were shot at everything that moved. They were [sometimes] fired at high angle over low buildings. . . .
>
> Multiple RPG rounds flying from different heights and directions limit a [Russian] vehicle commander's ability to respond. . . .
>
> Chechens chose firing positions high enough or low enough to stay out of the field of fire of tank and BMP weapons. . . . The Chechens used mobile tactics and "let the situation do the organizing," while the Russians relied more on brute strength.[95]
>
> — *Marine Corps Gazette,* April 2000

> [W]hat the Russians said was the most effective Chechen tactic: the 'hugging' of Russian units in the defense to neutralize Russian firepower.[96]
>
> — *Armed Forces Journal International,* May 2000

The Chechens had counted on Russian excess to sway public opinion. As the Russians were steadily bled by snipers and RPG gunners, they demonstrated less and less regard for civilian casualties and property damage. They finally captured the city on 26 February 1995.[97] In August 1996, the rebels counterattacked in a fashion reminiscent of Hue City.

> [T]he Chechen rebels infiltrated Grozny along three axes and attacked at dawn on 6 August 1996. The Chechens had conducted detailed reconnaissance of Russian garrisons and posts. . . . They used this information as they besieged every Russian position (manned by 12,000 Russian troops) and captured the city in a single day! . . .
>
> The Chechens conducted both tactical and operational maneuver consistent with Maoist revolutionary war strategy. The decisive aspect of this asymmetric approach was their extensive reliance on urban areas to "level the playing field." The Chechens recognized that employing a fixed defense based on urban strongpoints would allow the Russians to bring their firepower to bear, so instead, they used a mobile area defense strategy.[98]
>
> — *Marine Corps Gazette,* October 2001

After losing Grozny in August 1996, the Russians settled for a tenuous cease fire.[99] Their military campaign for the city had been an operational failure.[100] In January 2000 with more "human intelligence," a much larger force, and massive preparatory fire, the Russians retook the city. Their tanks stayed on its periphery this time.[101] Chechen tactics had started to suffer from overcontrol.

> Chechen tactics remained versatile and flexible. They boarded up first floor windows to slow Russian access to buildings, continued to "hug" Russian forces . . . and operated in a very centrally controlled fashion instead of in the "defenseless defense" or "let the situation do the organizing" mode of 1995. This was an obvious adjustment because the Russians refused to enter the city exposed and in mass formations. The Chechens used trenches more than in the first battle in order to move between buildings. They also positioned snipers in a "misdirection" tactic. . . . They constructed escape routes from their firing positions, and interconnected these positions.[102]
>
> — *Marine Corps Gazette,* April 2000

Part Four

The Winning Edge

The ordinary soldier has a surprisingly good nose for what is true and what is false. — Rommel

(Source: *A Dictionary of Military Quotations,* © 1989 by Trevor Royale, p. 60)

22 The Rising Value of the "Little Picture"

- *What effect does the little picture have on the big?*
- *Why must Eastern privates make strategic contributions?*

NEPALESE "GURKHA" CORPORAL, 1964-66

(Source: Courtesy of Cassell PLC, from *World Army Uniforms since 1939,* © 1975, 1980, 1981, 1983 by Blandford Press Ltd., Part II, Plate 85; *FM 21-76* (1957), p. 61)

"Big Pictures" Can Be Misleading

Every nation bases its battlefield style on assets available. One would think that countries with "low-tech" arsenals, non-Christian cultures, and huge populations might more freely expend their infantrymen. Lacking the firepower "surgically" to destroy their foe's strategic assets, they might resort to mass assaults. But saboteurs don't need "high-tech" equipment. And undervaluing a life does not necessarily put it at greater risk. Overprotecting a rifleman can be just as detrimental to his health. It saps him of the field skills and initiative he may need for the inevitable one-on-one encounter.

The well-endowed nation now has the weaponry to keep large numbers of people from overrunning its lines. For its adversary, the mass assault retains utility only as a feint. He must depend upon tiny contingents to close with those lines undetected. Subsequently to penetrate them, those contingents would need almost total surprise. As a result, the adversary often conducts his main attack with a single "sapper" team or "stormtrooper" squad. The former sneaks between enemy holes; the later assaults those holes with grenades and bayonets under an indirect-fire deception. Both can enter the position without alarming the majority of defenders. Given sufficient time, either could outsmart the latest surveillance equipment. If that sapper team or stormtrooper squad were additionally able to make its handiwork look like an accident, it could repeat the same penetration technique night after night.

On a mission this challenging, a tiny contingent will only be as effective as its weakest member. Because "low-tech" nations must necessarily depend on the strategic contributions of each rifleman, buddy team, fire team, and squad, they place more emphasis on lower-echelon training. Yet, this is not standardized training. Instead of having to memorize duty solutions from headquarters, these lower echelons get to refine their own methods. Most arrive at the same conclusions, and their parent units adopt similar small-unit techniques. Because "low-tech" armies value the collective opinions of their frontline fighters, they are more realistic about what happens at close range. In effect, that part of their "big picture" is based on a myriad of "firsthand accounts." This is not the case with an affluent army that is managed from the top down.

For U.S. Commanders to Do Their Sworn Duty

Big pictures can be influenced by everything from intraservice rivalries to budgetary projections. They will contain that part of truth that best supports the current political agenda. Political truth is different from ultimate truth. The former is often an overly optimistic view of some problem. To approach ultimate truth, one must consider the middle ground between a multitude of firsthand accounts. If too many firsthand accounts differ from the party line, then there is a good chance the party line is skewed.

Even without political subjectivity, the difference between the big picture and small would be one of scale and detail. Somewhere within the hodgepodge of frontline soldier accounts would lie the specifics of the enemy's main attack and preliminary deception—the secret maneuver behind his feint. Just as the U.S. battle chronicles are true, so too are the consensus opinions of the frontline soldiers. Those low-ranking Americans have been saying for a long time that their immediate opposition has been secretly doing more than pretended. They have told of tiny contingents of enemy soldiers covertly attacking and withdrawing. All the while, they have wondered why their commanders stuck to the large-unit, "high-diddle-diddle-right-up-the-middle" approach to war. Many have concluded that their Eastern counterparts were better prepared to survive. Both perspectives are true, but one has been largely ignored.

American commanders cannot do their sworn duty without the facts that most greatly affect their final casualty count. With their objectivity constantly subjugated to "command loyalty" within an "error-free" environment, only the most gifted has realized his oversight in time. Most have sent undertrained enlistees against soldiers of vastly superior short-range skill. Any number of U.S. privates may have already died unnecessarily since WWI. In essence, the U.S. military has no way of systematically capturing what its lowest ranks collectively learn in war. What PFCs Jack Wright, Win Scott, and Fred Davidson learned the hard way on the streets of Seoul in September of 1950 never entered the tactical memory of their parent service. The Western-style, "command-push" organization does not benefit from its lower-echelon mistakes as easily as the Eastern-style, "recon-pull" organization. To become as proficient at short-range combat, American units must develop a full portfolio of individual, fire team, and squad techniques from the bottom up.

This is a problem that Western armies have yet to correct through traditional means. The U.S. commander cannot rectify his short-range shortfall through martial arts instruction. Without drastic changes to his overall training method, he will lose just as many people to bullets and shrapnel as before. To save those people, he must remember the "big picture" and shift his focus to the small. What he now perceives as simplistic "basics," he must begin to see as intricate categories. What he currently thinks impossible, he must come to view as commonplace. To help his young enlistees to achieve their full fighting (and survival) potential, he must add "unconventional-warfare" battledrills to their daily curricula.

Only the "Big Picture" Is Complicated

War is not nearly as complicated at the squad level. The infantry squad spends most of its time standing guard duty, patrolling, or participating in a parent-unit operation. When the enemy shows up, its circumstances will vary less than those of a larger unit. Often, it will only experience differences in the amount of cover and direction of fire.

Saying too loudly that every country fights differently and that all war is chaos just gives U.S. commanders another excuse not to adequately prepare their lower echelons. The much-vaunted "fog of war" may be nothing more than failing to appreciate the short-range skill of one's enemies and small-unit advice of one's own front-line fighters.

Undeniably, the field skills of U.S. infantrymen have deteriorated over the last decade. Finally to acknowledge the tactical similarities between "low-tech" countries would accomplish the following: (1) enhance the U.S. rifleman's chances of survival by telling him (for the first time) what to expect from his enemy counterpart, and (2) identify any tactical innovations that U.S. forces may have failed to assimilate.

Eastern Armies Fight Differently Than Those in the West

There are pronounced trends in how America's adversaries have fought at close range since 1918. To ignore those trends could cut in half the survivability of the present generation of U.S. riflemen.

> When you are ignorant of the enemy, but know yourself, your chances of winning and losing are equal.[1]
> — Sun Tzu

Those trends cover the entire gamut of offensive and defensive operations. They are, in many instances, quite different from what Western forces might expect.

On offense, every one of America's foes has preferred to work at night. Their attacks have been closely reconnoitered and comprehensively rehearsed (often in life-size mock-ups of the objectives). All have relied heavily on covert infiltration to destroy U.S. materiel. All have used tiny, highly skilled elements to make their initial breakthroughs. To preserve the element of surprise, their assault elements have not been permitted to shoot their small arms. On defense, America's adversaries have dug innumerable trenches, gun pits, and tunnels. All have employed dummy positions and extensive camouflage. At times, this camouflage has been so exquisite as to conceal from view—at 50 yards—a major fortification. Most have resorted to the "gradually receding" strongpoint matrix—one in which the front row routinely withdraws to form a new back row.

All the while, U.S. forces have continued to hastily attack—in broad daylight—prepared enemy positions. Their reconnaissance has been from long range; their rehearsals have been cursory and over dissimilar ground. Most of their assaults have been of the old-fashioned, everyone-get-on-line, right-up-the-middle variety. On defense, American commanders have clung to linear formations and refused to withdraw under fire.

Details of the Difference

A technology- and materiel-poor country can only protect its strategic assets by hiding them. It can only threaten its foe's strategic assets by approaching them along the ground. To escape bombardment on both offense and defense, it must do more of the following: (1) extensive excavation, (2) frontline deception, (3) nighttime operations, (4) close-range combat, and (5) tactical withdrawal.

To generate sufficient surprise, the "low-tech" army must do many things differently. To best utilize the terrain, it must conduct "underground" or "maneuver" warfare at the squad level. (See Figure 22.1.) To become less predictable, it must decentralize control over training and operations. To get the most out of its greatest asset (people), it must require—of each small unit and individual soldier—a strategic contribution. Anything from shooting an enemy commander to blowing up opposition ammunition will qualify. To protect its greatest asset from piecemeal destruction, the "low-tech" army must give its soldiers and small units advanced field skills and commensurate leeway. For the penetration of enemy lines, they will have "sapper" and "stormtrooper" technique. To defend a piece of ground, they will hide below ground until the last moment

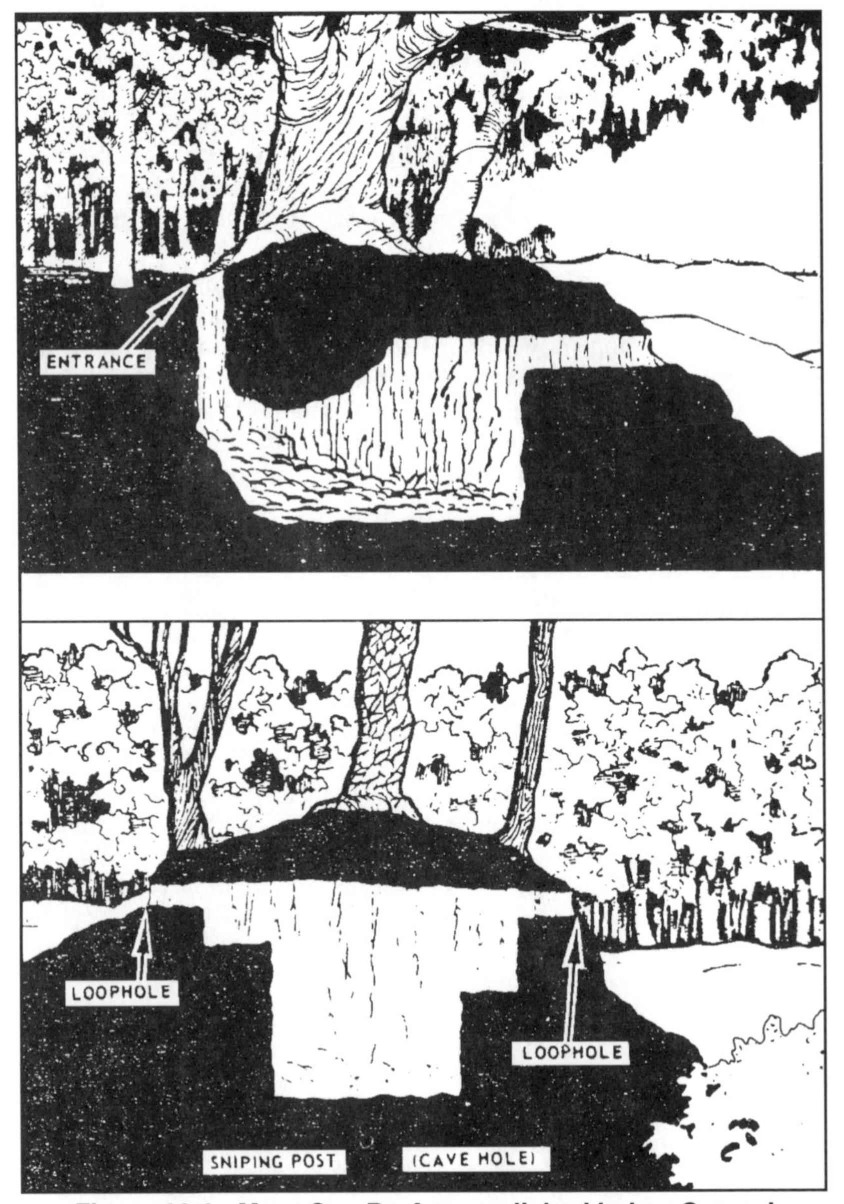

Figure 22.1: More Can Be Accomplished below Ground

(Source: *TC 23-14* (1988), p. 90)

and then tactically withdraw. To escape bombardment during a meeting engagement, they will hug or intermingle with their adversary. To do these kinds of things safely, they must be given advanced, short-range skills. Each will need more than just martial-arts training. He must be able to move through available cover to kill his Western counterpart with bayonet or grenade before getting shot himself. At 50 yards, such a soldier might have a considerable edge over a heavily equipped adversary.

Which Parts of the Eastern Way Should the U.S. Assimilate?

Where Eastern armies have lacked a bare minimum of wherewithal, they have not fared as well against someone equally proficient at maneuver warfare. In the Manchurian Nomonhan Incident, for example, the Russians had more tanks and artillery than the Japanese. Against a well-supported army practicing attrition warfare, the Japanese might still have won.

When Eastern armies have resorted to positional warfare, they also have done less well. The Chinese got embarrassed in their 1979 incursion into North Vietnam by failing to appreciate the power of the same underground defensive matrix that had worked so well for them in the Korean hill battles of 1953.

While the Russians have acquired much of their experience at the operational level, they may still possess fairly proficient advance elements. However, as their degree of control has ebbed and flowed, so too has their tactical expertise at the lowest echelons. Their distrust of NCOs played no small part in their defeat in Afghanistan. "Tactical initiative was not encouraged as it tended to upset operational timing."[2] The Russians rode into battle in BMPs and refused to fight a squad leaders' war. One could deduce that their reluctance to decentralize control and venture off road hurt them.

To become proficient at short range (expect minimal casualties from any ground war), the U.S. military may have to make—as its highest priority—the preparation of its lowest ranking members. It may also have to decentralize control over training and operations.

> The [Soviet] commander is expected to personally conduct all training. In armies with a professional NCO corps, such

> training and preparation is done by trained, seasoned sergeants who understand the unit missions and train their forces to meet them. . . . The Soviet system overburdened the company grade officers and limited individual training opportunities.[3]
>
> — DoD-published Soviet military academy study

If, as Germany did in WWII, the U.S. were to combine small-unit expertise with "high-tech" weaponry, it could win more wars at less cost in human life. While the V-1's and V-2's did the Germans little good,[4] their state-of-the-art tanks and planes were of great help to their infantry. It has been the "top-down" management style of the U.S., British, and French armies that has caused them to retain what amounts to premachinegun tactics at the squad level. Thanks in large part to the Gurkhas, British line infantrymen now appear to enjoy a slight edge in field skill over their American counterparts. However, it is the German, Russian, and Asian infantrymen who have the most experience with maneuver warfare. Therefore, it is their small-unit tactics that must be emulated. (See Figure 22.2.)

Figure 22.2: Smaller Teams Generate More Surprise

(Source: *FM 21-75* (1967), p. 122)

> I bought and read . . . *One More Bridge* . . . and *Phantom Soldier* some time ago. I'll have to let you know . . . which specific [Asian] techniques we found relevant, but we follow their general spirit closely.[5]
>
> — Lt.Col. Ian Thomas, 2d Royal Gurkha Rifles

The Two Possibilities

That such pronounced tactical trends are shared by the major Eastern armies can only mean two things. The first is that there is a growing body of knowledge on how to fight someone with more technology and firepower. If recent Western embarrassments in Vietnam, Somalia, Afghanistan, and Israel are any indication, this body of knowledge may have become fairly sophisticated.

The second possibility is that the Eastern armies have evolved tactically (at short range), while the Western ones have not. Either way, the army that now follows the Eastern trend toward small-unit skill will take fewer casualties than the one that doesn't.

America Is at a Moral Crossroads

As wars are fought on a moral as well as physical plain, they must necessarily entail some risk. Just as a soldier cannot shoot everyone who looks suspicious, neither can his unit bomb everything that might harbor an ambush. It must put highly skilled people on the ground to differentiate friend from foe. If it doesn't, it risks diluting its moral mandate. In the expulsion of the Taliban from Afghanistan and Saddam Hussein from Iraq, the U.S. strategy has been to obliterate from a distance the first hint of opposition. The Russians got little long-term satisfaction from a similar strategy in Afghanistan and Chechnya.

> Why did the Soviets fail to achieve military victory in Afghanistan? . . . They used the Afghan Peoples Army, Sarandoy, the Khad and militias . . . as cannon fodder. . . . Further, the Soviets conducted indiscriminate air and artillery attacks against the rural population . . . in order to dry up the *mujahideen* supply lines. Finally, the Soviets were reluctant to accept the casualties necessary for such a vic-

> tory and tried to substitute fire power for infantry close combat.[6]
>
> — DoD-published Soviet military academy study

> [T]he Russians could take the city [Grozny] back. It would take them half a year and they would have to destroy the town all over again. They could even take it in a month, but it would cost [them] ten to fifteen thousand men.[7]
>
> — Shamil Basayev, Chechen guerrilla leader

The Russians have since recaptured Grozny but failed to pacify the country. Suicide bombers and death squads are still active. According the *Christian Science Monitor* on 16 May 2003, about a dozen Russian soldiers die weekly in combat with the now mostly rural rebels.[8]

How the Tiger Is Born

- *Why is every private's self-confidence so important?*
- *How does the Eastern private get his?*

AMERICAN RANGER PRIVATE FIRST CLASS, 1970

(Source: Courtesy of Osprey Publishing Ltd. from *US Army Rangers & LRRP Units 1942-87*, Osprey Elite Series, © 1987 by Osprey, Plate F, No. 2; *FM 21-76* (1957), p. 99)

The *Ninja* Emulates a Tiger

The practitioner of *ninjutsu* (and its Mainland equivalents) enjoys "integrated mind-body awareness."[1] In strict compliance with the laws of the universe, he tries not to reach an accelerated state of aggression, but rather to avoid confrontation. As such, he is at one with nature and unafraid of any antagonist. He can sense that antagonist's presence, ascertain his intentions, and dodge his first blow. The *ninja* is convinced, and rightfully so, that he can single-handedly defeat many times his number of violence-bent pursuers.

How does the *ninja* attain this advanced state of readiness?

Under real-life conditions, he regularly rehearses any number of sensory-enhancement, individual-movement, and personal-concealment techniques. He spends long hours in the woods learning their every vibration. He trains his mind not to alter what his senses detect. He works over and over on mind-body coordination.

All the while, the *ninja* realizes that—to change with his circumstances—he must transcend technique.[2] So, by making his techniques almost instinctive, he is able to focus on his adversary. By constantly refining those techniques, he also maintains his proficiency at tactical-option decision making.

How Many Eastern Soldiers Know *Ninjutsu?*

Chapters 4 through 11 showed that Japanese privates have had *ninja* training since 1913. Chapter 2 proved that every nonrate in one Viet Cong security battalion practiced many hours of *ninja* ground-crossing technique each month. Chapter 10 revealed that VC units had "special-assignment squads" attached. Chapter 1 ascertained that each contemporary North Korean battalion has a training cadre of commandos with enough skill to cross the DMZ alone. One could easily conclude that every Oriental line infantryman now receives some training in *ninjutsu*. While that training may resemble gymnastics, it is really a series of individual battledrills in which soldiers compete against one another.

Within each Russian division are specially trained assault squads and sappers (footnotes 6, 32, and 33 of Chapter 19). Spetznaz commandos must help with their training. German line companies had stormtroopers as "teaching cadre" and dedicated half of every day to field-skill competitions (Appendix B). So Russian and German riflemen may receive *ninja*-like training as well.

A young American would much rather destroy a strategic target than a fellow human being's life. His "inner confidence" would be better served by nonconfrontational (or "unconventional") responses. To develop the tiger eye, the U.S. enlistee will need more than just martial-arts training. He will need *ninjutsu* training.

There Is Little Short-Range Capability without Technique

As every rifleman belongs to a team, he must be able to function as a part of that team. Practicing technique can do for a rifle squad what the *"kata"* sequence does for a *Tae Kwon Do* champion. While the *kata* produces muscle memory, technique practice produces team function memory. Without remembering how to work together on something similar, the members of a squad cannot respond to an enemy initiative quickly enough.

By late in World War I, every German line infantry squad was practicing "stormtrooper" technique.[3] At Camrai, many of those squads were able to transcend it.

> Painstaking preparation and repeated rehearsal had been replaced by rapid improvisation and the almost instinctive execution of battledrills.[4]

Not Just Any Battledrill Will Do

The U.S. infantry manuals contain one "duty solution" for each category of small-unit encounter. This "standard maneuver" is 60 years old and virtually devoid of surprise, but still published. Until the late 1980's, each squad leader required his people to reenact this "book solution" as a way of instilling discipline. Any deviation was met with "you screwed up so do it over." Against real-word resistance, his squad had about as much chance of moving forward as an overcontrolled football team with a few well-known plays.

Now the U.S. squad is so busy parading before VIP's, participating in large-unit exercises, and complying with bureaucratic requirements, that it seldom drills on anything but weapons' firing. While it masters the latest gadgetry, its "low-tech" counterpart refines its maneuvers. One remains disjointed and predictable; the other "works together" *(gung ho* in Chinese) to generate additional surprise. Whereas one has been forced to tailor "old-fashioned" technique to temperamental equipment, the other has been allowed to pursue tactical evolution. One will remain a clearly recognizable target, and the other will better blend into its surroundings. The probable winner of any close encounter is not difficult to predict.

How Other Nations Train

The Eastern armies have often been criticized for forcing their enlisted men to follow carefully scripted and mindless procedures in combat. In actuality, their soldiers, fire teams, and squads have being drilling on more than one maneuver technique for each category of enemy encounter. This gives them prerehearsed choices. It also gives them sufficient team-muscle memory to mix and match composite methods on the spur of the moment. To make matters worse, most Eastern technique better compensates for enemy machinegun fire than its Western equivalent. Most importantly, Eastern soldiers—unlike their Western counterparts—get to participate in battledrills that have been uniquely tailored to the very circumstances they will encounter in combat. The Eastern unit routinely reconnoiters its attack objective from the inside. It then builds a life-size mock-up that is accurate down to smallest defensive detail. Finally, its assault elements—through dynamic rehearsal—determine how best to circumvent those details.

All the while, the lower echelons of the U.S. Armed Forces have spent most of their field time rehearsing gun (and new equipment) drills. The published maneuvers lack enough surprise to even qualify as technique. To reenact them under differing circumstances would do more harm than good. At best, it would produce poor movement habits and total predictability.

The Germans

While Western headquarters tightly controlled all training during WWI, the German High Command let its companies train as they wanted. The 1917 German training day incorporated individual, buddy team, fire team, and squad competitions on tactical technique. Success was measured by more speed or stealth and fewer simulated casualties on successive tries. Between companies, differences in squad technique didn't matter.

> The training programs of the [German] Assault Battalions were quite different. . . . Close order drill . . . was largely, though not entirely, dispensed with. Its place was taken by exercises that cultivated rather than suppressed the initiative of the men. Half of each training day was usually devoted to sports. . . . [T]hese included obstacle courses and grenade-throwing contests.
>
> The other half of the day was spent in the practice of

> various battledrills—crossing "no man's land," breaching barbed-wire obstacles, clearing trenches, cooperating with flamethrowers, following closely behind a barrage, and the like were all practiced. These battledrills were often supplemented by live-fire exercises.[5]

To do well in combat, each echelon must have practice in matching its methods to unforeseen circumstances. Units didn't rigidly practice procedure so that its echelons could transcend technique.

> While the various drills in the use of . . . basic procedures were ingrained in the Imperial German Army through thorough training and repetition, the application of these techniques in the unique conditions of a battle was not done in a rigid fashion.[6]
>
> — U.S. Army, *Leavenworth Papers No. 4*

Just like the NVA and NKA would do 50 years later with their commando branches, each German field army had a storm battalion as teaching cadre.[7] That means every German line infantryman got to practice surreptitiously breaching enemy lines. Prior to WWII, the German squad also drilled on taking various targets alternately under fire with its light machinegun and riflemen.[8] Implicit was the unobserved forward motion that this shift facilitated.

The Japanese

The Nipponese knew the benefits of continual rehearsal. By 1913, their enlisted men were going through intensive drills on how to do the following at night: (1) march quietly, (2) participate in connecting files, (3) act as a messenger, (4) cross rough ground, (5) throw grenades, (6) bayonet fight, (7) entrench, (8) stand guard, (9) recognize friendly troops, (10) fire their weapons, (11) patrol, and (12) perform a hidden patrol.[9] These drills were conducted as experiments and often involved statistical feedback. Each man was given the opportunity to gauge his own progress and to share with the rest of the group any techniques he had discovered.[10] The Japanese practiced many of their battledrills with dummy rifles.[11]

There was competition between units. A 1935 photograph has inscribed "wall-scaling drill in a competition between the Imperial

Bodyguard and the Army's First Division."[12] Wall scaling is a *ninja* pastime. As with the Germans, these battledrills took up much of each day and were generally followed by a "free-play" exercise.

Before WWII, it was not unusual for a Japanese infantryman to be asked to work on his individual skills at night or during maneuvers. He trained 14 hours a day, six days a week. He often marched 25 miles a day and celebrated holidays with a "sham" battle.[13] The Japanese drills on individual know-how may have gone far beyond anything done by the Germans. Each so-called "basic" was subdivided and then subdivided again. During the hearing drill, Japanese privates had to distinguish between the sounds of "digging with a pick, driving a shovel strongly into the ground, pushing a spade into various kinds of ground, and a squad entrenching freely."[14] During messenger training, he not only learned advanced terrain association techniques, but also how to commit to memory what his route might look like after dark.[15] Just before WWII, a Japanese company arose at 4:00 A.M. to participate until dawn in a "bayonet-fighting tournament."[16]

The Chinese

Battledrills were nothing new to the Mainland Chinese. During the Korean War, those drills were very evident.

> The [Chinese] armies [entering Korea] had two phases of drills before going into battles. . . . The first phase focused on the exercises of small-group combat tactics, including shooting, throwing grenades, demolition, anti-aircraft training, and nighttime anti-air-raid practice. The second phase stressed courses of [larger] group-attack tactics . . . These exercises were aimed at the demolition of enemy defense works under enemy fire. . . .
>
> All armies offered classes in antitank training.[17]
>
> — *Mao's Generals Remember Korea*

The Russians

The individual battledrill is also common to the Russian unit. Soviet entry-level training includes "military sport games."[18]

> The Red Army had extensive experience in cities in World War II. . . . The effect of these experiences is apparent in Soviet training today. For instance, the standard obstacle course that all Soviet soldiers negotiate at least weekly as part of their physical training is designed to build skills specifically suited to combat in cities. It includes vaulting low obstacles, climbing walls, walking balance beams, diving under low doors, and jumping through windows.[19]

To do the most good, techniques must be improved in battle. In Manchuria in 1945, the Soviets not only refined pre-existing techniques, but also transcended them.

> The Soviets adjusted their . . . tactical techniques to the goal of . . . deep battle and rapid victory. Those techniques were aimed not only at defeating an enemy force, but also at mastering difficult terrain and beating the clock. . . . Consequently, while the Soviets used techniques developed during the war in the west, they molded those techniques to the unique theater requirements. Essentially, the Soviets emphasized maneuver at all levels.[20]
>
> — U.S. Army, *Leavenworth Papers No. 8*

It was technique (that had been recently practiced and altered) that made possible the capture of the highly sophisticated underground Japanese fortress of Hutou.

> The Soviet troops who participated in the reduction of the Hutou fortress complex were well trained and thoroughly rehearsed for the operation. . . . Their training exercises [were] conducted just before the invasion. . . .
>
> While the Soviet troops may have been well rehearsed, that did not mean their operations were stereotyped. Flexibility existed throughout the Manchurian operations. . . . Throughout the Hutou fighting, Soviet tactics were highly refined. . . . There were no massed frontal assaults or wave-type attacks. Soviet infantrymen flanked, enveloped, encircled, isolated, and then destroyed Japanese [fortified] strongpoints. . . . The infantry also showed its initiative by skillful infiltration tactics, usually conducted at night.[21]
>
> — U.S. Army, *Leavenworth Papers No. 8*

Of course, no battledrill will work well unless run at an appropriate time and place.

> The Soviets found that massed artillery and simple battledrills had little effect on the elusive guerrillas [in Afghanistan].[22]
>
> — DoD-published Soviet military academy study

The North Koreans

North Korean infantrymen depend heavily on battledrills as well. All dedicate almost two months a year to squad-level-and-below training.[23]

> Each [unit's biannual] cycle consists of approximately 760 hours of training, which progresses from individual . . . exercises to joint service maneuvers.[24]

The North Vietnamese

Within every NVA line company, small units diligently practiced many maneuvers. There were drills for almost every aspect of the offense and defense.

> Sapper units were also considered light infantry and were taught how to operate as an infantry unit. This training was the basis of specialized sapper training anyway, and sapper units trained hard so that they could do these small unit drills better than regular infantry units. Such tactics covered how to deploy on the battlefield for various types of operations (deliberate attack, hasty attack, meeting engagement, withdrawal, hasty and deliberate defense, regular patrolling, etc.).[25]

North Vietnamese soldiers who showed exceptional aptitude were subsequently sent to commando school. There they would drill on scouting and the more sophisticated aspects of short-range combat. As full-fledged commandos, they might return to the same line outfit as teaching cadre.

> Sappers recruited in North Vietnam . . . were given six to twelve month's training. The key subjects were:
>
> *Assault Techniques.* . . . A lot of these assault techniques were drill, to make the right moves automatic. . . .
>
> *Breaching Obstacles.* Sappers were carefully trained on the best techniques for getting through barbed wire, minefields, and floodlit areas. A lot of this was just practice, practice, practice. . . .
>
> *Camouflage.* . . . This was best accomplished by knowing how to hide yourself in plain sight. . . .
>
> *Exfiltration.* . . .
>
> *Explosives.* Sappers used explosives [and timing devices] to blow up obstacles and to destroy enemy fortifications and equipment. . . .
>
> *Infiltration.* . . . Sappers were known for their ability to sneak through the most formidable obstacles. . . .
>
> *Navigation.* . . . Sappers spent a lot of time in the field practicing. . . .
>
> *Planning.* . . . Sappers were trained on how U.S. and South Vietnamese bases were laid out and what [strategic] installations were found in all of them. So more reconnaissance was done until the model of the enemy base was complete. . . .
>
> *Reconnaissance.* . . .
>
> *Small Unit Tactics.* . . .
>
> *Special Equipment Training.* . . .
>
> *Weapons Training.* [26]

The Viet Cong

Every few months, VC rifleman received refresher "technique" training through a "mobile infantry school."[27] As techniques are best learned through battledrills, one can assume that this was the method of instruction. As the instructors taught several crawling methods,[28] they were probably commandos.

Even the village-defense-force members received training in sniper fire, trench warfare, use of grenades and mines, tunneling, and construction of homemade weapons. Meanwhile, the district maneuver force personnel received instruction in surprise raids, trench fighting, ambushing, and demolition work.[29] The VC who were to become sappers practiced working with demolitions and "observation and reconnaissance techniques."[30] Only a commando would have understood the latter well enough to teach it.

Afghan *Mujahideen*

Most of the videotapes of Al Qaeda training camps show lone soldiers running through live-fire obstacle courses. Those soldiers are doing so to build the hand-eye coordination and muscle memory required by combat.

That the *mujahideen* have since become hard to find in Afghanistan in no way refutes this assessment. A Western newsman might not have noticed the soldiers practicing *ninja* hiding techniques. Afghanistan does, after all, have a common border with China.

Battledrills Alone Will Not Produce a U.S. Tiger

Just as great plays can't ensure victory for a Western football team, neither can great battledrills promise a supremely confident American private. To emulate a tiger, the U.S. private must further be allowed to help design and then slightly deviate from the drills. Just as a halfback needs sometimes to change direction to avoid an enemy tackler, so too does a private need sometimes to alter his route to avoid an obstacle or danger area. As the halfback must have a straight-arm ready, so too must the private have a way to kill an unexpected assailant. If, to save his life, he must blow the element of surprise, then so be it. His unit's attack sequence should have optional steps with which to rebuild that surprise. Unfortunately, none of this will be possible until U.S. squads are allowed to operate semi-independently on offense and defense.

> Although combat experience should indicate otherwise, the rifle squad currently occupies a relatively minor place in . . . [U.S.] tactical thought. Squad level training and doctrine seem to suggest that the squad has little independent tactical value. The squad has been relegated to the role of subunit whose movements are closely controlled by the platoon commander. Considered in terms of maneuver war-

Figure 23.1: The Battledrill Instructors are Junior NCOs

(Source: *FM 100-5* (1994), p. 37)

fare, this attitude is disastrous . . . (maneuver warfare demands that the squad assume a primary tactical role).[31]

— William S. Lind
author of *The Maneuver Warfare Handbook*

To become maneuver warfare capable, American companies must learn to function under decentralized control. Every subunit and individual will need unpredictability and momentum. Like a winning football team, every squad will need two or three prerehearsed plays for each category of enemy encounter. Those plays will become the squad leader's "probable courses of action," greatly simplifying his decision process.

As companies now operate independently, there is no need for them to have identical squad procedures. Each company needs a way (through innovative training) to develop its own threat-dependent squad methods. The training must mirror the tactics—"recon pull" instead of "command push." Only through experimentation and research can the lower echelons of a tall organization stay up with the threat.

A training shortfall at any echelon—whether individual, fire team, or squad—can invalidate a commander's decision. Built into his unit training program must be a way to reassess subunit capabilities. Moreover, the program must give every rank practice at tactical decision making. There is a way to accomplish all of this. Commanders can choose worst-case combat scenarios and then let their NCOs develop ways to solve them at least cost (Figure 23.1). Group opinions won't guarantee worthwhile methods; only casualty assessment can do that. Company-grade officers can best help by doing the following: (1) shortstopping competing requirements, (2) arranging training support, (3) providing techniques from history, (4) monitoring casualty assessment, and (5) recording what has been learned. They must resist the temptation to overcontrol this process. Instead of the predictable book solution, they must require improvement on speed, stealth, deception, or casualties suffered. They must work through their SNCOs. The CO who was formerly enlisted can temporarily facilitate the learning, but should turn the responsibility over to his Gunnery Sergeant as soon as possible.[32] Many a well-seasoned Marine has seen the need for this "bottom-up" approach.

> Using football as an analogy, my players knew how to play the game, but had no team plays, much less any rehearsal of these plays. . . .
>
> " . . . My purpose is not to teach you anything but to pull our collective knowledge together, to get our purpose, our plays, our timing down to a razor edge."
>
> " . . . We will spend two hours each day here talking and walking ourselves through some offensive tactics that I feel are needed."[33]
>
> — 1st Lt. Wesley Fox
> upon assuming command of A/1/9 in Vietnam

What U.S. Enlistees Can Do to Help Themselves

The American private, private first class, lance corporal, or specialist must take it upon himself to rehearse—in its entirety—every aspect of his combat role. He should mentally run through any portion that cannot be physically practiced. Wherever he detects a lack of ability, he must personally correct the deficiency.

Then while in combat, like a *ninpo* practitioner, he should visualize every step of what he is about to do. For any portion that conflicts with his survival instincts, he should look for an acceptable alternative. Sometimes that alternative will require a leader's permission, but more often it won't.

24 Field Proficiency Has No Substitute

- *Can modern technology thwart every intrusion attempt?*
- *How might a ninja master outsmart electronic surveillance?*

AMERICAN SPECIAL FORCES SERGEANT, 1982

(Source: Courtesy of Cassell PLC, from *Uniforms of the Elite Forces,* ©1982 by Blandford Press Ltd., Plate 1, No. 2; *FM 5-20* (1968), p. 14)

Electronic Imagery Can Only Show So Much

Advanced satellite and aerial imaging helps the U.S. commander to see more of the battlefield's surface characteristics. It also helps him to see what his opponent wants him to see. As such, it may give an Eastern adversary additional opportunities for deception. He has only to flood the battlefield with bogus or nonstrategic units. Then whatever essential elements of (useful) information are retrieved by electronic imagery will get lost in its vast harvest of misinformation. Until aerial surveillance can distinguish operational tank from useless hulk, occupied bunker from dummy position, enemy soldier from noncombatant, and friendly IR (infrared) beacon from counterfeit tag, the U.S. commander will need close-in observers on the ground. Without them, his target acquisition will be poor, his friendly-fire losses will be bad, and his collateral damage will be terrible. Virtual reality can be manipulated. On 6 March 2003, ABC's *Nightly News* reported that Osama bin Laden had given away his position by making a call on his cell phone. It further noted that he had once before escaped by moving away from someone transmitting a recording of his voice.[1] If a homegrown political movement can so easily confound Western intelligence, imagine what a professional Asian army might do.

> [T]he [Chinese] PLA is fully aware of our intelligence gathering gadgets and are prepared to engage in countermeasures to deny us the level of battlefield information we often expect in training exercises.[2]
>
> — *Camp Lejeune (NC) Globe,* 8 March 2001

The Eastern unit is just as proficient at dispersion as deception. It can create so many tiny "blips" on the Western commander's computerized game board, that he can't tell which ones have strategic missions. In a major battle, there won't be enough planes, artillery pieces, and munitions to target every Eastern squad at once. To counter any further refinement to U.S. aerial surveillance, the "low-tech" opponent has only to disperse into fire teams.

As was apparent in Vietnam, tiny semi-autonomous elements can do considerable damage. They can take thousands of nibbles out of America's strategic assets. They can get U.S. forces to expend their strength against lightly manned objectives. They can capitalize on the inherent risks of over-the-horizon ordnance.

> "There are more beyond-visual-range engagements, since we no longer wait to see the whites of their eyes before we fire at the enemy," says John Pike, Director of GlobalSecurity.org in Washington.
>
> "So there are more chances for mistaken identity."[3]
>
> — *Christian Science Monitor,* 14 January 2003

As satellite- and airplane-mounted surveillance equipment cannot yet see below ground, the Easterner will shift almost exclusively to "underground warfare." Developed late in WWII, it enabled the Japanese to inflict tremendous losses on seasoned U.S. ground forces at Peleliu, Iwo Jima, and Okinawa. Used again in the Korean hill battles of 1953, it permitted the Chinese to curtail the advance of an opponent with vastly more wherewithal. Historians are just starting to realize that the Vietnam coastal plain was studded with Maoist, lowland "base areas."[4] These were not static, impenetrable lairs, but below-ground encampments that rotated among the ten or so fortified hamlets per commune. Many were part of a huge matrix of similar bastions. The best example is the Cu Chi tunnel complex. It extended all the way from the outskirts of Saigon to the Cambodian border. What an invisible maze of hardened strongpoints can accomplish was adequately demonstrated by the brief border clash between China and Vietnam in 1979. In that incident, Vietnamese home guard personnel thwarted some 17 fully supported Chinese divisions while their expeditionary brethren were off fighting the Khmer Rouge in Cambodia. The Vietnamese refer to this defensive marvel as their "Military Fortress" concept. Each composite strongpoint has a "vanish underground installation or hidden tunnel complex."[5]

Now that U.S. cordon maneuvers like Operation Anaconda have once again failed in Afghanistan, it's clear that the Taliban and al-Qaeda are well aware of the Eastern shift toward underground warfare. The Soviet Frunze Military Academy study—*The Bear Went over the Mountain*—freely admits to Afghan guerrillas routinely escaping Soviet encirclements through irrigation tunnels.[6] Oversized smart bombs can only do so much to solid granite. Any overpressure at the mouth of a tunnel can be sufficiently dissipated by blast doors, air shafts, and rear entrances. In an operation south of Gardez in May of 2002, the British discovered tunnels with metal doors.[7]

The Latest Technology Doesn't Go to the Frontline Soldier

Whether one is a fan of technology or not, all can agree that a small unit's technique should dictate its equipment and not the other way around. In the vastly successful U.S. infantry manual supplement—*The Last Hundred Yards*—is a technique through which a platoon or squad can secretly assault a prepared enemy position in the daytime. Its most crucial step is causing the defenders to think that they have spotter (as opposed to screening) smoke in their wire. As spotter smoke rounds are no longer available to U.S. artillery and mortar batteries, the technique is no longer viable. If a unit were to substitute white phosphorous (WP) for smoke, it would destroy itself. Nor can a U.S. infantry leader place a tight enough sheaf of artillery or mortar rounds onto a machinegun bunker to kill it. While still in the official manuals, the "precision fire" method of adjustment (in which only one gun shoots) is viewed by most batteries as violating their "area target" mandate. They are further unwilling to put enough rounds into their computer-generated 100-meter sheaf to ensure a direct hit on a hardened, pinpoint target. Anyone who has played ring toss at a county fair knows why. While the U.S. squad or platoon leader could direct a smart shell onto the bunker himself with a hand-held laser designator, he is seldom allowed that much equipment, ordnance, or authority.

A modern vision enhancement device helps its wearer to detect a stationary, upright adversary at night. It in no way conceals that wearer. While fidgeting around with his uncomfortable head gear, he creates more of a motion signature than he had before. Of course, the U.S. Army is not totally oblivious to this problem. It has given the Massachusetts Institute of Technology (MIT) 50 million dollars to develop a totally reflective uniform that will help American privates to "more easily blend in with their surroundings."[8] Hopefully, those privates won't try to do so after the sun, moon, or stars come out.

In March 2002, the U.S. Army sent experts to Afghanistan from its Natick Soldier Center, a Massachusetts research facility dedicated to meeting the soldiers' needs. They wanted to hear firsthand from recent combat veterans how their gear performed.

> Helmets were criticized by 85 percent of the surveyed GIs as way too heavy, especially when combined with night-vision equipment. Serious head and neck aches were a constant companion.[9]
>
> — Scripps Howard News Service, 6 October 2002

The Advantages of a Closer Look at Enemy Targets

By reconnoitering each objective from the inside, the Eastern commander greatly increases his odds of attacking that objective successfully. By discovering its defensive details, he can do two things differently: (1) make a foolproof plan, and (2) have his subordinates rehearse every step of that plan in a life-size mock-up. With enough surprise, they are almost ensured of winning. Even the often-maligned Russians follow this procedure.

> Soviets conducted a systematic tactical exercise on terrain similar to that for an [Afghan] ambush site.[10]
>
> — DoD-published Soviet military academy study

An Eastern Unit Can Operate on Autopilot

The Eastern way of war depends more on the initiative of front-line fighters than on the direction of headquarters. In the final assault on Saigon in 1975, the North Vietnamese high command issued only the most general of mission-type orders to its main-force units. The rural VC had become so proficient at controlling their respective TAORs that they had little trouble detaining Allied reinforcements.

Western military planners have been slow to realize that such a unit depends little on the advice or health of its commander. It cannot be stopped by cutting off its head.

The Eastern Mix of Art and Science

The six other nations in this book had something in common. As they often faced greater firepower, they learned to practice war as an art as well as a science.

> In developing doctrine, the [WWI] Germans always considered another critical factor, the enemy. Unlike [French Gen.] Nivelle, who unfortunately acted as if the success of his plans were utterly independent of the existence of his enemy, the Germans respected their enemies. The German consideration of influential factors made the application of their doctrine an art rather than [only] a science.[11]
>
> — U.S. Army, *Leavenworth Papers No. 4*

Modern Soviet tactical doctrine calls for more aggressiveness, flexibility, and initiative from commanders than Western analysts like to concede. Russian commanders have two habits that many Western counterparts lack. They study their foe's tactics (as opposed to operations) and prefer human battlefield intelligence.[12] Theoretically based on collective wisdom and objective knowledge, the Soviet offensive algorithm could be formidable.[13] While war cannot be conducted by formula, it can be waged in a way that increases one's odds of winning. Prerequisite, of course, would be periodically admitting to a tactical defeat and then asking frontline fighters what went wrong. The U.S. military is not famous for either.

An Eastern army prefers to disseminate its tactical wisdom as guidelines rather than doctrine. It does so to give its lowest-echelons the leeway to cope with the ever changing circumstances of battle.

> The tactical principles were guides for the exercise of good judgment in unique situations, not formulas to eliminate the need for good judgment. While the various drills . . . and basic procedures were ingrained in the Imperial German Army through thorough training and repetition, the *application* of these techniques in the unique conditions of a battle was not done in a rigid fashion.[14]
>
> — U.S. Army, *Leavenworth Papers No. 4*

Today's U.S. infantry leader may be more tactically sophisticated than his forefathers, but he has left his riflemen, fire teams, and squads virtually bereft of movement technique. A great tactical decision must still be executed.

Learning Takes Admitting to Error

From the time the American enlistee enters boot camp, he is told that pride in his unit will take precedence over almost every other consideration. As he acquires rank, he comes to believe that openly criticizing any aspect of his unit's (or his own) readiness constitutes disloyalty to his commander. As a result, it is difficult to get anyone within a U.S. unit to acknowledge a deficiency. Unfortunately, the unit that is perfect has no reason to improve. As the lower ranks absorb most of the casualties in combat, it makes perfect sense to let them identify their own deficiencies.

Eastern nations have placed more emphasis on the collective opinions of each unit's frontline fighters than on that of its commander. The North Vietnamese used *kiem thao* sessions to acquire this knowledge.[15] The Soviets expected their commanders to monitor the collective opinions of their subordinates. Unfortunately, the Soviet version put more stress on the enforcement of rules than on common sense. Often those rules only masqueraded as collective opinion. In other words, the North Vietnamese allowed the criticism of combat performance to improve their tactics while the Soviets manipulated peer pressure to keep everybody in line. As a result the NVA privates liked the collective process,[16] while the Soviet privates didn't.[17]

To profit from tactical errors, an infantry unit must first admit to them and then be given enough leeway to make the requisite changes.

> The success of [the] Soviet night operations was in large part due to intensive training and the ability to profit from mistakes and failures. . . . Although the Soviet combat leaders were acutely aware that "the success of any battle is determined . . . by the extent of preparation," *training standards for night operations were not uniform* [italics added].[18]
>
> — U.S. Army, *Leavenworth Papers No. 6*

A "zero-defect" environment inhibits learning. American infantrymen have been struggling with just such an environment for the better part of a century. To their utter dismay, the problem has worsened over the last decade.

The Shocking Truth—Asymmetric Warfare Isn't

For every category of combat, this book has revealed the emerging body of tactical technique with which an army can compensate for deficits in technology and firepower. It has shown how these techniques have been progressively refined by the WWI Germans, WWII Germans, WWII Japanese, Chinese, North Koreans, and

North Vietnamese. It has demonstrated that these evolutionary trends in tactics are well known to the Southwest-Asian armies, *mujahideen,* and Al Qaeda.

> Several combat principles lay at the heart of the *mujahideen's* [Soviet-Afghan War] tactics. First, they avoided direct contact with superior might. . . . Second, the *mujahideen* practically never conducted positional warfare and, when threatened with encirclement, would abandon their positions. Third, in all forms of combat, the *mujahideen* always strove to achieve surprise. Fourth, the *mujahideen* used examples from the Basmachi movement (Muslims who resisted Red rule in Central Asia from 1918 to 1933).[19]
>
> — DoD-published Soviet military academy study

To have the best odds of surviving any future conflict, America's lowest-ranks must be apprised of the Eastern methods. Yet, more than their survival may be at stake. Without better small-unit technique, America's ground forces may have trouble prevailing in close terrain at any cost. The only way to defeat this emerging style of warfare is with frontline combatants of commensurate skill and self-assurance. American privates must begin to see themselves not as pawns in the "big picture," but rather as skilled warriors with the opportunity to make a strategic contribution.

The Potential Quagmire

After Mogadishu, "Third-World" militants knew where to confront a U.S. expeditionary force. In urban terrain, they can hide—beneath rooftops and among civilians—from the most advanced electronic surveillance. In a city, they can snipe U.S. soldiers and then benefit from reprisal bombardment. If the growing number of casualties didn't undermine America's resolve, then the collateral damage would sway its public opinion. Make no mistake, the former leader of Iraq had figured this out.

> Mr. Hussein has recently moved batteries of surface-to-air missiles from the desert into Baghdad, a signal that if attacked he plans to fight U.S. troops in the city instead of the open desert, said defense intelligence officials. His elite

Figure 24.1: Iraqi Special Forces Trooper, 1982

(Source: Courtesy of Cassell PLC, from *Uniforms of the Elite Forces*, ©1982 by Blandford Press Ltd., Plate 26, No. 76)

Republican Guard, which before the Gulf War trained mostly in open terrain, also has stepped up its urban training, defense officials say.[20]

— *Wall Street Journal,* 22 August 2002

While badly exposed in the desert, Saddam Hussein did have a few advantages. Greater Baghdad is vast—fully 2,000 square miles. That's over three times the size of 1945 Berlin. Within Baghdad, U.S. firepower would have been of little use against swarming bands

Figure 24.2: Muslim Irregular, 1982

(Source: Courtesy of Cassell PLC, from *World Army Uniforms since 1939*, © 1975, 1980, 1981, 1983 by Blandford Press Ltd., Part II, Plate 115)

of determined guerrillas (Figures 24.1 and 24.2). Only the skill and initiative of the individual U.S. soldier or Marine could have gotten the job done without setting the stage for future problems in the region. Precise munitions require precise intelligence and stationary targets. Too many replays of the Amiriya bomb shelter tragedy of 1991 could have compromised world opinion. Fortunately for the U.S., the Iraqi army had been overcontrolled for so long that it could not field squad-sized maneuver elements with enough skill and motivation to operate semi-independently. According to *Newsweek,* even the Fedayeen had difficulty dispersing.[21] Still, it was a close call for U.S. troops.

On 24 March 2003, ABC's *Nightly News* reported that British Marines had encountered urban guerrillas in Umm Qasr on the Persian Gulf.[22] On 25 March 2003, the lead story in the *Christian Science Monitor* was "Guerrilla Tactics vs. U.S. War Plan."[23] On the ABC's *Nightly News* of 27 March 2003, Saddam Hussein was quoted as saying that U.S. troops would be confronted on every street in Baghdad and all along their tenuous resupply routes.[24] Implied, of course, was the method of confrontation. On 28 March 2003, ABC's *Morning News* reported that "U.S. troops are coming under attack wherever they go from roving bands of militia."[25] On 30 March 2003, ABC's *Nightly News* reported Coalition forces fighting guerrillas in Basra.[26] On 3 April 2003, National Public Radio (NPR) said that every street corner in Baghdad was being manned by soldiers ready to wage guerrilla war.[27] The 7 April 2003 issue of *Newsweek* indicated that "Saddam has built an extensive underground tunnel system [beneath Baghdad] that could be used to ambush forces or as a quick getaway."[28] Only by the grace of God, was tragedy averted in the Iraqi capital during its initial capture. Only time will tell if U.S. occupied Iraq will now become the preferred destination for every Islamic *jihad* guerrilla organization in the region. There is already abundant evidence of their presence. According to the Associated Press (AP) on 13 April 2003, as many as 5000 foreign fighters may have been bypassed in Kut alone.[29] On 18 May 2003, a full six weeks after the declared end of hostilities, AP indicated that 25,000 U.S. troops were having trouble restoring security to Baghdad.[30] In Afghanistan, Coalition forces don't even try to provide security outside Kabul.[31]

The "Maneuver Warfare" Heritage of Southwest Asia

As Attila the Hun's raiders had been Mongolian nomads from the Caspian Steppes, they were in small part Turkish.[32] From 618 A.D. to 906 A.D., the Chinese T'ang Dynasty controlled everything from northern India to Turkestan.[33] Part of Turkestan abutted the Caspian Sea and became the Turkmen Soviet Socialist Republic.[34] However, other parts extended south. Semi-nomadic Turkomans frequented what is now northern Iran.[35] To this day, the Turkmen are Iraq's third largest ethnic group.[36] For this reason, the ancient Turks provide an excellent example of how Far-Eastern military thought may have affected the tactics of the region. During the 10th century, the Turks preferred to operate in small, highly mobile bands:

> The Turks posed a different problem: innumerable bands of light horsemen armed with javelins and scimitars, but, like the Huns, relying on the bow and arrow, . . . given to ambushes and stratagems of every sort. . . . [I]n battle they advanced not in one mass, but in small scattered bands, which swept along the enemy's front and around his flanks, pouring in flights of arrows, and executing partial charges if they saw a good opportunity.[37]

During the first Crusade from 1095 A.D. to 1099 A.D., the Turks practiced tactical withdrawal just as expertly as their distant in-laws:

> Their [the Turks'] tactics skillfully blended surprise, mobility and firepower. . . .
>
> The Turk was a master of feigned retreat. On occasion, he led Frankish horse [cavalry] a chase lasting days, on other occasions, he lured them into prepared ambush. When attacking marching columns, he concentrated on separating the components, usually striking the rear. And, just as disconcerting, if things went wrong for the attackers, they did not hesitate in breaking off action and disappearing.[38]

This was no fluke. At the turn of the first millennium, even the East-Roman (Byzantine) occupants of Constantinople (present-day Istanbul) followed a decidedly Asian military philosophy.

> The generals of the East considered a campaign brought to a successful issue without a great battle as the cheapest and most satisfactory consummation in war. They considered it absurd to expend stores, money, and the valuable lives of veteran soldiers in achieving by force an end that could equally well be obtained by skill. . . . They had a strong predilection for stratagems, ambushes, and simulated re-

> treats. For the officer who fought without having first secured all the advantages for his own side, they had the greatest contempt (Oman, *A History of War in the Middle Ages,* 1924, 1).[39]

To make matters worse, what is modern-day Iraq and Iran then became part of the Mongol Empire for 250 years.[40] While traveling through the region around 1270 A.D., Marco Polo observed, firsthand, Tartar tactics. For those familiar with Japanese "retreat combat" and Chinese "mobile warfare," his description is chilling.

> When these Tartars come to engage in battle, they never mix with the enemy, but keep hovering about him, discharging their arrows first from one side and then from another, occasionally pretending to fly [run away], and during their flight shooting arrows backwards at their pursuers, killing men and horses, as if they were combating face to face. In this sort of warfare, the adversary imagines he has gained a victory, when he has in fact lost the battle; for the Tartars, observing the mischief they have done him, wheel about, and renewing the fight, overpower his remaining troops.[41]
>
> — Marco Polo

So, one could say that southwest Asia has been exposed for 1500 years to the forerunner of "maneuver warfare." The Mongols had to know of Sun Tzu's implicit "soft defense" and "soft-spot offense." From the 16th century to the end of WWI, Iraq was part of the Constantinople-based Ottoman Empire.[42] Then, as a Soviet client state, it gained even more exposure to the alternative way of war.

> Through the Mongol-Tartars, Sun Tzu's ideas were transmitted to Russia and became a substantial part of her Oriental [military] heritage.[43]
>
> — Brig.Gen. Samuel B. Griffith USMC (Ret.)

The Persian Gulf Has Its Own *Ninja* Equivalent

The Near East sits astride the ancient overland and oceanic trade routes from the Far East. Chinese *moshuh nanren* traveled these routes, offering their skills to the highest bidder. The word "assassin" comes from *hashishin*—a hashish-eating cult of medieval night stalkers who spread terror throughout the Middle East for over 200 years. Founded in the late 11th century by Hasan-ibn-Sabah (dubbed "The Old Man of the Mountain" by Marco Polo), it was a radical offshoot of the Shiite branch of Islam. Though the cult's mountain stronghold in Western Persia was eventually overrun by the Mongols, many of its traditions survive to this day. All modern Islamic terrorist groups trace their roots to this cult of killers.[44] Their specialty was bypassing sentries.

> The *hashishin's* favorite technique was to sneak past all the security a king or potentate could muster and, in the dead of night, while the official slept, place a dagger beside his head. For those few who wouldn't take the hint, *hashishin fidavis* ("daggermen") would make a return visit.[45]

It is one thing to worry about someone who knows *ninjutsu*. What if there were a higher art form—one that could produce some of the more amazing feats of the *ninja* legend? How much of a chance would the American scout or sentry have against its practitioner?

Ninjutsu's Higher Form Comes from Edge of Southwest Asia

Whereas *ninjutsu* is the art of stealth through accomplishment, *ninpo* is the way of invisibility through enlightenment (utilization of universal laws). It is a higher and more spiritual form of *ninjutsu* for all circumstances and levels of human activity.[46] Through spiritual study, the *ninpo* practitioner attempts to capitalize on those elements of life that are normally referred to as "luck," "coincidence," and "fate."[47] For the intended doer of good, it must be practiced carefully. In essence, *ninpo* combines *ninjutsu* with *yojutsu* (mystical) and *genjutsu* (illusion)[48]—metaphysical disciplines from the Indian subcontinent.[49] Based on Himalayan Tantric spiritual teachings,[50] *ninpo mykko* helps its practitioner to better focus through meditation.[51] The psychological techniques of *ninpo* run the gamut of possible applications—from visualizing future actions to self-hypnosis. (See Table 24.1.)

The ancient Hindu *Bhagavad-Gita* reputedly contains a step-by-step procedure on how to appear not to exist.[52] Within the art of making oneself invisible *(onshinjutsu),*[53] the Eastern warrior creates shapes and images with which to confuse and delude his foe. As his first target is that foe's mind, many of his techniques border on hypnosis. According to the *ninpo* grandmaster, *fudo kanashibari no jutsu* paralyzes the victim with a spell.[54] Other techniques do what any self-respecting maneuver warfighter might do—create an opening where none previously existed. Of particular interest might be the scouting techniques of *sekkojutsu*.[55]

While these early forms of psychological warfare are well beyond the scope of this book, every U.S. private must know of them. They are more than legend. Hypnosis has long been recognized as a remedy for fear and pain.[56] Marco Polo warned of a daytime "darkness" being somehow conjured up by the caravan-raiding *Karaunas* of Upper Persia.[57] While retracing Marco Polo's steps, both Col. Henry Yule and Maj. Sykes encountered a "dry mist" in Upper India.[58]

The *ninpo* techniques for clouding a quarry's mind fall under the broad category of *saiminjitsu*.[59] Among these techniques is hazing. Monotonous motion has often been associated with hypnosis. Why couldn't any repetitive sight or sound have the same effect on a sentry? Only deep hypnosis would require that he be willing. Shallow hypnosis has been described as "an altered state of intense but narrow concentration."[60]

> A popular myth is that hypnosis is a form of sleep. . . . In fact, EEG [electroencephalogram] studies show that the hypnotic state is not a form of sleep at all. It is a form of focused alertness, with increased attention in one area and decreased or absent focus on other events. . . . Hypnosis is perhaps best described as a daydream so intense that one temporarily believes one is in it, and is oblivious to anything else going on in the surroundings.[61]

Even the Russians will create a diversion in front of a sentry post while a few flankers crawl to within assaulting distance.[62] Couldn't an Eastern infiltrator's accomplice sufficiently distract a U.S. watchstander to permit his partner's unobserved passage? By cooing like a dove, a *ninja* master might create his own monotonous diversion. At 3:00 A.M., a young American would be highly susceptible to dreams of home. *Ninjas* are, after all, expert at narrowing their own focus. An accomplished student of psychology and Eastern religions had this to say about the possibilities:

> Perhaps some form of meditation is involved in which the outer form or appearance is somehow altered so as to render the practitioner invisible to the eye of the beholder [quarry].
>
> When I have stalked animals, I have been able to get within a few feet of deer, and woodchucks and whatever, even when they have been looking right at me. Needless to say I moved very slowly, and tried to blend in with the surroundings. Keeping an inner stillness seemed to help.
>
> Most of us have an inner dialogue going, with one part of ourselves debating with another part . . . about the best way to do something or other. One result . . . is that we are not processing what our outer perceptions are telling us: e.g., windshifts, noises (that stamp of the hoof when a deer is beginning to get spooked, or a snort that says he/she doesn't like what the winds or motion are telling him/her). So by being empty (no inner dialogue), the hunter stays alert to what his sensory perceptions are telling him and is also alert to premonitions, intuitions, about the prey. He can sometimes feel its eyes on him and know when to freeze and when to move.
>
> Also, . . . stopping the inner dialogue . . . makes one invisible because one is more a part of nature at that moment and not a preoccupied human being. You can't even want to kill the prey—they will pick that up.[63]
>
> — Dr. David H. Reinke

Several well-respected *ninpo* masters admit to being able to hypnotize an adversary in broad daylight.[64] One Chinese-oriented author shows how. Just to be on the safe side, any U.S. sentry who is approached by a figure with strangely clasped hands, should not look at the hands.

> *Kuji Kiri* is the technique of performing hypnotic movements with the hands. These magical in-signs created by knitting the fingers together may be used to . . . hypnotize an adversary into inaction or temporary paralysis. Each is a key or psychological trigger to a specific center of power in the [human] body. . . . These are keyed to the twelve meridiens *[sic]* of acupuncture.[65]

- Achieving harmony with universal forces [66]
  - Protecting one's own body, mind, and spirit [67]
    - Developing a benevolent heart *(jihi no kokoro)* [68]
  - Secret spiritual power *(mykko, tantra,* and *ekkyo)* [69]
    - Viewing reality from the outside *(kongokai mandala)* [70]
    - Viewing reality from the inside *(taizokai mandala)* [71]
  - Ascetic power *(shugendo)* [72]
  - An intuitive knowledge of fate [73]

- Accomplishing one's mission through indirect influence *(bo-ryaku)* [74]

- Psychological strategies (from *yojutsu* and *genjutsu*) [75]
  - Spiritual refinement (seishin *teki kyoyo)* [76]
    - Achieving a relaxed mental state [77]
    - Channeling one's energy through chants & finger weaving *(kuji-in)* [78]
  - Mind control *(saiminjitsu)* [79]
    - Harnessing one's subconscious through self-hypnosis [80]
      - Maintaining a "can-do" attitude
        - Gaining inner strength through meditation [81]
        - Transcending fear or pain [82]
      - Enhancing one's external awareness [83]
        - Concentrating on sensory impressions [84]
        - Paying attention to detail [85]
        - Reading the thoughts of others [86]
        - Perceiving danger [87]
        - Making decisions [88]
        - Visualizing the task to be accomplished [89]
    - Clouding a sentry's mind [90]
      - Hazing the sentry [91]
      - Distracting the sentry through hypnotic suggestion [92]
      - Paralyzing the sentry with a spell *(fudo kanashibari no jutsu)* [93]
  - Threatening appearance or shouts as a weapon (kiaijutsu) [94]
  - Scouting methods *(sekkojutsu)* [95]

- Penetrating a fortress
  - Falling in behind a sentry or incoming patrol [96]

- Hiding *(inpo)* [97]
  - Making oneself invisible *(onshinjutsu)* [98]

- Escaping *(tonpo)* [99]
  - Establishing a false exit (e.g., opening an outside door without leaving) [100]
  - Leaving a recognizable trail and then retracing one's steps [101]
  - Doubling back after the foe has bypassed one's concealed position [102]
  - Abruptly terminating one's trail [103]

- Realm devoid of specific recognizable manifestation *(ku no seikai)* [104]

Table 24.1: A Few of the Additional Capabilities of *Ninpo*

Fourth-Generation-Warfare Implications

The Oriental's fascination with hypnotism may have far more sinister implications on the world stage. What if an Asian nation were to use it as a part of its expansionist agenda? Some years ago, there was a movie entitled "The Manchurian Candidate." It was about a man who had been somehow programmed to snipe a Western leader. Within many a story lies a grain of truth. According to *Newsweek,* 17-year-old Lee Malvo had received fatherly advice, reading material, marksmanship training, and "as many as 70 pills a day" from his self-professed-Muslim handler for several months before going on his shooting spree around Washington, D.C. in November of 2002.[105] It was during this shooting spree that Congress approved the war on Iraq, and it was shortly thereafter that North Korea felt threatened by U.S. foreign policy. On ABC's *Good Morning America,* Todd Petit—Malvo's court-appointed guardian — later produced the teenager's political manifesto. He further speculated that Malvo had been "brainwashed."[106]

There are those who believe that war has evolved beyond battlefield confrontation. That which is "Fourth Generation" takes place on one's own soil. It covers the full spectrum of electronic, psychological, and guerrilla warfare.[107] While Islamic movements have been credited with recent attacks on America, Far Eastern nations are much better suited to conduct this alternative style of war. They have long depended on deception, dispersion, and sabotage on the battlefield and could easily shift their area of operations. High-ranking members of the Chinese Army recently pointed out how the electronic age had increased its opportunities.

> *[Unrestricted Warfare]* advocates a chilling asymmetric warfare strategy that includes the use of terrorism, cyber-warfare, propaganda and the use of unconventional weapons.[108]
>
> — Brig.Gen. Robert M. Shea in Congress testimony

Of late, China has been active from Mayanmar to Afghanistan. Thailand's biggest newspaper reported a "third party" responsible for the Mayanmar border trouble in the summer of 2002.[109] "Intelligence sources indicate that Chinese nuclear technicians, under PLA protection were sent into Afghanistan to remove evidence of radiological materials that could have been traced back to Beijing."[110]

China's obvious interest in the west may portend more subtle intentions in the east. To take back Taiwan, it would need an excuse and U.S. forces fully committed at two other locations.

> [T]he [U.S.] administration's strike-first military strategy could embolden other powers to do likewise.
> . . . China could assert such rights to "reclaim" Taiwan.[111]
> — Associated Press, 19 March 2003

Homeland Security

The security measures for many of the sensitive installations within the continental United States were designed to stop a determined thief, not a professional infiltrator. Counterespionage measures would take considerably more effort. Here is a recent example of what can happen under existing security procedures.

> Life at the Army's largest stockpile of chemical weapons was back to normal Friday, despite the failure of police and soldiers to find the person who infiltrated the property a day earlier. The Army ended the search for the intruder Thursday night following a "thorough sweep" of the Deseret *[sic]* Chemical Depot, about 20 miles south of Tooele, said depot spokeswoman Alaine Southworth. The sweep involved a large but undisclosed number of soldiers, Tooele County sheriff's deputies and police dogs. They were assisted by a state police helicopter. The FBI, which dispatched two agents from the agency's Homeland Security division and Joint Terrorism Task Force, also assisted [in] the investigation. Covering virtually every foot of the depot, the search revealed no clues about the intruder's whereabouts. Southworth said the intruder apparently escaped. "We know he's not on the installation," she said.
>
> Much, however, remains unknown about what is believed to be the most serious breach of security in the depot's 60-year history. Four soldiers in two separate patrol vehicles spotted the intruder about 9:20 A.M. Thursday. He—or she—was on foot about 100 yards inside the northern fence of the 19,000-square-foot depot and about one mile

> from where millions of pounds of nerve and mustard gas are stored in dozens of underground bunkers.[112]
> — *The Salt Lake Tribune,* 7 September 2002

In a 9 December 2002 interview on ABC's *Good Morning America,* Foster Zeh—a security manager at the Indian Point Nuclear Power Facility near New York City—blew the whistle on its security procedures.[113] He claimed that his superiors and the National Regulatory Commission had taken no corrective action after the plant had been penetrated in 36 seconds during a recent test. Those so-called "infiltrators" probably had little training in such matters. To the accomplished *ninja,* the most sophisticated of Western security precautions would pose little challenge. If the outer perimeter had too many electronic gadgets, he would simply tunnel under it or glide over it.

At the root of the problem is the "top-down" way in which most, if not all, U.S. government agencies operate. The more productive "bottom-up" approach only occasionally appears in Western business. To admit that there might be a better way of running a military, police, or security agency would be to commit political suicide. It would bring to light so many past mistakes as to jeopardize national resolve.

Few Western Installations Are Totally Secure

To enter a Western installation, the Eastern soldier need not always sneak through its perimeter defenses. With a workable disguise and forged identity card, a novice intruder could walk right in. By hiding in a shipment of materiel or beneath a vehicle, he could bypass all sensors, barriers, and sentries. "Personnel subordinate to the [North Korean] Light Infantry Training Guidance Bureau receive [one month of] 'assimilation' training to allow them [to] pose as ROKA [South Korean] soldiers."[114]

> Some military analysts say that North Korea may have some 20 [undiscovered] tunnels coursing under the DMZ. They say it may be North Korean Special Operations units—at 80,000 men, the largest in the world—that would cause the greatest military challenge, should those forces infiltrate

> through those tunnels and come up behind the U.S. forces in the event of conflict.[115]
>
> — *Christian Science Monitor,* 11 March 2003

It takes someone with slightly more skill to outwit electronic surveillance or detection equipment. While the latest electronic warning devices may appear foolproof, all have limitations. Even with adequate power, calibration, and maintenance, they still require human follow-up. Machines can't think, and the guards who answer their alarms are susceptible to deception. Because of a *ninja* heritage and fewer restrictions, Eastern commandos have had more opportunity to explore their potential.

> Any perimeter, no matter how well guarded or supported by what modern monitoring equipment, can be penetrated, given sufficient time.[116]
>
> — Dr. Lung

As those electronic gadgets have increased in sensitivity, they have generated more false alarms. They can now be set off by wind, fog, rain, snow, ice, dust, smoke, electrical storm, bright light, voltage surge, seismic tremor, vehicular vibrations, wild animals, and temperature change.

> Statistics indicate that 95 percent of all activated perimeter alarms are *false* alarms. . . . [A] fair-sized . . . installation can expect four or five false alarms a day, every day.[117]

Too many false alarms might lessen the thoroughness with which each incident was investigated. In fact, there appears to be an inverse relationship between the amount of "high-tech" security equipment and the motivation of "low-paid" security personnel.[118]

Of course, the infiltrator need not always go through the electronic warning system, he can also go over or under it. To go over, he has only to use a hang glider, high-altitude/low-opening parachute, or elevated utility line. To go under, he has only to crawl through a water pipe, feeder sewer, utility conduit, ventilation shaft, or maintenance tunnel. Or that infiltrator can just dig a new passageway.

Electromagnetic energy takes many forms. In order of increasing wave length, it covers the following spectrum: (1) gamma, (2) X-ray, (3) ultraviolet, (4) visible, (5) infrared, (6) microwaves, and (7) radio waves.[119] The newest surveillance devices fall into three categories: (1) those that magnify reflected, visible light; (2) those that sense natural or reflected infrared light; and (3) those that detect reflected microwaves. All three categories can penetrate darkness but experience varying degrees of difficulty with obstacles, rain, fog, mist, smoke, dust, and haze.[120] To make good sense of the new surveillance equipment, one must only realize that any heat source will emit infrared light.[121]

Of little concern to the skilled infiltrator is his initial approach. He knows that thermal imaging will not do any of the following: (1) work in heavy fog or rain, (2) see through dense vegetation or anything else, (3) distinguish a human being from the desert floor right after sunset, (4) detect a swimmer for a few seconds after emerging from a pond, and (5) easily tell the difference between a person and a cow. He also knows that night vision goggles will be useless while aimed in the general direction of a distant light source. More disturbingly, these exceptions appear to be common knowledge to every paramilitary organization and dissident faction in the Eastern world.

> "At night, when these groups heard a Predator or AC-130 coming, they pulled a blanket over themselves to disappear from the night-vision screen," Maj.Gen. Franklin L. Hagenbeck, who led U.S. forces in Afghanistan, told the Army's *Field Artillery Magazine*. "They used low tech to beat high tech."[122]
>
> — *Washington Times,* 15 January 2003

The skilled infiltrator also knows the limitations of antipersonnel radars, electronic trip wires, pressure mats, listening devices, seismic detectors, motion sensors, and video memory. Most early warning devices fit into two categories—motion-triggered alarms and energy fields.

Some motion devices are attached to fences. While the "taut wire" is supposed to detect a cut fence, it can be fooled by temperature changes.[123] The vial of mercury that triggers the "balanced-level-capacitance device" can be jiggled by wind or wild animals.[124] Microwave antipersonnel radar is only effective at short range.[125]

Seismic intrusion devices can be set off by lightning strikes and vehicular vibrations. Video motion detectors are vulnerable to electrical storms and any sudden change in illumination.[126]

In the energy field category are beams of infrared and microwave light. Any interruption to the beam triggers an alarm. Of course, infrared beams are clearly visible through infrared-sensing goggles.[127] Just as water vapor absorbs infrared rays,[128] rain or fog will easily disrupt them. Volumetric electric fields are vulnerable to animals and electrical interruption.[129] Directional listening devices are at a loss during heavy wind or thunder.

Moreover, all of this can happen without the help of the infiltrator himself. All he has to do is wait. Of course, he has a few tricks of his own—like shutting off perimeter lighting by aiming an invisible beam at the photoelectric cell that detects dawn, or broadcasting static to confuse a sound monitor.[130] Just by releasing his pet cat, he could deceive the average U.S. security force. To stop a *ninja,* it takes a *ninja.*

True Preparedness Takes Greater Individual Skill

The U.S. Customs agents with the best record for apprehending drug traffickers in the American Southwest are not the ones with the best equipment, but the ones with the most skill. Detecting and stopping an adversary are not one in the same. Those agents were "desert-wise" Native Americans.

> Rather than relying solely on high-tech gadgetry—night vision goggles or motion sensors buried in the ground, members of this [21-man-strong] unit "cut for sign." Sign is physical evidence—footprints, a dangling thread, a broken twig, a discarded piece of clothing, or tire tracks. "Cutting" is searching or analyzing it once it's found. . . .
>
> . . . Between October 2001 and October 2002, the [Navajo] Shadow Wolves seized 108,000 pounds of illegal drugs, nearly half of all the drugs intercepted by U.S. Customs in Arizona.[131]
>
> — *Smithsonian,* January 2003

For far too long, U.S. infantry commanders have thought their short-range capabilities as good as their long (Figure 24.3). They

Figure 24.3: The Cost of Inferior Skill at Close Range
(Source: *FM 100-5* (1994), p. 39)

have taken—at face value—the claims of arms and equipment manufacturers. They have assumed their tactical doctrine to be a reflection of the ongoing, combined experience of frontline fighters. It isn't. Over the years, their doctrine has been changed only slightly to accommodate the latest American technology. One's equipment should not dictate what he does at short range, but the other way around. That American commanders can find little in the official battle chronicles to indicate any U.S. shortfall in individual or small-unit skill can be easily explained. Those chronicles have been based on random accounts and command chronologies. The former came mostly from high-ranking, standoff observers. The latter were hastily assembled in the heat of battle. Compiled by headquarters personnel, command chronologies tend to put whatever happened in the most favorable light. U.S. battle chronicles focus more on the big picture, than the small. They describe large-unit operations, not individual firefights. They record little of what occurred at 75 yards or less. They have failed to save for posterity the detail of U.S. small-unit technique, much less that of the enemy. A "big picture" will draw attention to what the largest concentration of enemy troops appears to be doing. With a highly deceptive enemy, that big picture can be extremely misleading. Without polling hundreds of frontline American soldiers, one would have little chance of discovering the details of an Eastern army's main attack. For it is by tiny "extraordinary" elements—sappers and stormtrooper squads—that his main attack is normally conducted.

Within Each U.S. Line Company Lies Its Own Salvation

This book has produced compelling evidence of a chronic short-range combat deficiency within the U.S. military, police, and security establishment. It has done so not to undermine their self-confidence, but to strengthen it. A problem of this magnitude and tenure may seem insurmountable, but it isn't. While it would help to attach two or three *ninja*-trained commandos to each U.S. rifle company as training cadre, it is not absolutely essential. In fact, this problem could be solved without any bureaucratic reform whatsoever. Because it has sprung from ignoring what happens at the lowest echelons of the U.S. military, it can be easily fixed there by the people who live there. Then, while organizational doctrine may remain the same, the tactical proficiency of each company will drastically improve. As noninfantry companies now guard most of America's strategic assets, they too should follow this training regimen.

A company commander need not know the answer to every question, only where to look for it. On the subject of short-range combat, those answers lie more in the realm of appreciating "alternate methods" than arriving at a "duty solution." For, once the enemy knows what that duty solution is, it will no longer work.

Within the average company are about 40 NCO billet holders. When allowed to work in the field on their own perceived deficiencies, they will collectively discover more about short-range combat than is contained in any manual. To instill initiative and tactical decision making at every echelon within the company without compounding its predictability, its commander has only to do one thing. He has only to allow his junior enlisted men to *collectively* design and conduct their own tactical training up to squad level. Those

200 or so nonrates must—through constant experimentation—arrive at situationally appropriate techniques that can be statistically proven to be the most rapid, stealthy, deceptive, or safe. Promising techniques can be initially identified by a show of hands. Then, those techniques must be continually tested against stop watch, human eye, or casualty simulation. Finally, the techniques must be compulsorily practiced by every member of the company. That the individual, fire team, and squad "plays" will slightly differ between companies makes little difference on the modern battlefield—where predictability and mass can be lethal. For the parent battalion, the increase in learning dynamics will more than compensate for any loss in standardization.

Daily, aboard any military base in America, one must see thousands of privates and hundreds of squads (from every type of unit) doing a wide-assortment of movement-oriented battledrills. They will be rehearsing their infantry "plays" in all types of terrain during the time normally allotted to physical fitness. Then, and only then, will the United States be ready to take on another Eastern opponent at close quarters.

By U.S. admission, the North Korean army prepares its privates to loosely follow orders, outwit enemy technology, and take on many times their number.[132] In contrast, the American military prepares its privates to strictly follow orders, master their own technology, and seek a 3 to 1 advantage.[133] One encourages its lowest ranks to think and act decisively, and the other doesn't. It's not difficult to see why America's foes routinely have the edge at short range. Their approach to war is more humanistic and tactically intuitive. Until the U.S. military follows suit, it will continue to lose too many of its privates to enemy contact. It will also have trouble winning any terrorist, guerrilla, or "4th-generation" war.

Appendix A
Casualty Comparisons

	Chinese	NKPA	Total Reds	U.S.	ROK	Total UN
Participants	**3.0** [1]	**NA**	**NA**	**2.0** [2] 1.32 [3]	**NA**	**NA**
Killed in Action	***0.15+** [4]	**NA**	**NA**	***0.04** [5]	**0.41+** [6] *0.26 [7]	**(0.45+)** *0.29+ [8]
Overall Losses	**<1.01>** [9] *0.36 [13]	**NA**	**NA**	**0.54+** [10] *0.14 [14]	**{0.90+}** [11] 0.84+ [15]	**[1.47+]** [12] {1.17} [16]

Note: All entries in millions (NA means not available)
Non-bold print — lowest legitimate estimate
Parentheses — sum or remainder of other entries
Asterisk — confirmed from more than one source
Hatch marks — Chinese military archival research
Curly brackets — South Korea Defense Dept. statistics
Brackets — Associated Press estimate of 23 October 1953

Table A:1: Participant's Highest Estimate of Own Losses in Korea

	Chinese	NKPA	Total Reds	U.S.	ROK	Total UN
Participants	**2.3** [17]	**NA**	**NA**	**NA**	**NA**	**NA**
Killed in Action	**{0.18+}** [18] 0.5+ [20]	**{0.29+}** [19]	**(0.48)**	**NA**	**NA**	**NA**
Overall Losses	**0.7** [21] *1.0+ [24]	**0.6** [22] (0.5+)	**(1.3)** *1.5+ [25]	**NA** [0.41] [26]	**0.72** [23] [1.1] [27]	**NA** [1.54] [28]

Note: All entries in millions (NA means not available)
Non-bold print — highest legitimate estimate
Parentheses — sum or remainder of other entries
Asterisk — confirmed from more than one source
Curly Brackets — ROK statistics
Brackets — Statistics from North Korean War Museum in Pyongyang

Table A.2: Participant's Lowest Estimate of Foe Losses in Korea

	NVA/VC	U.S.	ARVN	All Allies
Participants	***67,000** [29]	**NA**	**NA**	**NA**
Casualties	**18,560+** [30]	**[24,013]** [31]	**[20,977]** [32]	**(45,000)**

Note: Actual numbers shown (NA means not available)
Parentheses — sum of other entries
Brackets — Hanoi's interpretation of Saigon statistics
Asterisk — North Vietnamese interpretation of U.S. statistics

Table A.3: Participant's Highest Est. of Own Losses for 60 Days of Tet

	NVA/VC	U.S.	ARVN	All Allies
Participants	**80,000** [33]	**492,000** [34]	**626,000** [35]	**1,179,000** [36]
Casualties	**40,000** [37]			**101,400+** [38]
		{43,000}[39]	{104,000}[40]	{147,000}[41]

Note: Actual numbers shown (NA means not available)
Non-bold print — highest legitimate estimate
Curly brackets — North Vietnamese Military History Institute statistics

Table A.4: Participant's Lowest Est. of Foe Losses for 60 Days of Tet

Appendix B
Enemy Entry-Level Training

German

German training programs throughout WWI strongly emphasized individual training.[1] Recruit training behind the front stressed individual training for the new German Hutier (infiltration) and Stormtroop (assault squad tactics).[2] "The Germans recognized the considerable training effort that their tactical changes required."[3] Their teaching cadre came from the storm battalion within each German field army.[4] The WWII-era German infantryman was highly trained as well.

> A considerable part of the military training in Germany is . . . pre-Army training by other military and auxiliary organizations.[5]
> — U.S. War Dept., *TM-E 30-451*

> The basic training . . . in infantry training units normally is planned for 16 weeks. . . . This period may be followed by an indefinite period of advanced training . . . , lasting up to the time of transfer of the recruits to a field unit. The basic training usually is divided into three parts, the first of which is devoted to individual training, the second to the training of the individual recruit within the framework of the squad, and the third to the training of the squad within the framework of the platoon. . . . The basic training components, listed in order of the importance attributed to them, are: combat training, firing, lectures, drilling, sports.[6]
> — U.S. War Dept., *TM-E 30-451*

The WWII Germans also saw a need for experienced lower ranks. They had—in order of increasing rank—the following: senior private, private first class (ordinary), senior private first class, and chief private first class.[7]

Japanese

In Manchuria, the Japanese trained their enlistees extensively during their 1939 conflict with the Russians:

> [T]he *new conscripts* of the 7th [Japanese] Division underwent their advanced infantry training on the dusty plains near Tsitsihar [in Manchuria]. . . . Most of the riflemen were just completing their first year of company training, a year that contained *thirty-eight weeks of night combat instruction*. . . . The men studied night attacks in various echelons, obstacle clearing, concealment, noise prevention, orientation at night, patrolling, and security [italics added].[8]
>
> — U.S. Army, *Leavenworth Papers No. 2*

Yet, this in-theater training was not their first. From the age of eight, Japanese boys had received preconscription training from Regular Army officers throughout the WWII era. Then their conscript training lasted a full two years.

> In peacetime this amounts to 2 or more hours per week with 4-6 days of actual maneuvers, but recently [in wartime] the amount of time . . . has been greatly increased.[9]
>
> — U.S. War Dept., *TM-E 30-480*

> In peacetime the training of [Japanese] men assigned to active service . . . . covers a period of two years. . . .

January to May.	Recruit training. This includes general instruction, squad (section) training, bayonet training, and target practice. In February a march of 5 days, with bivouacking at night, is held to train men in endurance of cold.
June and July.	Target practice, field works, platoon and company training, and bayonet training. Marching 20 miles a day.
August.	Company and battalion training, field work, combat firing, swimming, and bayonet fighting. Marching 25 miles per day.
October and November.	Battalion and regimental training. Combat firing. Autumn maneuvers. . . .

> Conscripts may often receive the bulk of their training in operational areas. The Japanese are known to have used the Chinese Theater for training purposes, where men perform garrison du-

> ties and sometimes get actual combat experience during their period of training.[10]
>
> — U.S. War Dept., *TM-E 30-480*

To further season those who must do the fighting, the Japanese slowed their accumulation of rank. There were four different categories of the lowest grade: second-class private, first-class private, superior private, and leading private.[11]

Russian

In 1982, Soviet entry-level training lasted 6 months with the first month dedicated to individual skills, the second to squad skills, and the third to platoon skills.[12] Few receive training in map reading.[13] Soviet entry-level training is very strict and tries to instill discipline.[14]

Chinese

While no data on Chinese entry level training is available, the ability of their individual soldier in Korea speaks for itself.

> Stealth and infiltration were not the only skills of Chinese light infantrymen. In the absence of significant combat support, CCF tactical success required that Chinese soldiers possess many highly developed skills and attributes.[15]
>
> — U.S. Army, *Leavenworth Research Survey No. 6*

North Korean

Each North Korean conscript receives entry-level instruction that is far more comprehensive than anything in the West. He graduates with a full portfolio of conventional- and nonconventional-warfare skills.

> The KPA training system is designed to produce tough, disciplined, and politically well indoctrinated soldiers, who, by dint of their superior ideological training, physical training, and superior skills in guerrilla warfare, can defeat a *numerically* and technologically superior enemy [italics added].[16]
>
> — U.S. Army, *FC 100-2-99*

Mountain and night skills are taught, and the soldiers are trained in both conventional and unconventional warfare.[17]

North Vietnamese

At boot camp, North Vietnamese soldiers were imbued with more field skills than were their U.S. counterparts. While U.S. recruits were getting accustomed to functioning in a group, the North Vietnamese recruits were learning how to survive on their own. Unlike American enlistees, the North Vietnamese were shown how to generate surprise.

> The typical North Vietnamese regular fighting in the South was twenty-three years old—four years older than his U.S. counterpart. . . . [H]e had already logged three years of compulsory service, undergoing military training and also instructing local militia.[18]

> An NVA PFC who rallied in South Vietnam in May 1967 stated that during his two months of basic training in the North, he learned about infantry combat tactics, "that is to say, squad assault, platoon air-defense activities; we also learned how to roll, to creep, to crawl, to shoot, to use the bayonet, to throw grenades *(RAND Vietnam Interviews,* Series SX, 1972, 2:1)." A corporal squad leader captured in June 1967 described his training . . . "walking in step with others, saluting, shooting at targets, crawling, rolling, attacking, defending, digging foxholes, shooting with heavy machineguns *(RAND Vietnam Interviews,* Series SX, 1972, 5:1)."[19]

> Two months of basic training was pretty standard [for North Vietnamese soldiers], followed by one or more months of specialist training (even for infantrymen)
>
> Once sent to their units, training continued. Most of the time, Viet Cong and North Vietnamese units were not involved in combat operations. During these more relaxed times, the Communist units followed a strict schedule. Up at 0500 (5:00 A.M.), two hours of preparing and eating breakfast and cleaning up the camp. Then came several hours of political lectures, training, or working on fortifications. Then came the traditional midday siesta, during the hottest part of the day. Then another five hours of work. Then another two to three hours for the evening meal and rest. Often there followed several hours of night training.[20]

After boot camp, North Vietnamese infantrymen received specialized training before heading south.[21] As suggested by the first excerpt above, many may have had to instruct militia before joining a frontline outfit. This would have given them valuable experience at exercising initiative and making tactical decisions. To improve their courses of instruction, they would have had to field-test, and then choose between, technique options. The entry-level training for VC differed between locations. Most received training on bayonet fighting and grenade throwing [22]—the two means by which WWI German stormtroopers assaulted without compromising surprise.

Appendix C
Advised U.S. Battledrills

Individual Battledrills

Within the traditional categories of practical knowledge ("shoot," "move," and "communicate"), the procedures that *do not* telegraph intentions are the most important to modern war. To them must be added techniques associated with the new "basics" in Part Two of this book. Without regularly practicing this wide assortment of conventional and unconventional warfare skills, American troops will lack the "muscle memory" to fend for themselves in combat. Below are listed some field-tested examples.

a. "Discreet-force-at-close-range" exercises:

(1) Bayonet fighting with the blades covered by scabbards

(2) Grenade toss

(a) Place a row of wastebaskets 30 yards from, and parallel to, a row of students with low cover
(b) Have the students toss practice grenades from the prone position for accuracy points
(c) Every "clang" registers a hit
(d) Keep score

b. "Nondetectable-movement" exercises:

(1) Crawl race

(a) Establish parallel start and finish lines
(b) With students on line, hold crawl race
(c) So as not to harass, instructors must participate

(2) Silent movement

(a) Send columns of students off in different directions

(b) Each group moves 100 meters and then back along the same route
(c) Each man counts the noises he makes on each repetition

(3) Crossing of rough ground

(a) Students charge in column through all sorts of rough ground, uphill and downhill, day and night
(b) Individuals count the times they stumble

(4) Urban fun run

(a) Warn troops that walking a street is too risky in war but a good way to learn how to avoid the first bullet
(b) List for them the most probable ways of getting shot in the city

1 By bullets deflected along walls
2 From machineguns at the ends of the street
3 From upper stories to the front or rear
4 From spaces between buildings
5 From windows and open doors
6 From tiny wall apertures (like inlet vents)
7 Through closed doors or thin walls

(c) Place aggressors next to the walls on both sides of the street near its end
(d) Have aggressors assess their kills by three-second sight pictures of fully exposed human beings
(e) Send troops, two at a time, up opposite sides of the street working on the first threat
(f) Have aggressors record and apprise their respective runners of how many times they got killed
(g) Repeat the drill and ask students to beat their own previous scores
(h) Have aggressors move to provide the second threat
(i) Send troops, two at a time, up opposite sides of the street to work on the second threat
(j) Repeat the drill and ask students to beat their previous scores
(k) Follow the same procedure for each of the remaining threats
(l) Then have the troops work on two threats at a time, then three, then four, then five, then six, then seven

c. "Guarded communication" exercises:

(1) Muted hand-and-arm signal conversation

(a) Students get an assortment of written messages
(b) They form into two rows and choose partners
(c) One row moves 100 meters away from the other
(d) Sets of partners take turns sending and recording messages with a minimum of ostensible movement

(2) Tug-rope talk

(a) Students pair off
(b) Each pair has a string and a list of ambush messages to convert to tug code

1 Enemy on the trail
2 Enemy on the inside track (next to the treeline)
3 Enemy coming from the front
4 Enemy coming from the back
5 Emplace claymores
6 Collect claymores
7 Pull out to the left
8 Pull out to the right
9 Pull out to the back
10 Wait, not ready

(c) One partner describes a written scenario in tug talk
(d) The other records the scenario in writing
(e) Compare notes and then switch roles and scenarios

d. "Combat deception" exercises:

(1) Students separately practice pretending to be shot and then crawling behind cover during an individual rush

(2) Distracting a sentry

(a) Students pair up
(b) Each pair goes to a different place in the woods
(c) One partner will be victim, and the other stalker
(d) With sights and sounds, the stalker tries to confuse the victim as to his approach route
(e) Then the roles are reversed

e. "Seeing" exercises:

(1) Students separately practice defocusing drills to enhance peripheral vision[1]

(2) Gurkha trail

(a) Silhouette targets placed in the brush along trail
(b) Extended column move up trail
(c) Each student quietly counts the targets he sees

f. "Not-being-seen" exercises:

(1) Students pair up and take turns hiding within plain sight of each other[2]

(2) Stalk walk

(a) Students form an extended column of two's
(b) Double column moves into the woods and stops
(c) Pairs turn back to back and face outboard
(d) One row moves 100 yards out and begins to stalk
(e) Other row turns to see how close stalkers can get to their partners without being detected
(f) Reverse the roles

g. "Hearing" exercises:

(1) Ambush trail

(a) Instructors are positioned along a trail to make pre-ambush sounds

1 M-16 coming off safe
2 Twig snapping
3 Handguard bumping on ground
4 Whispering

(b) Students move up the trail in extended column
(c) Students quietly record what they hear

(2) Direction and distance

(a) Students form a tight circle and face outboard

(b) Assistant instructors make various sounds at different directions and distances

1 Noises associated with digging in
2 Talking and whispering
3 Coughs and sneezes
4 Sounds of jangling gear

(c) Students write down the direction and distance of each noise
(d) Instructor announces actual directions and distances

h. "Not-being-heard" exercises:

(1) Students separately practice crawling quietly through dry leaves or swamp

(2) Sneak

(a) Students pair off in heavy brush
(b) They separate and then see how close they can approach each other without being heard
(c) Students repeat the exercise for a better score

i. "Microterrain appreciation" exercises:

(1) Students split up and separately crawl in a 50-foot circle through the lowest ground available

(2) Take cover

(a) Instructor tells what constitutes cover
(b) Students form an extended column
(c) Column moves through rural terrain
(d) Instructor shouts direction of enemy
(e) How long it takes the slowest student to take cover is the time to beat
(f) Repeat the exercise

j. "Night familiarity" exercises:

(1) Troops practice various night-seeing techniques on their own

(a) Creating natural horizons from the prone in low ground
(b) Locating artificial horizons

(c) Cupped-hand technique
(d) Indirect-vision technique

(2) Point to him[3]

(a) Students pair up and go to separate locations
(b) Students take turns being victim and stalker
(c) Victim shuts his eyes and memorizes all sensations
(d) Victim imagines eight cone-shaped areas of awareness
(e) Stalker approaches from the direction of his choice
(f) Stalker tries to get within arms' reach of the victim undetected
(g) Victim points to stalker as soon as he can
(h) Students reverse roles

k. "One-on-one decision-making" exercises:

(1) Discerning civilian from combatant

(a) Instructor cuts from magazines several sets of pictures of civilians and obvious guerrillas
(b) Students must instantly identify each

(2) React trail

(a) Students sent up a trail one at a time
(b) Instructors pop up to test reactions
(c) Instructors do not dictate proper behavior but rather learn from the most common response

Small-Unit Battledrills

As in football, each buddy team, fire team, and squad must develop *several* "plays" with which to deal with each expected category of combat situation. If they do not continually rehearse those plays, they will not have the teamwork to use them in combat.

a. "Buddy team" drills:

(1) Point

(a) Have students pair off
(b) Tell each pair to walk in a staggered column
(c) First man looks short while second looks long

(d) Send each pair up a Gurkha trail

(2) Fire and move

(a) Divide unit into three equal-sized groups
(b) Make lanes with string (one for every six men)
(c) Put two umpires at each end of the course
(d) Put two groups at one end, one at other
(e) Groups alternately attack, defend, and watch
(f) Attackers must jump up, run, and jump down within three seconds, and then crawl to a new spot
(g) Those in the first group move randomly up lanes guiding on simulated squad leader in the center
(h) Instructor walks with attackers to promote alignment, disqualify crawling, and stop them short of the end
(i) Each defender counts a hit every time he acquires a three-second sight picture of someone in his lane
(j) The defenders of each lane debrief their respective attackers as to the number of hits
(k) Then the group that was defending attacks and the group that was watching from the other end defends
(l) The entire evolution is repeated until all three groups have attacked twice
(m) Each man tries to get killed fewer times during his next try

b. "Fire team" drills:

(1) Counterambush or hasty flanking attack

(a) Students file past a simulated ambush site
(b) The instructor specifies the kill zone
(c) Those in the kill zone take cover and return fire
(d) Those outside kill zone do the following:

1 Crawl back down the trail
2 Crawl into the brush occupied by the foe
3 Rear man automatically becomes the leader
4 He stays next to trail within full view of base of fire
5 Rest guide on leader and move forward on line

(e) The maneuver element first "covers and stalks" and then "fires and moves" against the enemy
(f) The students who were in the kill zone cease fire when the maneuver element leader comes abreast of them

(2) Speed assault

(a) Use a hilltop with two slightly separated roles of concertina wire as the objective
(b) Find a defiladed attack position at the hill bottom
(c) Put twelve men on line at final coordination line (FCL)
(d) Have them run in column through the hole in the wire and then realign without stopping

1 One fire team goes left and then obliques to right
2 One fire team goes right and then obliques to left
3 The last fire team spreads out in the middle

(e) The squad should cross the finish line in a lazy "V" without ever stopping to align
(f) At no time should anyone mask another's fire
(g) Grade the event on time and alignment
(h) Every squad runs it twice

c. "Squad" drills:

(1) Ambush

(a) Instructor has two tug ropes and acts as squad leader
(b) As squad enters an ambush site from the rear, he attaches one end of each rope to a twig and dispatches flank security teams with other ends
(c) All personnel advance along the tug rope and lie on it alternately facing forward and backward
(d) Flank security teams and assigned personnel put out simulated claymores with backblast protection
(e) Instructor signals with tug rope for claymore retrieval and exit to the right or left instead of rear
(f) Squad executes the command

(3) Daylight attack

(a) Use same training site as for speed assault
(b) Put squad on line at FCL
(c) Call for simulated artillery beyond danger close range
(d) Walk simulated smoke rounds in to mask the wire
(e) Emplace bangalore
(f) Call for repeat smoke
(g) When simulated artillery round heard coming in, do the following:

1 Blast machinegun bunker with SMAW (bunker buster)
2 Blow bangalore to breach wire

(h) Shift artillery and ask for precision, continuous fire
(i) Run speed assault with these extra steps:

1 Mask breach with smoke grenades in same place
2 Toss concussion grenades
3 Cut claymore wires

Free-Play Segment

All individual and small-unit skills must be honed in simulated combat. As tactical excellence is inversely proportional to casualties suffered, free play offers the best medium for practical application.

a. Opposing sides are assigned defensive positions 100-300 yards apart to man with one-third their strength (no outposts)

b. Sides are required to attack (with two-thirds their strength) each other's position twice

(1) Termination time is established and flour grenades issued

(2) One side turns its shirts inside out

(3) Umpires and sides dispatched to the defensive positions

(4) Sides place flags on ground inside defensive positions

(5) Weaponry is flour grenades and rubber rifles

(6) Players record three-second sight pictures of upright foes and grenades landing within 10 feet of themselves

(7) Casualties must return to defensive position to reenter play but can't disclose enemy locations when they do

(8) No vehicles and impersonating other side's players

(9) *Secretly* capturing flag nets 30 point bonus

(10) Bodily contact, skipping required attacks, or divulging information earn 10 demerits for each occurrence

(11) Umpires separate sides to add kills and bonuses and subtract times killed and demerits for each

(12) The side with the highest score wins

(13) Winner is gets time off, cleanup reprieve, or food prize

Notes

SOURCE NOTES

Illustrations

Pictures on pages 12, 14, 37, 75, 321, and 329 reproduced after written assurance from Cassell PLC, London, that the copyright holders for *UNIFORMS OF THE ELITE FORCES,* text by Leroy Thompson, color plates by Michael Chappell, could no longer be contacted. The illustrations are from Plate 15 (No. 43, Soviet Airborne Private), Plate 28 (No. 82, CRCC Paratrooper), Plate 15 (No. 45, Soviet Airborne Sniper), Plate 17 (No. 49, Soviet Naval Infantry Commando), Plate 1 (No. 2, Special Forces Sergeant), Plate 26 (No. 76, Iraqi Special Forces Trooper), respectively. Copyright © 1982 by Blandford Press Ltd. All rights reserved.

Pictures on pages 19, 34, 47, 57, 93, 111, 131, 143, 185, 237, 279, 299, and 330 reproduced after written assurance from Cassell PLC, London, that the copyright holder for *WORLD ARMY UNIFORMS SINCE 1939,* text by Andrew Mollo and Digby Smith, color plates by Malcolm McGregor and Michael Chappell, could no longer be contacted. The illustrations are from Part I (Plate 103), Part I (Plate 155), Part I (Plate 156), Part II (Plate 79), Part II (Plate 9), Part II (Plate 2), Part II (Plate 4), Part I (Plate 201), Part II (Plate 1), Part II (Plate 104), Part I (Plate 193), Part II (Plate 85), and Part II (Plate 115), respectively. Copyright © 1975, 1980, 1981, and 1983 by Blandford Press Ltd. All rights reserved.

Pictures on pages 23, 54, 91, 180, 216, 239, 242, and 282 reproduced with permission of Dr. Anatol Taras, Minsk, Belarus, from *PODGOTOVKA RAZVEGCHIKA: SISTEMA SPETSNAZA GRU,* by A.E. Taras and F.D. Zaruz. The illustrations are from pp. 330, 144/145/152, 150, 374/375, 173, 142, 278/279, and 147, respectively. Copyright © 1998 by A.E. Taras and F.D. Zaruz. All rights reserved.

Picture on page 35 reproduced with permission of Allen and Unwin Pty. Ltd., St. Leonards, NSW Australia, from *CONSCRIPTS AND REGULARS,* by Michael O'Brien. The illustration is from Table 11.2, "C2 D445 Training Program, 13-27 August 1970," on page 214. Copyright © 1995 by The Seventh Battalion, The Royal Australian Regiment Association, Inc. All rights reserved.

Pictures on pages 69, 101, 193, and 203 reproduced with permission of Osprey Publishing Ltd., London, from *ARMIES OF THE VIETNAM WAR 1962-75,* Men-at-Arms Series, Issue 104, text by Philip Katcher and Lee E. Russell, color plates by Michael Chappell. The illustrations are from Plate C (Nos. 1 & 2, "Viet Cong," 1960's-70's), Plate H (No. 1, Enlisted Man, North Vietnamese Army, 1975), and Plate C (No. 3, "Khmer Rouge," 1960's-70's), respectively. Copyright © 1980 by Osprey Publishing Ltd. All rights reserved.

Pictures on pages 80, 81, and 82 reproduced with permission of The Gioi Publishers, Hanoi, from *THE HO CHI MINH TRAIL,* by Hoang Khoi. The illustrations are from page 26 (Truong Son Network of Strategic Routes 1965-1968), page 70 (Truong Son Network of Strategic Routes 1969-1973), and page 96 (Truong Son Network of Strategic Routes 1973-1975), respectively. Copyright © 2001 by Hoang Khoi and The Gioi. All rights reserved.

Picture on page 149 reproduced with permission of Presidio Press, Novato, CA, from *SOVIET AIRLAND BATTLE TACTICS,* by William P. Baxter, page 137. Copyright © 1986 by Presidio Press. All rights reserved.

Pictures on pages 164, 165, and 166 reproduced after asking the permission of Polygon Publishers, St. Petersburg, Russia, from *VOINA V KOREE: 1950-1953,* by A.A. Kuryacheba, Voenno-Estorecheskaya Biblioteka. The illustrations are from pp. 601, 613, 582, 590, 585, 590, and 614, respectively. Copyright © 2002 by Polygon. All rights reserved.

Picture on page 169 reproduced with permission of The Wylie Agency, Inc., from *BROTHER ENEMY: THE WAR AFTER THE WAR,* by Nayan Chanda. The illustration is from "Chinese Invasion of Vietnam," page 337. Copyright © 1986 by Nyan Chanda. All rights reserved.

Pictures on pages 177, 221, 229, and 267 reproduced with permission of Osprey Publishing Ltd., London, from *THE KOREAN WAR 1950-53,* Men-at-Arms Series, Issue 174, text by Nigel Thomas and Peter Abbott, color plates by Michael Chappell. The illustrations are from Plate H (No. 1, Chinese Infantryman Winter Field Dress 1950), Plate A (No. 1, North Korean Sergeant Summer Field Dress 1950), Plate H (No. 2, Chinese Infantryman Summer Field Dress 1951), and Plate A (No. 2, North Korean Private Winter Field Dress 1950), respectively. Copyright © 1986 by Osprey Publishing Ltd. All rights reserved.

Pictures on page 264, 273, and 274 reproduced with permission of The Gioi Publishers, Hanoi, from *THE TET MAU THAN 1968 EVENT IN SOUTH VIETNAM* by Ho Khang. The illustrations are from page 74 (The Road No. 9—Khe Sanh Campaign in 1968), page 82 (Tet Mau Than General Offensive and Uprising in Tri Thien—Hue 1968), and page 88 (First Stage of Tet Mau Than General Offensive and Uprising in Saigon from 31 Jan. to 28 Feb. 1968), respectively. Copyright © 2001 by The Gioi Publishers. All rights reserved.

Pictures on page 212, 213, and 214 reproduced with permission of Paladin Press, Boulder, CO, from *TACTICAL TRACKING OPERATIONS,* by David Scott-Donelan. The illustrations are from page 42 (The Effects of Age and Weather on Spoor) and page 57 (Counting the Tracks—The Average Pace Method). Copyright © 1998 by David Scott-Donelan. All rights reserved.

Picture on page 259 reproduced with permission of The Gioi Publishers, Hanoi, from *DIEN BIEN PHU,* by Gen. Vo Nguyen Giap. The illustration is from the insert "Development of the Battle of Dien Bien Phu." Copyright © 1999 by The Gioi Publishers. All rights reserved.

Picture on page 309 reproduced with permission of Osprey Publishing Ltd., London, from *U.S. ARMY RANGERS & LRRP Units 1942-1987,* Osprey Elite Series, text by Gordon L. Rottman, color plates by Ron Volstad. The illustration is from Plate F (No. 2, PFC, Ranger Co., 75th Infantry, Vietnam, 1970). Copyright © 1987 by Osprey Publishing Ltd. All rights reserved.

Text

Reprinted with permission of the Associated Press, from "Military Visits Urban Tactics against Baghdad," by John J. Lumpkin; "War with Iraq Has Wild Cards," by Tom Raum; and "Urban Battle in Nasiriyah Hurts Troops from Lejeune," as published in the *JACKSONVILLE (NC) DAILY NEWS* on 23 September 2002, 19 March 2003, and 27 March 2003, respectively. Copyrights © 2002 and 2003 by Associated Press. All rights reserved.

Reprinted with permission of Tuttle Publishing, North Clarendon, VT, from *THE NINJA AND THEIR SECRET FIGHTING ART,* by Stephen K. Hayes. Copyright © 1981 by Charles E. Tuttle, Co., Inc. All rights reserved.

Reprinted with permission of Alfred A. Knopf, Inc., from *CROSSING THE THRESHOLD OF HOPE,* by His Holiness John Paul II. Original copyright © 1994 by Arnoldo Mondadori Editore. Translation copyright © 1994 by Alfred A. Knopf, Inc. All rights reserved.

Reprinted with permission of Osprey Publications Ltd., from *JAPANESE ARMY OF WORLD WAR II,* by Philip Warner, vol. 20, Men-at-Arms Series. Copyright © 1972 by Osprey Publications Ltd. All rights reserved.

Reprinted with permission of Kensington Publishing Corp., from *SECRETS OF THE NINJA,* by Ashida Kim. Copyright © 1981 by Ashida Kim. All rights reserved.

Reprinted with permission of the Copyright Clearance Center, from "Street Smarts: As Threats Evolve, Marines Should Learn Skill of Combat in Cities," by Greg Jaffe, as published in the *WALL STREET JOURNAL* on 22 August 2002. Copyright © 2002 by Dow Jones. & Co., Inc. All rights reserved.

Reprinted with permission of the Copyright Clearance Center, from "Depot Stumped over Intruder: Army Doesn't Know How He Got Away," by Brent Israelsen, as published in *SALT LAKE TRIBUNE ON LINE* on 7 September 2002. Copyright © 2002 by Salt Lake Tribune. All rights reserved.

Reprinted with permission of *THE WASHINGTON TIMES,* from "Al-Qaeda Adapts to Pursuit Tactics," by Rowan Scarborough, as published in the 15 January 2003 issue. Copyright © 2003 by the Washington Times. All rights reserved.

Reprinted with permission of Greenwood Publishing Group, Inc., from *THE BATTLE OF AP BAC, VIETNAM,* by David M. Toczek. Copyright © 2001 by David M. Toczek. All rights reserved.

Reprinted with permission of Paladin Press, from *KNIGHTS OF DARKNESS: SECRETS OF THE WORLD'S DEADLIEST NIGHT FIGHTERS,* by Dr. Haha Lung. Copyright © 1998 by Dr. Haha Lung. All rights reserved.

Reprinted with permission of the Greenwood Publishing Group, Inc., Westport, CT, from *STORMTROOP TACTICS—INNOVATION IN THE GERMAN ARMY 1914-1918,* by Bruce I. Gudmundsson. Copyright © 1989 by Bruce I. Gudmundsson. All rights reserved.

Reprinted with permission of the *MARINE CORPS GAZETTE,* from the following articles: "'Urban Warrior'—A View from North Vietnam, by Lt.Col. Robert W. Lamont, April 1999; "Seventeen Days at Iwo Jima," by Merritt E. Benson, February 2000; "Russian Lessons Learned from the Battles for Grozny," by Lt.Col. Timothy L. Thomas and Lester W. Grau, April 2000; "Russia's 1994-96 Campaign for Chechnya: A Failure in Shaping the Battlespace," by Maj. Norman L. Cooling, October 2001; and "Night Patrol on Iwo Jima," by Col. David E. Severance, March 2003. Copyrights © 1999, 2000, 2001, and 2003 by the Marine Corps Association. All rights reserved.

Reprinted with permission of United Media, from "Among Troops' Problems in Afghanistan—Their Own Stuff," by Lisa Hoffman, as published in the *JACKSONVILLE (NC) DAILY NEWS* on 6 October 2002. Copyright © 2002 by Scripps Howard News Service. All rights reserved.

Reprinted with permission of Boston Publishing Company, Inc., Newton, MA, from *VIETNAM EXPERIENCE: A CONTAGION OF WAR,* by Terrence Maitland and Peter McInerney. Copyright © 1983 by Boston Publishing Company. All rights reserved.

Reprinted with permission of Frank Cass and Company, Ltd., c/o International Specialized Book Services, Inc., Portland, OR, from *THE SOVIET CONDUCT OF TACTICAL MANEUVER: SPEARHEAD OF THE OFFENSIVE,* by David M. Glantz. Copyright © 1991 by David M. Glantz. All rights reserved.

Reprinted with permission of Random House, Inc., from *PAVN: PEOPLE'S ARMY OF VIETNAM,* by Douglas Pike. Copyright © 1986 by Presidio Press. All rights reserved.

Reprinted with permission of Random House, Inc., from *COMMON SENSE TRAINING: A WORKING PHILOSOPHY FOR LEADERS,* by Lt.Gen. Arthur S. Collins, Jr. Copyright © 1978 by Presidio Press. All rights reserved.

Reprinted with permission of Random House, Inc., from *CHOSIN: HEROIC ORDEAL OF THE KOREAN WAR,* by Eric M. Hammel. Copyright © 1981 by Eric M. Hammel. All rights reserved.

Reprinted with permission of Random House, Inc., from *INSIDE THE VC AND THE NVA: THE REAL STORY OF NORTH VIETNAM'S ARMED FORCES,* by Michael Lee Lanning and Dan Cragg. Copyright © 1992 by Michael Lee Lanning and Dan Cragg. All rights reserved.

Reprinted with permission of Random House, Inc., from *OPERATION BUFFALO: USMC FIGHT FOR THE DMZ,* by Keith William Nolan. Copyright © 1991 by Keith William Nolan. All rights reserved.

Reprinted with permission of Random House, Inc., from *PORTRAIT OF THE ENEMY,* by David Chanoff and Doan Van Toai. Copyright © 1986 by David Chanoff and Doan Van Toai. All rights reserved.

Reprinted with permission of Random House, Inc., from *IWO JIMA: LEGACY OF VALOR,* by Bill D. Ross. Copyrights © 1885 and 1986 by Bill D. Ross. All rights reserved.

Reprinted with permission of Doubleday, a division of Random House, Inc., from *A POPULAR HISTORY OF THE MIDDLE AGES,* by Joseph Dahmus. Copyright © 1968 by Joseph Dahmus. All rights reserved.

Reprinted without objection from Random House, Inc., or DeAgostini, London, from *THE KOREAN WAR: HISTORY AND TACTICS,* edited by David Rees. Copyright © 1984 by Orbis Publishing, London. All rights reserved.

Reprinted without objection from Dell Publishing, a division of Random House, Inc., or F. Allin Elford, Firenze, Italy, from *DEVIL'S GUARD,* by George Robert Elford. Copyright © 1971 by George Robert Elford. All rights reserved.

Reprinted with permission of The Random House Group Ltd., London, from *THE FALL OF BERLIN,* by Anthony Read and David Fischer, and published by Hutchinson. Copyright © 1992 by Anthony Read and David Fischer. All rights reserved.

Reprinted with permission of Stanford Univ. Press, Palo Alto, CA, from *NOMONHAN: JAPAN AGAINST RUSSIA, 1939,* by Alvin D. Coox. Copyright © 1985 by the Board of Trustees of the Leland Stanford Junior University. All rights reserved.

Reprinted with permission of *THE CHRISTIAN SCIENCE MONITOR,* from the following articles: "In Era of High-Tech Warfare, 'Friendly Fire' Risk Grows," by Brad Knickerbocker, 14 January 2003; "GIs in South Korea Unfazed by Crisis," by Robert Marquand, 11 March 2003; and "How the U.S. Plans to Take Control of Baghdad," by Seth Stern, 7 April 2003. Copyright © 2003 by The Christian Science Monitor. All rights reserved.

Reprinted with permission of Brassey's, Inc., from *THIS KIND OF WAR: THE CLASSIC KOREAN WAR HISTORY,* by T.R. Fehrenbach. Copyright © 1963 by T.R. Fehrenbach. All rights reserved.

Reprinted with permission of The Wylie Agency, Inc., from *BROTHER ENEMY: THE WAR AFTER THE WAR,* by Nayan Chanda. Copyright © 1986 by Nyan Chanda. All rights reserved.

Reprinted with permission of the University Press of Kansas, Lawrence, KS, from *MAO'S GENERALS REMEMBER KOREA,* edited and translated by Xiaobing Li, Allan R. Millett, and Bin Yu. Copyright © 2001 by the University Press of Kansas. All rights reserved.

Reprinted with permission of Presidio Press, Novato, CA, from *SOVIET AIRLAND BATTLE TACTICS,* by William P. Baxter. Copyright © 1986 by Presidio Press. All rights reserved.

Reprinted with permission of Scribner, an imprint of Simon & Schuster Adult Publishing Group, from *INSIDE THE SOVIET UNION,* by Viktor Suvorov. Copyright © 1982 by Viktor Suvorov. All rights reserved.

Reprinted with permission of Simon & Schuster Adult Publishing Group, from *THE KOREAN WAR,* by Max Hastings. Copyright © 1987 by Romadata Ltd.. All rights reserved.

Reprinted with permission of Brassey's, Inc., from *MARINE RIFLEMAN: FORTY-THREE YEARS IN THE CORPS,* by Col. Wesley L. Fox. Copyright © 2002 by Wesley L. Fox. All rights reserved.

Reprinted with permission of Naval Institute Press, from *NORTH KOREAN SPECIAL FORCES,* by Joseph S. Bermudez, Jr.. Copyright © 1998 by Joseph S. Bermudez, Jr. All rights reserved.

Reprinted with permission of The Gale Group, from *ENCYCLOPEDIA OF THE VIETNAM WAR,* edited by Stanley I. Kutler. Copyright © 1996 by Charles Scribner's Sons. All rights reserved.

Reprinted with permission of The Gioi Publishers, Hanoi, from *DIEN BIEN PHU,* by Gen. Vo Nguyen Giap. Copyright © 1999 by The Gioi Publishers. All rights reserved.

Reprinted with permission of The Gioi Publishers, Hanoi, from *THE TET MAU THAN 1968 EVENT IN SOUTH VIETNAM,* by Ho Khang. Copyright © 2001 by The Gioi Publishers. All rights reserved.

Reprinted with permission of The Gioi Publishers, Hanoi, from *HOW SOUTH VIETNAM WAS LIBERATED,* by Gen. Hoang Van Thai. Copyright © 1996 by The Gioi Publishers. All rights reserved.

Reprinted with permission of The Gioi Publishers, Hanoi, from *THE 30-YEAR WAR: 1945-1975,* volume II. Copyright © 2001 by The Gioi Publishers. All rights reserved.

Reprinted with permission of the publishers of *THE PHILADELPHIA INQUIRER ON LINE,* 16 November 1997, from "Blackhawk Down," by Mark Bowden. Copyright © 1997 by The Philadelphia Inquirer. All rights reserved.

Reprinted with permission of Harcourt, Inc., from *THE KOREAN WAR: PUSAN TO CHOSIN, AN ORAL HISTORY,* by Donald Knox. Copyright © 1985 by Donald Knox. All rights reserved.

Reprinted with permission of International Publishers Co., Inc., from *MAO TSE-TUNG: AN ANTHOLOGY OF HIS WRITINGS,* edited by Anne Fremantle. Copyrights © 1954 and 1956 by International Publishers Co., Inc. Copyright © 1962 by Anne Fremantle. All rights reserved.

Reprinted with permission of Allen and Unwin Pty. Ltd., St. Leonards, NSW Australia, from *CONSCRIPTS AND REGULARS,* by Michael O'Brien. Copyright © 1995 by The Seventh Battalion, The Royal Australian Regiment Association, Inc. All rights reserved.

Reprinted with permission of Monthly Review Foundation, from *THE MILITARY ART OF PEOPLE'S WAR,* by Gen. Vo Nguyen Giap. Copyright © 1970 by Monthly Review Press. All rights reserved.

Reprinted with permission of Westview Press, Boulder, CO, from *THE MANEUVER WARFARE HANDBOOK,* by William S. Lind. Copyright © 1985 by Westview Press, Inc. All rights reserved.

Reprinted with permission of W.W. Norton & Co., from *THE TRAVELS OF MARCO POLO,* revised from Marsden's translation, edited and with introduction by Manuel Komroff. Copyright © 1926 by Boni & Liveright, Inc., © 1930 by Horace Liveright, Inc., and © 1953 renewed by Manuel Komroff. All rights reserved.

Reprinted with permission of Texas A&M Univ. Press, College Station, TX, from *SAPPERS IN THE WIRE: THE LIFE AND DEATH OF FIREBASE MARY ANN,* by Keith William Nolan. Copyright © 1995 by Keith William Nolan. All rights reserved.

Reprinted after written assurance from McGraw-Hill Education that, with the acquisition of Contemporary Books, it inherited no rights to *THE ESSENCE OF NINJUTSU,* by Dr. Masaaki Hatsumi. Copyright © 1988 by Dr. Masaaki Hatsumi. All rights reserved.

Reprinted after written assurance from McGraw-Hill Education that, with the acquisition of Contemporary Books, it inherited no rights to *THE GRANDMASTER'S BOOK OF NINJA TRAINING,* by Dr. Masaaki Hatsumi. Copyright © 1988 by Dr. Masaaki Hatsumi. All rights reserved.

Reprinted after written assurance from McGraw-Hill Education that, with the acquisition of Contemporary Books, it inherited no rights to *MIND OF THE NINJA: EXPLORING THE INNER POWER,* by Kirtland C. Peterson. Copyright © 1986 by Kirtland C. Peterson. All rights reserved.

Reprinted with permission of the *NEWSWEEK,* from the following articles: "The Kosovo Cover-Up," by John Barry and Evan Thomas, 15 May 2000; "Father, Where Art Thou," by Kevin Peraino and Evan Thomas, 27 January 2003; "A Plan under Attack," by Evan Thomas and John Barry, 7 April 2003; and "Counting Civilians: Iraq's War Dead," by Suzanne Smalley, 7 April 2003. Copyrights © 2000 and 2003 by Newsweek, Inc. All rights reserved.

Reprinted with permission of Tribune Media Services International, from "U.S. Hopes to Take Baghdad Quickly," by Knight Ridder, as published in the *JACKSONVILLE (NC) DAILY NEWS* on 26 March 2003. Copyright © 2003 by Knight Ridder. All rights reserved.

Reprinted with permission of John Wiley & Sons, Inc., from *THE BATTLE FOR OKINAWA,* by Col. Hiromichi Yahara, translated by Roger Pineau and Masatoshi Uehara. Copyright © 1955 by Pacific Basin Institute. All rights reserved.

Reprinted with permission of Stackpole Books, Harrisburg, PA, from *EXPLORING THE OUTDOORS WITH INDIAN SECRETS,* by Allan A. Macfarlan. Copyright © 1971 by Allan A. Macfarlan. All rights reserved.

Reprinted with permission of Oxford Univ. Press, India, from *JUNGLE LORE,* by James Corbett. Copyright © 1953 by Oxford University Press, Inc. All rights reserved.

Reprinted with permission of Oxford Univ. Press, New York, from *THE ART OF WAR,* by Sun Tzu, trans. Samuel B. Griffith. Copyright © 1963 by Oxford University Press, Inc. All rights reserved.

Reprinted with permission of Oxford Univ. Press, India, from *JIM CORBETT'S INDIA,* by James Corbett, edited by R.E. Hawkins. Selections and introduction copyright © 1978 by Oxford University Press, India, Inc. All rights reserved.

Reprinted with permission of Harper Collins Publishers, from *POINT MAN,* by James Watson and Kevin Dockery. Copyright © 1993 by James Watson and Kevin Dockery. All rights reserved.

Reprinted with permission of Gene Duncan Books, Boonville, MO, from *GREEN SIDE OUT,* by Maj. H.G. Duncan and Capt. W.T. Moore, Jr. Copyright © 1980 by H.G. Duncan. All rights reserved.

Reprinted with permission of Reader's Digest Assoc., Inc., from *THE COMPLETE MANUAL OF FITNESS AND WELL-BEING.* Copyright © 1984 by Marshall Editions Ltd., London. Revised edition copyright © 1988 by Reader's Digest, Assoc., Inc. All rights reserved.

Reprinted with permission of St. Martin's Press, from *DIRTY LITTLE SECRETS OF THE VIETNAM WAR,* by James F. Dunnigan and Albert A. Nofi. Copyright © 1998 by James F. Dunnigan and Albert A. Nofi. All rights reserved.

Reprinted after an unsuccessful attempt by Doubleday & Co. to reach the author in Cadiz, Spain, from *WAR IN THE SHADOWS,* by Robert B. Asprey. Copyright © 1975 by Robert B. Asprey. All rights reserved.

Reprinted with permission of Mark Wheeler, Canoga Park, CA, from "Shadow Wolves" in the January 2003 issue of *THE SMITHSONIAN*. Copyright © 2003 by Mark Wheeler. All rights reserved.

Reprinted with permission of Penguin Books, Ltd., London, from *THE PENGUIN HISTORICAL ATLAS OF RUSSIA,* by John Channon and Robert Hudson. Copyright © 1995 by John Channon and Robert Hudson. All rights reserved.

Reprinted with permission of Army Times Publishing Co., from the following *ARMED FORCES JOURNAL INTERNATIONAL* articles: "The World Turned Upside Down: Military Lessons of the Chechen War," by Anatol Lieven, August 1998; and "Weapons of Restraint: Emphasis on Limiting Non-Combatant Casualties Is Good Reason for Developing New Non-Lethal Weapons," by Paul Churchill Hutton III, May 2000. Copyrights © 1998 and 2000 by Armed Forces Journal International, Inc. All rights reserved.

ENDNOTES

Introduction

1. Lt.Gen. Arthur S. Collins, Jr., U.S. Army (Ret.), *Common Sense Training: A Working Philosophy for Leaders* (Novato, CA: Presidio Press, 1978), p. 214.

2. George Robert Elford, *Devil's Guard* (New York: Dell Publishing, 1971), p. 68.

3. Sun Tzu, *The Art of War,* trans. and with intro. by Samuel B. Griffith, foreword by B.H. Liddell Hart (New York: Oxford Univ. Press, 1963), p. 84.

4. Robert B. Asprey, *War in the Shadows* (Garden City, NY: Doubleday & Co., 1975), pp. 36-38.

5. John Channon and Robert Hudson, *The Penguin Historical Atlas of Russia* (London: Penguin Books, 1996), p. 20; *Random House Encyclopedia,* electronic ed., s.v. "Hun."

6. Joseph Dahmus, *A Popular History of the Middle Ages* (New York: Barnes & Noble Books, 1968), pp. 84-90.

7. Ibid., pp. 365, 366.

8. *Journey into China,* ed. Kenneth C. Danforth (Washington, D.C.: Nat. Geographic Society, n.d.), pp. 87, 88.

9. *Random House Encyclopedia,* electronic ed., s.v. "Kublai Khan."

10. Dahmus, *A Popular History of the Middle Ages,* p. 367.

11. Channon and Hudson, *The Penguin Historical Atlas of Russia,* p. 28.

12. Abraham Resnick, *Russia: A History to 1917* (Chicago: Children's Press, 1983), p. 26.

13. *Random House Encyclopedia,* electronic ed., s.v. "Batu Khan."

14. Dahmus, *A Popular History of the Middle Ages,* p. 368.

15. Resnick, *Russia,* p. 25.

16. *Random House Encyclopedia,* electronic ed., s.v. "Tamerlane."

17. Dahmus, *A Popular History of the Middle Ages,* p. 369.

18. *Random House Encyclopedia,* electronic ed., s.v. "Tatars."

19. Channon and Hudson, *The Penguin Historical Atlas of Russia,* p. 34.

20. Bruce I. Gudmundsson, *Stormtroop Tactics—Innovation in the German Army 1914-1918* (New York: Praeger, 1989), p. 21. (This work will henceforth be cited as Gudmundsson, *Stormtroop Tactics.)*

21. Alvin D. Coox, *Nomonhan: Japan against Russia, 1939* (Stanford, CA: Stanford Univ. Press, 1985), p. 1007.

22. His Holiness John Paul II, *Crossing the Threshold of Hope* (New York: Alfred A. Knopf, 1995), pp. 205, 206.

23. *The Bear Went over the Mountain: Soviet Combat Tactics in Afghanistan,* trans. and ed. Lester W. Grau, Foreign Mil. Studies Office, U.S. Dept. of Defense (Soviet Union: Frunze Mil. Academy, n.d.; reprint Washington, D.C.: Nat. Defense Univ. Press, 1996), p. 52. (This work will henceforth be cited as *The Bear Went over the Mountain.)*

24. Translator's preface, from *Night Movements*, trans. and preface by C. Burnett (Tokyo: Imperial Japanese Army, 1913; reprint Port Townsend, WA: Loompanics Unlimited, n.d.), p. 4.

25. Edward J. Drea, "Nomonhan: Japanese—Soviet Tactical Combat, 1939," *Leavenworth Papers No. 2* (Ft. Leavenworth, KS: Combat Studies Inst., U.S. Army Cmd. & Gen. Staff College, 1981), p. 88. (This work will henceforth be cited as Drea, "Nomonhan," *Leavenworth Papers No. 2.)*

Chapter 1: *American Units Must Further Disperse*

1. (Lt.Col.) Lester W. Grau, "Technology and the Second Chechen Campaign," www.da.mod.uk/CSRC/Home/Caucasus/P31 (Sandhurst).

2. Arkady Shipunov and Gennady Filimonov, "Field Artillery to be Replaced with Shmel Infantry Flamethrower," *Military Parade (Moscow),* Issue 29, September 1998, www.milparade.com/security/29.

3. Lt.Col. Timothy L. Thomas and Lester W. Grau, "Russian Lessons Learned from the Battles for Grozny," *Marine Corps Gazette,* April 2000, p. 48.

4. *Soviet Combat Regulations of November 1942* (Moscow: [Stalin], 1942), republished as *Soviet Infantry Tactics in World War II: Red Army Infantry Tactics from Squad to Rifle Company from the Combat Regulations,* with trans., intro., and notes by Charles C. Sharp (West Chester, OH: George Nafziger, 1998), p. 3. (This work will henceforth be cited as *Soviet Combat Regulations of November 1942.)*

5. Ibid., pp. 13-17.

6. Steven L. Canby, "Classic Light Infantry and New Technology," produced under contract MDA 903-81-C-0207 for the Defense Advanced Research Project Agency (Potomac, MD: C&L Associates, 1983), p. ii.

7. Ibid., p. iv.

8. Ibid., pp. 109, 110.

9. *The Bear Went over the Mountain,* p. 205.

10. Joseph S. Bermudez, Jr., *North Korean Special Forces* (Annapolis, MD: Naval Inst. Press, 1998), p. 155. (This work will henceforth be cited as Bermudez, *North Korean Special Forces.)*

11. Ibid., p. 169.

12. Ibid., pp. 172, 173.

13. Ibid., p. 169.

14. Ibid., p. 219.

15. U.S. Army, "OPFOR: North Korea," *TC-30-37,* draft of January 1979, p. 3-1, from Bermudez, *North Korean Special Forces,* p. 227.
16. Bermudez, *North Korean Special Forces,* p. 155.
17. Ibid., p. 171.
18. *Inside North Korea: Three Decades of Duplicity* (Seoul: Inst. of Internal and External Affairs, 1975), p. 74, from Bermudez, *North Korean Special Forces,* p. 233.
19. *The Strategic Advantage: Sun Zi & Western Approaches to War,* ed. Cao Shan (Beijing: New World Press, 1997), p. 55.
20. "Handbook on German Military Forces," *TM-E 30-451* (Washington, D.C.: U.S. War Dept., 1945; reprint Baton Rouge, LA: LSU Press, 1990), pp. 222-226.
21. "Handbook On U.S.S.R. Military Forces," *TM 30-340* (Washington, D.C.: U.S. War Dept., 1945), republished as *Soviet Tactical Doctrine in WWII,* with foreword by Shawn Caza (West Chester, OH: G.F. Nafziger, 1997), p. V-2.
22. Harriet Fast Scott and William F. Scott, *The Armed Forces of the U.S.S.R.* (Boulder, CO: Westview Press, 1979), p. 143.
23. Viktor Suvorov, *Inside the Soviet Army* (New York: Berkley Books, 1983), p. 275.
24. *Soviet Combat Regulations of November 1942,* p. 9.
25. Memo for the record by H.J. Poole.
26. "Handbook On U.S.S.R. Military Forces," *TM 30-340,* p. V-19.
27. Ibid., p. V-39.
28. *Soviet Combat Regulations of November 1942,* p. 22.
29. Suvorov, *Inside the Soviet Army,* p. 280.
30. Canby, "Classic Light Infantry and New Technology," p. ii.
31. Bermudez, *North Korean Special Forces,* p. 172.
32. H. John Poole, *Phantom Soldier: The Enemy's Answer to U.S. Firepower* (Emerald Isle, NC: Posterity Press, 2001), p. 93. (This work will henceforth be cited as Poole, *Phantom Soldier.)*
33. "Handbook On U.S.S.R. Military Forces," *TM 30-340,* pp. V-138, V-143.
34. T. R. Fehrenbach, *This Kind of War: The Classic Korean War History* (Dulles, VA: Brassey's, 1963), p. 138. (This work will henceforth be cited as Fehrenbach, *This Kind of War.*)
35. *Mao's Generals Remember Korea,* trans. and ed. Xiaobing Li, Allan R. Millett, and Bin Yu (Lawrence, KS: Univ. Press of Kansas, 2001), p. 115. (This work will henceforth be cited as *Mao's Generals Remember Korea.)*
36. *Sun Bin's Art of War: World's Greatest Military Treatise,* trans. Sui Yun (Singapore: Chung Printing, 1999), p. 190.
37. Mao Tse-tung, as quoted in "Handbook on the Chinese Communist Army," *DA Pamphlet 30-51* (Washington, D.C.: Hdqts. Dept. of the Army, 7 December 1960), p. 19.

38. "Handbook On U.S.S.R. Military Forces," *TM 30-340,* p. V-1.
39. *Soviet Combat Regulations of November 1942,* p. 7.
40. Ibid.
41. Ibid., p. 5.
42. H.J. Poole, *The Last Hundred Yards: The NCOs Contribution to Warfare* (Emerald Isle, NC: Posterity Press, 1997), p. 272.
43. *Soviet Combat Regulations of November 1942,* pp. 4, 5.
44. Ibid., p. 4.
45. Timothy L. Lupfer, "The Dynamics of Doctrine: The Changes in German Tactical Doctrine during the First World War," *Leavenworth Papers No. 4* (Ft. Leavenworth, KS: Combat Studies Inst., U.S. Army Cmd. & Gen. Staff College, 1981), p. 8. (This work will henceforth be cited as Lupfer, "The Dynamics of Doctrine," *Leavenworth Papers No. 4.)*
46. Ibid.
47. Ibid., p. 57.
48. Ibid., pp. 57, 58.
49. *Soviet Combat Regulations of November 1942,* p. 4.
50. Ibid., pp. 13, 14.
51. Ibid., p. 8.
52. Ibid., p. 10.
53. William P. Baxter, *Soviet Airland Battle Tactics* (Novato, CA: Presidio Press, 1986), p. 70.
54. A. Babenko, *Soviet Officers* (Moscow: Progress Publishers, 1976), p. 36, as quoted in *Soviet Airland Battle Tactics,* by William P. Baxter (Novato, CA: Presidio Press, 1986), p. 70.
55. V.V. Shelyag, A.D. Glotochkin, and K.K. Platinov, *Military Psychology: A Soviet View* (Moscow, 1972), p. 320, from *Cohesion: The Human Element in Combat,* by Wm. Darryl Henderson (Washington, D.C.: Nat. Defense Univ. Press, 1985), p. 67.
56. Baxter, *Soviet Airland Battle Tactics,* p. 65.
57. *Vietnam Interviews,* Interview "K-5" (Santa Monica, CA: Rand Corporation, K Series, n.d.), pp. 32, 33, from *Cohesion: The Human Element in Combat,* by Wm. Darryl Henderson (Washington, D.C.: Nat. Defense Univ. Press, 1985), p. 28.

Chapter 2: *Orphaned Squads Are at Greater Risk*

1. Ron Fournier, Associated Press, "Bush: Russian Firms Aiding Iraqi War Effort," *Jacksonville (NC) Daily News,* 25 March 2003, p. 5A.
2. Suzanne Smalley, "Counting Civilians: Iraq's War Dead," *Newsweek,* 7 April 2003, p. 11.
3. *Mao Tse-tung: An Anthology of His Writings,* ed. Anne Fremantle (New York: Mentor, 1962), pp. 132, 133.

4. Ibid., p. 139.
5. *The Bear Went over the Mountain,* p. 197.
6. Ibid., p. 204.
7. *Night Movements*, trans. and preface by C. Burnett (Tokyo: Imperial Japanese Army, 1913; reprint Port Townsend, WA: Loompanics Unlimited, n.d.), p. 109. (This work will henceforth be cited as *Night Movements.)*
8. "Handbook On U.S.S.R. Military Forces," *TM 30-340,* p. V-42.
9. Ibid., p. V-43.
10. Ibid.
11. *Night Movements*, pp. 109-111.
12. Gudmundsson, *Stormtroop Tactics,* pp. 147-149.
13. "Handbook On U.S.S.R. Military Forces," *TM 30-340,* pp. V-43, V-44.
14. *Soviet Combat Regulations of November 1942,* p. 23.
15. Gudmundsson, *Stormtroop Tactics,* p. 94.
16. "Handbook On U.S.S.R. Military Forces," *TM 30-340,* p. V-2.
17. Ibid., V-46.
18. Ibid., V-26.
19. Stephen K. Hayes, *The Ninja and Their Secret Fighting Art* (Rutland, VT: Charles E. Tuttle Company, 1981), p. 40.
20. Ashida Kim, *The Invisible Ninja: Ancient Secrets of Surprise* (New York: Citadel Press, 1983), p. 151.
21. Dr. Haha Lung, *Knights of Darkness: Secrets of the World's Deadliest Night Fighters* (Boulder, CO: Paladin Press, 1998), p. 9.
22. Hayes, *The Ninja and Their Secret Fighting Art,* pp. 15, 16.
23. Ibid., p. 18.
24. Stephen K. Hayes, *The Mystic Arts of the Ninja: Hypnotism, Invisibility, and Weaponry* (Chicago: Contemporary Books, 1985), p. 101.
25. Ashida Kim, *Secrets of the Ninja* (New York: Citadel Press, 1981), p. x.
26. Lung, *Knights of Darkness,* pp. 8, 9.
27. Kirkland C Peterson, *Mind of the Ninja: Exploring the Inner Power* (Chicago: Contemporary Books, 1986), p. 26.
28. Stephen K. Hayes, *Legacy of the Night Warrior* (Santa Clarita, CA: Ohara Publications, 1985), p. 16.
29. Ibid.
30. Ibid.
31. Ibid., p. 17.
32. Stephen K. Hayes, *Warrior Path of Togakure* (Santa Clarita, CA: Ohara Publications, 1983), author's foreword page.
33. Hayes, *Legacy of the Night Warrior,* p. 20.
34. Joe Bernard *(ninjutsu* student), in e-mails from JBernard29 to author, 27 November 2002 and 17 January 2003.

35. Dr. Masaaki Hatsumi, *The Essence of Ninjutsu* (Chicago: Contemporary Books, 1988), p. 12; Hayes, *The Ninja and Their Secret Fighting Art,* p. 155.

36. Dr. Masaaki Hatsumi, *The Grandmaster's Book of Ninja Training,* trans. Chris W. P. Reynolds (Chicago: Contemporary Books, 1988), pp. 1, 2.

37. Hatsumi, *The Essence of Ninjutsu,* p. 66.

38. Hayes, *The Ninja and Their Secret Fighting Art,* pp. 15, 16.

39. Hatsumi, *The Essence of Ninjutsu,* p. 12.

40. Hatsumi, *The Grandmaster's Book of Ninja Training,* p. 2.

41. Hatsumi, *The Essence of Ninjutsu,* p. 15.

42. Ibid, p. 15.

43. Hayes, *The Mystic Arts of the Ninja,* p. 114; Kim, *Secrets of the Ninja,* p. 3.

44. Hayes, *The Mystic Arts of the Ninja,* p. 114.

45. Hayes, *The Mystic Arts of the Ninja,* p. 114; Kim, *Secrets of the Ninja,* p. 3.

46. Hatsumi, *The Essence of Ninjutsu,* p. 37.

47. Lung, *Knights of Darkness,* p. 56.

48. Hatsumi, *The Essence of Ninjutsu,* p. 15.

49. Lung, *Knights of Darkness,* p. 71.

50. Hayes, *The Ninja and Their Secret Fighting Art,* p. 99.

51. Hatsumi, *The Grandmaster's Book of Ninja Training,* p. 42.

52. Lung, *Knights of Darkness,* pp. 75, 76.

53. Ibid., p. 75.

54. Dr. Masaaki Hatsumi, *Ninjutsu: History and Tradition* (Burbank, CA: Unique Publications, 1981), p. 17.

55. Ibid.

56. Kim, *Secrets of the Ninja,* p. 69.

57. Hayes, *Legacy of the Night Warrior,* p. 27.

58. Kim, *Secrets of the Ninja,* p. 55.

59. Ibid., p. 49.

60. Peterson, *Mind of the Ninja,* pp. 118, 119; *Night Movements,* pp. 40-42.

61. Lung, *Knights of Darkness,* pp. 134-136.

62. Hayes, *The Mystic Arts of the Ninja,* p. 114; Kim, *Secrets of the Ninja,* p. 3.

63. Kim, *Secrets of the Ninja,* p. 47.

64. Hayes, *The Mystic Arts of the Ninja,* p. 114.

65. Kim, *Secrets of the Ninja,* p. 79.

66. Hayes, *The Mystic Arts of the Ninja,* p. 114.

67. Ibid., pp. 38-41.

68. Hatsumi, *The Essence of Ninjutsu,* pp. 37-39; Hatsumi, *Ninjutsu: History and Tradition,* p. 16.

69. Hatsumi, *The Essence of Ninjutsu,* p. 39.

70. Kim, *Secrets of the Ninja,* p. 37.
71. Ibid.
72. Hatsumi, *The Grandmaster's Book of Ninja Training,* p. 98.
73. Lung, *Knights of Darkness,* p. 137.
74. Kim, *Secrets of the Ninja,* pp. 39, 44.
75. Hayes, *The Mystic Arts of the Ninja,* p. 114.
76. Kim, *Secrets of the Ninja,* p. 57.
77. Kim, *The Invisible Ninja,* pp. 22, 23.
78. Ibid., p. 33.
79. Kim, *Secrets of the Ninja,* p. 73.
80. Ibid., p. 75.
81. Kim, *Secrets of the Ninja,* p. 3.
82. Ibid., p. 75.
83. Ibid., p. 89.
84. Ibid.
85. Hayes, *The Mystic Arts of the Ninja,* pp. 31-54.
86. Lung, *Knights of Darkness,* p. 134.
87. Hayes, *The Ninja and Their Secret Fighting Art,* p. 104; Hayes, *The Mystic Arts of the Ninja,* pp. 33-35.
88. Hayes, *The Mystic Arts of the Ninja,* pp. 42-43.
89. Hayes, *The Ninja and Their Secret Fighting Art,* p. 102.
90. Hayes, *The Mystic Arts of the Ninja,* pp. 36-37.
91. Kim, *Secrets of the Ninja,* p. 54.
92. Hayes, *The Ninja and Their Secret Fighting Art,* p. 103; Kim, *Secrets of the Ninja,* p. 40.
93. Lung, *Knights of Darkness,* pp. 112, 134; Hatsumi, *Ninjutsu: History and Tradition,* p. 230.
94. Hayes, *The Mystic Arts of the Ninja,* pp. 50-53.
95. Ibid.
96. Kim, *Secrets of the Ninja,* p. 45.
97. Ibid., p. 46.
98. Hayes, *The Ninja and Their Secret Fighting Art,* p. 130; Kim, *The Invisible Ninja,* pp. 1, 41.
99. Hayes, *The Ninja and Their Secret Fighting Art,* p. 100.
100. Kim, *The Invisible Ninja,* p. 6.
101. Ibid., p. 14.
102. Ibid., p. 18.
103. Hayes, *The Ninja and Their Secret Fighting Art,* p. 130; Hatsumi, *The Essence of Ninjutsu,* p. 102.
104. Hayes, *The Ninja and Their Secret Fighting Art,* p. 100.
105. Ibid.
106. Hayes, *The Ninja and Their Secret Fighting Art,* p. 100; Lung, *Knights of Darkness,* p. 70.
107. Hayes, *Warrior Path of Togakure,* p. 52; Bernard e-mails.
108. Kim, *The Invisible Ninja,* p. 18.

109. Hatsumi, *The Essence of Ninjutsu,* p. 115.
110. Hatsumi, *The Essence of Ninjutsu,* p. 55; Hayes, *Legacy of the Night Warrior,* p. 154.
111. Hayes, *The Ninja and Their Secret Fighting Art,* p. 130; Kim, *The Invisible Ninja,* p. 2; Hayes, *The Mystic Arts of the Ninja,* p. 114.
112. Kim, *The Invisible Ninja,* pp. 7, 15.
113. Lung, *Knights of Darkness,* p. 131.
114. Ibid., p. 100.
115. Kim, *The Invisible Ninja,* p. 29.
116. Ibid., p. 20.
117. Ibid., pp. 25-29.
118. Ibid., p. 30.
119. Hatsumi, *The Essence of Ninjutsu,* p. 39; Hayes, *The Mystic Arts of the Ninja,* p. 101.
120. Hatsumi, *The Essence of Ninjutsu,* p. 40.
121. Hatsumi, *The Essence of Ninjutsu,* p. 40; Kim, *Secrets of the Ninja,* p. 39.
122. Hatsumi, *The Essence of Ninjutsu,* p. 111; Hayes, *Legacy of the Night Warrior,* p. 19; Hatsumi, *Ninjutsu: History and Tradition,* p. 16.
123. Hayes, *Warrior Path of Togakure,* p. 54.
124. Hayes, *The Mystic Arts of the Ninja,* p. 103.
125. Hatsumi, *The Essence of Ninjutsu,* p. 21.
126. Kim, *Secrets of the Ninja,* p. 57.
127. Ibid.
128. Hatsumi, *The Essence of Ninjutsu,* p. 21.
129. Hayes, *Warrior Path of Togakure,* p. 52.
130. Kim, *Secrets of the Ninja,* p. 3.
131. Ibid., p. 74.
132. Ibid., p. 75.
133. Ibid., p. 34.
134. Ibid.
135. Ibid., p. 75.
136. Hayes, *Legacy of the Night Warrior,* p. 19.
137. Hayes, *The Mystic Arts of the Ninja,* p. 114.
138. Ibid.
139. Hayes, *The Ninja and Their Secret Fighting Art,* p. 25.
140. Lung, *Knights of Darkness,* p. 123.
141. Kim, *Secrets of the Ninja,* p. 77.
142. Hayes, *Legacy of the Night Warrior,* pp. 42, 43.
143. Hayes, *The Ninja and Their Secret Fighting Art,* p. 107.
144. Ibid., p. 26.
145. Hayes, *Legacy of the Night Warrior,* p. 19.
146. Hayes, *The Ninja and Their Secret Fighting Art,* p. 108.
147. Lung, *Knights of Darkness,* p. 132; Kim, *Secrets of the Ninja,* p. 51.

148. Hayes, *The Ninja and Their Secret Fighting Art,* p. 26.

149. Hayes, *The Ninja and Their Secret Fighting Art,* p. 25; Hatsumi, *Ninjutsu: History and Tradition,* pp. 211, 212.

150. Hayes, *The Ninja and Their Secret Fighting Art,* p. 128; Hatsumi, *Ninjutsu: History and Tradition,* pp. 214, 215.

151. Hayes, *The Mystic Arts of the Ninja,* pp. 106, 107.

152. Hayes, *The Ninja and Their Secret Fighting Art,* p. 25.

153. Hayes, *Legacy of the Night Warrior,* pp. 36, 37.

154. Hayes, *The Mystic Arts of the Ninja,* pp. 110, 111; Hatsumi, *Ninjutsu: History and Tradition,* p. 9.

155. Kim, *Secrets of the Ninja,* p. 51.

156. Hayes, *The Mystic Arts of the Ninja,* pp. 112, 113.

157. Hayes, *The Ninja and Their Secret Fighting Art,* p. 131; Hatsumi, *Ninjutsu: History and Tradition,* p. 211.

158. Hatsumi, *Ninjutsu: History and Tradition,* p. 16.

159. Kim, *Secrets of the Ninja,* p. 78.

160. Ibid., p. 57.

161. Ibid., pp. 80, 81.

162. Ibid., p. 33.

163. Kim, *The Invisible Ninja,* p. 12.

164. Kim, *Secrets of the Ninja,* p. 79.

165. Ibid., p. ix.

166. Hayes, *The Mystic Arts of the Ninja,* p. 116.

167. Ibid.

168. Hayes, *Legacy of the Night Warrior,* p. 19.

169. Hayes, *The Ninja and Their Secret Fighting Art,* p. 130.

170. Hayes, *Legacy of the Night Warrior,* p. 27; Hayes, *The Ninja and Their Secret Fighting Art,* p. 111.

171. Hayes, *The Mystic Arts of the Ninja,* p. 116.

172. Ibid., pp. 104, 105.

173. Hatsumi, *Ninjutsu: History and Tradition,* p. 204.

174. Hayes, *The Mystic Arts of the Ninja,* p. 116.

175. Hayes, *The Mystic Arts of the Ninja,* p. 116; Kim, *Secrets of the Ninja,* p. 33.

176. Hayes, *Legacy of the Night Warrior,* pp. 29, 34, 35; Hatsumi, *Ninjutsu: History and Tradition,* p. 216.

177. Kim, *Secrets of the Ninja,* p. 33.

178. Hatsumi, *The Essence of Ninjutsu,* p. 54; Hayes, *The Mystic Arts of the Ninja,* p. 116.

179. Hayes, *The Mystic Arts of the Ninja,* p. 116; Hatsumi, *Ninjutsu: History and Tradition,* p. 183.

180. Hatsumi, *The Essence of Ninjutsu,* p. 165.

181. Kim, *Secrets of the Ninja,* p. 33.

182. Hayes, *The Mystic Arts of the Ninja,* p. 116.

183. Ibid., p. 117.

184. Hayes, *Legacy of the Night Warrior,* pp. 19, 28; Hatsumi, *Ninjutsu: History and Tradition,* p. 190.
185. Hayes, *The Ninja and Their Secret Fighting Art,* p. 111; Kim, *Secrets of the Ninja,* p. 145.
186. Lung, *Knights of Darkness,* p. 72.
187. Ibid., p. 133.
188. Kim, *The Invisible Ninja,* p. 19.
189. Ibid.
190. Hayes, *The Mystic Arts of the Ninja,* p. 114.
191. Kim, *The Invisible Ninja,* p. 46; Kim, *Secrets of the Ninja,* pp. 81, 82.
192. Hayes, *The Mystic Arts of the Ninja,* pp. 108, 109.
193. Hatsumi, *Ninjutsu: History and Tradition,* p. 13.
194. Ibid., p. 52.
195. Ibid.
196. Kim, *The Invisible Ninja,* p. 53.
197. Hayes, *Legacy of the Night Warrior,* p. 19.
198. Kim, *Secrets of the Ninja,* p. 143.
199. Ibid., p. 146.
200. Hatsumi, *Ninjutsu: History and Tradition,* p. 52.
201. Kim, *The Invisible Ninja,* p. 50; Hayes, *The Mystic Arts of the Ninja,* p. 48.
202. Hayes, *The Mystic Arts of the Ninja,* p. 44.
203. Ibid., p. 46.
204. Hayes, *Legacy of the Night Warrior,* p. 14.
205. Hayes, *The Ninja and Their Secret Fighting Art,* p. 15.
206. *Night Movements*, pp. 23-32.
207. Hayes, *The Ninja and Their Secret Fighting Art,* p. 105.
208. Lung, *Knights of Darkness,* p. 12.
209. Bermudez, *North Korean Special Forces,* p. 171.
210. Michael O'Brien, *Conscripts and Regulars: With the Seventh Battalion in Vietnam* (St. Leonards, Australia: Allen & Unwin, 1995), p. 215.
211. Memo for the record by H.J. Poole.
212. Kim, *The Invisible Ninja,* pp. 13-16.
213. Hayes, *The Ninja and Their Secret Fighting Art,* p. 46.
214. Ibid., p. 135.
215. Ibid., p. 37.
216. Ibid., pp. 136-143.
217. Ibid., pp. 146-148.
218. Ibid., p. 145.
219. Ibid., p. 94.
220. Ibid., p. 111.
221. Ibid., p. 22.
222. Hatsumi, *The Essence of Ninjutsu,* p. 15.

223. Hatsumi, *The Grandmaster's Book of Ninja Training,* pp. 43, 44.

Chapter 3: *U.S. Riflemen Will Need More Skill*

1. Maj.Gen. Vandegrift, as quoted in "Fighting on Guadalcanal," *FMFRP 12-110* (Washington, D.C.: U.S.A. War Office, 1942), p. v.
2. Col. Merritt A. Edson, as quoted in "Fighting on Guadalcanal," *FMFRP 12-110* (Washington, D.C.: U.S.A. War Office, 1942), pp. 14-19.
3. Dutch Marine exchange officer, in conversation with author, 16 January 2002.
4. *Encyclopedia of the Vietnam War,* ed. Stanley I. Kutler (New York: Macmillan Reference, 1996), p. 392.
5. Ibid.
6. O'Brien, *Conscripts and Regulars,* p. 29.
7. Ibid., p. 30.
8. Coox, *Nomonhan,* p. 1003.
9. *Soviet Combat Regulations of November 1942,* p. 13.
10. Fehrenbach, *This Kind of War,* p. 200.
11. Excerpt from Chinese Communist Forces pamphlet, in Fehrenbach, *This Kind of War,* p. 200.
12. Excerpt from *Primary Conclusions of Battle Experience at Unsan* (Chinese Communist Forces, n.d.), in Fehrenbach, *This Kind of War,* p. 200; and *The Korean War: An Encyclopedia,* ed. Stanley Sandler (New York: Garland Publishing, 1995), p. 266.

Chapter 4: *Microterrain Appreciation*

1. Gudmundsson, *Stormtroop Tactics,* pp. xi, xii.
2. "Historical Study—Night Combat," *DA Pamphlet 20-236* (Washington, D.C.: Hdqts. Dept. of the Army, 1953), pp. 19-21.
3. A.E. Taras and F.D. Zaruz, *Podgotovka Razvegchika: Sistema Spetsnaza GRU* (Minsk, Belarus: AST Publishing, 2002), p. 149.
4. Maj. Scott R. McMichael, "The Chinese Communist Forces in Korea," chapt. 2 of "A Historical Perspective on Light Infantry," *Leavenworth Research Survey No. 6* (Ft. Leavenworth, KS: Combat Studies Inst., U.S. Army Cmd. & Gen. Staff College, 1987), p. 60. (This work will henceforth be cited as McMichael, "The Chinese Communist Forces in Korea," *Leavenworth Research Survey No. 6.)*
5. H. John Poole, *One More Bridge to Cross* (Emerald Isle, NC: Posterity Press, 1999), pp. 35, 36.
6. "Handbook on German Military Forces," *TM-E 30-451,* p. 231.

7. Merritt E. Benson, "Seventeen Days at Iwo Jima," *Marine Corps Gazette,* February 2000, p. 32.
8. Memo for the record by H.J. Poole.
9. Hayes, *Legacy of the Night Warrior,* pp. 42, 43.
10. Hayes, *The Ninja and Their Secret Fighting Art,* p. 107.
11. Ibid.
12. Hayes, *The Mystic Arts of the Ninja,* p. 117.
13. Elford, *Devil's Guard,* p. 254.
14. Taras and Zaruz, *Podgotovka Razvedcheka,* p. 333.
15. John Barry and Evan Thomas, "The Kosovo Cover-Up," *Newsweek,* 15 May 2000, p. 23.

Chapter 5: *Harnessing the Senses*

1. Hatsumi, *The Grandmaster's Book of Ninja Training,* p. 86.
2. Spencer Chapman, *The Jungle Is Neutral* (London: Chatto & Windus, 1952), p. 27.
3. *Soviet Combat Regulations of November 1942,* p. 13.
4. Col. Merritt A. Edson, as quoted in "Fighting on Guadalcanal," *FMFRP 12-110* (Washington, D.C.: U.S.A. War Office, 1942), p. 14.
5. *Night Movements*, p. 23.
6. Hatsumi, *The Essence of Ninjutsu,* p. 15.
7. Ibid., p. 37.
8. Lung, *Knights of Darkness,* p. 54.
9. Korean War veteran, in conversation with author, 1990.
10. Lung, *Knights of Darkness,* p. 55.
11. Ibid., pp. 53-55.
12. Ibid., p. 56.
13. Allan A. Macfarlan, *Exploring the Outdoors with Indian Secrets* (Harrisburg, PA: Stackpole Books, 1971), p. 93.
14. Chapman, *The Jungle Is Neutral,* p. 73.
15. *Patrolling and Tracking* (Boulder, CO: Paladin Press, n.d.), p. 46.
16. O'Brien, *Conscripts and Regulars,* p. 31.
17. *Night Movements*, p. 17.
18. Lung, *Knights of Darkness,* p. 70.
19. *U.S. Special Forces Reconnaissance Manual* (n.p., n.d.; reprint, Sims, AR: Lancer Militaria, 1982), p. 51.
20. *Night Movements*, pp. 30-34.
21. Ibid., p. 31.
22. Hatsumi, *The Essence of Ninjutsu,* p. 15.
23. *Night Movements*, p. 27.
24. Ibid.

25. Trihn Duc, as quoted in *Portrait of the Enemy,* by David Chanoff and Doan Van Toai (New York: Random House, 1986), p. 21.
26. James Corbett, *Jungle Lore* (London: Oxford Univ. Press, 1953), p. 65.
27. Ibid., pp. 52, 53.
28. Ibid., p. 53.
29. Ibid., p. 54.
30. Ibid., pp. 54, 55.
31. Ibid., pp. 65.
32. Hayes, *The Ninja and Their Secret Fighting Art,* p. 99.
33. Hatsumi, *The Grandmaster's Book of Ninja Training,* pp. 43, 44.
34. Chapman, *The Jungle Is Neutral,* p. 73.
35. *Night Movements*, p. 107.
36. McMichael, "The Chinese Communist Forces in Korea," *Leavenworth Research Survey No. 6,* p. 33.
37. *U.S. Special Forces Reconnaissance Manual,* p. 48.
38. James F. Dunnigan and Albert A. Nofi, *Dirty Little Secrets of the Vietnam War* (New York: Thomas Dunne Books, 1999), p. 278.
39. Lung, *Knights of Darkness,* p. 75.
40. Hatsumi, *The Grandmaster's Book of Ninja Training,* p. 42.
41. Lung, *Knights of Darkness,* p. 75.
42. Ibid., p. 76.
43. Ibid., p. 75.
44. David Scott-Donelan, *Tactical Tracking Operations: The Essential Guide for Military and Police Trackers* (Boulder, CO: Paladin Press, 1998), p. 16.
45. Hatsumi, *Ninjutsu: History and Tradition,* p. 12.
46. Hatsumi, *The Essence of Ninjutsu,* p. 66.
47. Corbett, *Jungle Lore,* p. 172.
48. Lung, *Knights of Darkness,* p. 46.
49. Bermudez, *North Korean Special Forces,* p. 247.
50. David M. Glantz, "August Storm: Soviet Tactical and Operational Combat in Manchuria, 1945," *Leavenworth Papers No. 8* (Ft. Leavenworth, KS: Combat Studies Inst., U.S. Army Cmd. & Gen. Staff College, 1983), p. 1. (This work will henceforth be cited as Glantz, "August Storm," *Leavenworth Papers No. 8.)*
51. Corbett, *Jungle Lore,* p. 34.

Chapter 6: *Night Familiarity*

1. "Urban Battle in Nasiriyah Hurts Troops from Lejeune," Associated Press, *Jacksonville (NC) Daily News,* 27 March 2003, p. 2A.

2. Col. Dave E. Severance, "Night Patrol on Iwo Jima," *Marine Corps Gazette,* March 2003, p. 70.
3. Col. Wesley L. Fox, *Marine Rifleman: Forty-Three Years in the Corps* (Dulles, VA: Brassey's, 2002), p. 42.
4. Memo for the record by H.J. Poole.
5. Fox, *Marine Rifleman,* p. 275.
6. *Night Movements,* pp. 23-29.
7. *Mao's Generals Remember Korea,* p. 15.
8. "Primary Conclusions of Battle Experience at Unsan" (Korea: Communist Chinese Forces, 1950), from *This Kind of War: A Study in Unpreparedness,* by T.R. Fehrenbach (New York: Macmillan, 1963), pp. 300, 301, as cited in McMichael, "The Chinese Communist Forces in Korea," *Leavenworth Research Survey No. 6,* p. 69.
9. "Handbook on the Chinese Communist Army," *DA Pamphlet 30-51* (Washington, D.C.: Hdqts. Dept. of the Army, 1960), p. 54.
10. "Handbook On U.S.S.R. Military Forces," *TM 30-340,* p. V-104.
11. John Erickson, "Soviet Combined-Arms: Theory and Practice," photocopy of transcript (Edinburgh, Scotland: Univ. of Edinburgh, September 1979), p. 51, as quoted in "Soviet Night Operations in World War II," *Leavenworth Papers No. 6,* by Maj. Claude R. Sasso (Ft. Leavenworth, KS: Combat Studies Inst., U.S. Army Cmd. & Gen. Staff College, 1982), p. 14.
12. Lt.Col. David M. Glantz, Curriculum Supervisor, in foreword to "Soviet Night Operations in World War II," by Maj. Claude R. Sasso, *Leavenworth Papers No. 6* (Ft. Leavenworth, KS: Combat Studies Inst., U.S. Army Cmd. & Gen. Staff College, 1982), p. viii.
13. Republisher's intro., *Soviet Combat Regulations of November 1942,* p. 3.
14. Maj. Claude R. Sasso, "Soviet Night Operations in World War II," *Leavenworth Papers No. 6* (Ft. Leavenworth, KS: Combat Studies Inst., U.S. Army Cmd. & Gen. Staff College, 1982), p. 35. (This work will henceforth be cited as Sasso, "Soviet Night Operations in World War II," *Leavenworth Papers No. 6.)*
15. *The Bear Went over the Mountain,* p. 38.
16. Editor's commentary, *The Bear Went over the Mountain,* p. 174.
17. *The Bear Went over the Mountain,* p. 197.
18. Ibid., p. 24.
19. Lung, *Knights of Darkness,* pp. 43-50.
20. Ibid., p. 56.
21. Hayes, *Warrior Path of Togakure,* p. 52.
22. Lung, *Knights of Darkness,* p. 59.
23. Ibid., p. 64.
24. *Night Movements,* p. 26.
25. Ibid., p. 110.

26. Ibid., p. 55.

Chapter 7: *Nondetectable Movement*

1. Hatsumi, *The Essence of Ninjutsu,* p. 12.
2. Stephen K. Hayes, *Ninjutsu: The Art of the Invisible Warrior* (Chicago: Contemporary Books, 1984), p. 153.
3. Hatsumi, *The Grandmaster's Book of Ninja Training,* p. 2.
4. Ibid., p. 28.
5. Ibid., p. 8.
6. Ibid., p. 4.
7. Ibid., p. 83.
8. Kim, *The Invisible Ninja,* p. 1.
9. Ibid., p. 23.
10. Hatsumi, *The Grandmaster's Book of Ninja Training,* p. 30.
11. Ibid.
12. Hatsumi, *Ninjutsu: History and Tradition,* author's preface page.
13. Dr. David H. Reinke (expert on parapsychology and Eastern religions), in e-mails from DRDAVIDHREINKE to author, 17 December 2002 and 28 January 2003.
14. McMichael, "The Chinese Communist Forces in Korea," *Leavenworth Research Survey No. 6,* p. 56.
15. Ibid., pp. 56, 57.
16. Glantz, "August Storm," *Leavenworth Papers No. 8,* p. 1.
17. Ibid., p. 51.
18. Ibid., p. 17.
19. Ibid., p. 64.
20. Bruce Cumings, in "The Battle of the Minds" segment of *Korea—the Unknown War Series* (London: Thames TV in assoc. with WGBH Boston, 1990), NC Public TV.
21. Elford, *Devil's Guard,* pp. 148-166.
22. Ibid., p. 200.
23. Ibid., p. 129.
24. Ibid., pp. 312-329.
25. Ibid., p. 313.
26. *The Battle of Dien Bien Phu,* Visions of War Series, vol. 10 (New Star Video, 1988), 50 min., videocassette #4010.
27. Ibid.
28. Ibid.
29. Hoang Khoi, *The Ho Chi Minh Trail* (Hanoi: The Gioi Publishers, 2001), p. 45.
30. Ibid., p. 48.
31. Ibid., p. 47.

32. Ibid., p. 52.
33. Ibid., p. 73.
34. Ibid., p. 61.
35. Memo for the record by H.J. Poole.
36. Ibid.
37. Charles "Tag" Guthrie (Vietnam-era infantryman and intell. analyst), in e-mails from tagguthrie@cox.net to author, 18 August 2002, 4 -18 January 2003, 7 March 2003, and 9 April 2003.
38. Ibid.
39. Ibid.
40. Ibid.
41. Gen. Van Tien Dung, *Our Great Spring Victory: An Account of the Liberation of South Vietnam,* trans. John Spragens, Jr. (Hanoi: The Gioi Publishers, 2001), p. 14.
42. Khoi, *The Ho Chi Minh Trail,* pp. 53, 54.
43. Poole, *Phantom Soldier,* p. 194.
44. "Handbook On U.S.S.R. Military Forces," *TM 30-340,* p. V-3.
45. Coox, *Nomonhan,* p. 998.
46. Gen. Hoang Van Thai, *How South Vietnam Was Liberated* (Hanoi: The Gioi Publishers, 1996), pp. 235, 236.
47. *Soviet Combat Regulations of November 1942,* p. 26.
48. *Night Movements,* p. 36.
49. Ibid., pp. 79-82.
50. Ibid., p. 85.
51. Ibid., pp. 86, 87.
52. Chapman, *The Jungle Is Neutral,* p. 73.
53. Elford, *Devil's Guard,* p. 273.
54. "Handbook on German Military Forces," *TM-E 30-451,* p. 227.
55. Hoang Tat Hong, as quoted in *Portrait of the Enemy,* by David Chanoff and Doan Van Toai (New York: Random House, 1986), p. 165.
56. Memo for the record from H.J. Poole.
57. McMichael, "The Chinese Communist Forces in Korea," *Leavenworth Research Survey No. 6,* p. 59.
58. *The Bear Went over the Mountain,* p. 197.
59. *Soviet Combat Regulations of November 1942,* p. 22.
60. "Handbook On U.S.S.R. Military Forces," *TM 30-340,* p. V-105.
61. *The German Squad in Combat*, trans. and ed. U.S. Mil. Intell. Service from a German manual (n.p., 1943); republished as *German Squad Tactics in WWII,* by Matthew Gajkowski (West Chester, OH: Nafziger, 1995), p. 14. (This work will henceforth be cited as *The German Squad in Combat.)*
62. *Night Movements,* pp. 37-59.
63. Ibid., p. 40.
64. O'Brien, *Conscripts and Regulars,* p. 214.
65. Ibid., p. 215.

66. Nguyen Van Mo, as quoted in *Portrait of the Enemy,* by David Chanoff and Doan Van Toai (New York: Random House, 1986), pp. 161, 162.

67. "Light Infantry Platoon/Squad," *FM 7-70* (Washington, D.C.: Hdqts. Dept. of the Army, 1986). p. D-14.

68. "Handbook On U.S.S.R. Military Forces," *TM 30-340,* p. V-112.

69. "Handbook on German Military Forces," *TM-E 30-451,* p. 253.

70. *Soviet Combat Regulations of November 1942,* p. 51.

Chapter 8: *Guarded Communication*

1. Robert Burns, Associated Press, "Retired General: Exercise Rigged," *Jacksonville (North Carolina) Daily News,* 17 August 2002, p. 3A.

2. Glantz, "August Storm," *Leavenworth Papers No. 8,* p. 185.

3. *Night Movements,* p. 46.

4. Poole, *One More Bridge to Cross,* p. 74.

5. Unidentified NCO to Chesty Puller, as quoted in *Fighting on Guadalcanal* (Washington, D.C.: U.S.A. War Office, 1942), p. 35.

6. Nguyen Van Mo, as quoted in *Portrait of the Enemy,* by David Chanoff and Doan Van Toai (New York: Random House, 1986), p. 163.

7. *The Bear Went over the Mountain,* p. 128.

8. Chapman, *The Jungle Is Neutral,* p. 73.

9. *Night Movements,* pp. 44-46.

10. Ibid. p. 93.

11. Ibid., pp. 91, 92.

12. Eric M. Hammel, *Chosin: Heroic Ordeal of the Korean War* (Novato, CA: Presidio Press, 1981), p. 55.

13. Ibid., pp. 71, 155, 156.

14. Maj. Donald H. Campbell USMC (Ret.) (platoon sergeant during Korean hill battles of 1952-53), in conversation with author, 26 March 2001.

15. Hammel, *Chosin,* p. 71.

16. Joseph C. Goulden, *Korea: The Untold Story of the War* (New York: Times Books, 1982), p. 356.

17. "Fight at Ia Drang," in *Seven Firefights in Vietnam,* by John A. Cash, John Albright, and Allan W. Sandstrum (Washington, D.C.: Center of Mil. Hist., U.S. Army, 1985), p. 13.

18. Unidentified member of 3d Battalion, 27th Marines, in conversation with author, June 1968.

19. Keith William Nolan, *Operation Buffalo: USMC Fight for the DMZ* (New York: Dell Publishing, 1991), p. 49.

20. *Night Movements,* pp. 91, 92.

21. Memo for the record by H.J. Poole.
22. Maj. Norman L. Cooling, "Russia's 1994-96 Campaign for Chechnya: A Failure in Shaping the Battlespace," *Marine Corps Gazette,* October 2001, p. 64.
23. Rowan Scarborough, "Al-Qaeda Adapts to Pursuit Tactics," *Washington Times,* 15 January 2003, p. 10.
24. Glantz, "August Storm," *Leavenworth Papers No. 8,* p. 27.
25. Memo for the record by H.J. Poole.
26. *Night Movements,* pp. 48-54.

Chapter 9: *Discreet Force at Close Range*

1. Sun Tzu, *The Art of War,* trans. and with intro. by Samuel B. Griffith, p. vii.
2. Hatsumi, *The Essence of Ninjutsu,* p. 23.
3. Peterson, *Mind of the Ninja,* p. 123.
4. Maj.Gen. Hoang Min Thao, *The Victorious Tay Nguyen Campaign* (Hanoi: Foreign Languages Publishing House, 1979), pp. 102, 103.
5. Maj. Frank D. Pelli, "Insurgency, Counterinsurgency, and the Marines in Vietnam," as extracted on 10 November 2002 from http://www.ehistory.com/vietnam/essays/insurgency/index.cfm, p. 3. (This work will henceforth be cited as Pelli, "Insurgency, Counterinsurgency, and the Marines in Vietnam.")
6. *Mao's Generals Remember Korea,* p. 68.
7. *Vietnam Interviews,* Interview "K-5" (Santa Monica, CA: Rand Corporation, K Series, n.d.), pp. 7-15, from *Cohesion: The Human Element in Combat,* by Wm. Darryl Henderson (Washington, D.C.: Nat. Defense Univ. Press, 1985), pp. 122, 123.
8. *The Bear Went over the Mountain,* p. 46.
9. *Night Movements,* p. 124.
10. *Mao Tse-tung,* ed. Fremantle, p. 71.
11. "Handbook on the Chinese Communist Army," *DA Pamphlet 30-51,* p. 5.
12. *Soviet Combat Regulations of November 1942,* p. 4.
13. Ibid., p. 26.
14. Editor's commentary, *The Bear Went over the Mountain,* p. 162.
15. *Soviet Combat Regulations of November 1942,* pp. 18, 50, 70.
16. Taras and Zaruz, *Podgotovka Razvedcheka,* pp. 529-536.
17. "Handbook on Japanese Military Forces," *TM-E 30-480* (Washington, D.C.: U.S. War Dept., 1944; reprint, Baton Rouge, LA: LSU Press, 1991), pp. 210-212.
18. "Handbook on German Military Forces," *TM-E 30-451,* pp. 402, 403.

19. Ibid., p. 403.
20. Ibid.
21. Poole, *Phantom Soldier,* p. 101.
22. Ibid., p. 230.
23. Memo for the record by H.J. Poole.
24. *The Bear Went over the Mountain,* p. 97.
25. *Soviet Combat Regulations of November 1942,* p. 112.
26. Memo for the record by H.J. Poole.
27. Taras and Zaruz, *Podgotovka Razvedcheka,* pp. 90-93.
28. *Night Movements,* p. 95.
29. Ibid., p. 112.
30. Ibid., p. 125.
31. Ibid., p. 115.
32. Gudmundsson, *Stormtroop Tactics,* pp. 49, 191.
33. Poole, *One More Bridge to Cross,* p. 30.
34. *The German Squad in Combat,* p. 14.
35. Ibid., p. 23.
36. *Soviet Combat Regulations of November 1942,* p. 22.
37. "Handbook on Japanese Military Forces," *TM-E 30-480,* p. 98.
38. Hammel, *Chosin,* p. 71.
39. Ibid.
40. Goulden, *Korea,* p. 358.
41. Ibid., p. 356.
42. *NVA-VC Small Unit Tactics & Techniques Study,* ed. Thomas Pike, part I (n.p.: U.S.A.R.V., n.d.) p. VI-5.
43. *Soviet Combat Regulations of November 1942,* p. 5.
44. *The German Squad in Combat,* p. 21.
45. Ibid., p. 23.
46. Unidentified veteran, as quoted in "Fighting on Guadalcanal," *FMFRP 12-110* (Washington, D.C.: U.S.A. War Office, 1942), p. 63.
47. *Mao's Generals Remember Korea,* p. 33.
48. *The German Squad in Combat,* p. 15.
49. Campbell conversation.
50. *The Bear Went over the Mountain,* p. 97.
51. *Soviet Combat Regulations of November 1942,* p. 14.
52. Unidentified veteran, as quoted in "Fighting on Guadalcanal," *FMFRP 12-110* (Washington, D.C.: U.S.A. War Office, 1942), pp. 66, 67.
53. *Night Movements,* p. 99.
54. *The Bear Went over the Mountain,* p. 174.
55. Mark Bowden, "Blackhawk Down," *Philadelphia (Inquirer) On Line,* http://www3.phillynews.com/packages/somalia/ncv16/rang16.asp, 16 November 1997.

56. Patrick Peterson, Knight Ridder, "Marines Shoot at Anything that Moves," *Jacksonville (NC) Daily News,* 26 March 2003, p. 1.

57. Ellen Knickmeyer and Ravi Nes, Associated Press, "A Cargo Plane Lands at Airport outside Capital," *Jacksonville (NC) Daily News,* 7 April 2003, p. 4A.

Chapter 10: *Combat Deception*

1. Sun Tzu, as quoted in *Sun Tzu's Art of War: The Modern Chinese Interpretation,* by Gen. Tao Hanzhang, trans. Yuan Shibing (New York: Sterling Publishing, 1990), p. 95.
2. *Soviet Combat Regulations of November 1942,* p. 9.
3. Poole, *Phantom Soldier,* p. 31.
4. Hatsumi, *The Essence of Ninjutsu,* pp. 32, 115.
5. Gudmundsson, *Stormtroop Tactics,* p. 21.
6. *Sun Bin's Art of War,* p. 9.
7. Poole, *Phantom Soldier,* pp. 27, 28.
8. *The 30-Year War: 1945-1975,* vol. II (Hanoi: The Gioi Publishers, 2001), p. 175.
9. Ho Khang, *The Tet Mau Than 1968 Event in South Vietnam* (Hanoi: The Gioi Publishers, 2001), p. 53. (This work will henceforth be cited as Khang, *The Tet Mau Than 1968 Event in South Vietnam.)*
10. Nguyen Khac Can and Pham Viet Thuc, *The War 1858 - 1975 in Vietnam* (Hanoi: Nha Xuat Ban Van Hoa Dan Toc, n.d.), no. 544; *The 30-Year War,* p. 173; Khang, *The Tet Mau Than 1968 Event in South Vietnam,* p. 94.
11. Bob O'Bday (Marine machinegunner at Khe Sanh), in conversation with author, 30 July 2000; *The 30-Year War*, p. 177.
12. Ibid.
13. Gen. Vo Nguyen Giap, *Dien Bien Phu,* 6th ed. supplemented (Hanoi: The Gioi Publishers, 1999), "Development of the Battle" insert.
14. Can and Thuc, *The War 1858 - 1975 in Vietnam,* no. 540.
15. *The 30-Year War,* p. 176.
16. Ibid., p. 177.
17. Ibid.
18. Khang, *The Tet Mau Than 1968 Event in South Vietnam,* p. 10.
19. Ibid., p. 131.
20. "Handbook On U.S.S.R. Military Forces," *TM 30-340,* p. V-2; *The Bear Went over the Mountain,* p. 18.
21. Khang, *The Tet Mau Than 1968 Event in South Vietnam*, p. 37.
22. Foreword to *Mao Tse-Tung On Guerrilla Warfare*, trans. Brig.Gen. Samuel B. Griffith USMC (New York: Frederick A. Praeger, Publisher, 1961), from Pelli, "Insurgency, Counterinsurgency, and the Marines in Vietnam," p. 3.

23. Khang, *The Tet Mau Than 1968 Event in South Vietnam*, p. 9.
24. Dung, *Our Great Spring Victory*, p. 113.
25. Ibid., p. 11.
26. Pelli, "Insurgency, Counterinsurgency, and the Marines in Vietnam," p. 2.
27. Khang, *The Tet Mau Than 1968 Event in South Vietnam*, pp. 14-16.
28. Pelli, "Insurgency, Counterinsurgency, and the Marines in Vietnam," p. 2.
29. *The 30-Year War*, p. 81.
30. Pelli, "Insurgency, Counterinsurgency, and the Marines in Vietnam," p. 2.
31. Poole, *Phantom Soldier*, p. 143.
32. Khang, *The Tet Mau Than 1968 Event in South Vietnam*, p. 130.
33. Ibid., p. 46.
34. Ibid., p. 62.
35. Allan R. Millett and Peter Maslowski, *For the Common Defense: A Military History of the United States of America* (New York: The Free Press, 1984), p. 464.
36. Thao, *The Victorious Tay Nguyen Campaign*, p. 68.
37. Khang, *The Tet Mau Than 1968 Event in South Vietnam*, p. 18.
38. Ibid., p. 34.
39. Douglas Pike, *PAVN: People's Army of Vietnam* (Novato, CA: Presidio Press, 1986), p. 268.
40. Khang, *The Tet Mau Than 1968 Event in South Vietnam*, pp. 15, 16.
41. *Mao Tse-tung*, ed. Fremantle, pp. 128, 129.
42. Excerpts from the November 1961 Resolution for the Central Board for South Vietnam, from *The 30-Year War*, p. 94.
43. Khang, *The Tet Mau Than 1968 Event in South Vietnam*, p. 10.
44. Ibid., p. 18.
45. Ibid., p. 10.
46. Dunnigan and Nofi, *Dirty Little Secrets of the Vietnam War*, pp. 265, 266.
47. Khang, *The Tet Mau Than 1968 Event in South Vietnam*, p. 15.
48. Ibid.
49. Ibid., p. 19.
50. Ibid., p. 47.
51. *The 30-Year War*, p. 165.
52. Thai, *How South Vietnam Was Liberated*, p. 17.
53. Ibid., p. 75.
54. Khang, *The Tet Mau Than 1968 Event in South Vietnam*, pp. 7-84.
55. Ibid., p. 128.

56. "Handbook on Japanese Military Forces," *TM-E 30-480,* p. 98.
57. Gudmundsson, *Stormtroop Tactics,* p. 49.
58. *Soviet Combat Regulations of November 1942,* p. 22.
59. Hammel, Chosin, p. 71.
60. NVA-VC Small Unit Tactics & Techniques Study, p. VI-5.
61. Unidentified NCO to Chesty Puller, as quoted in "Fighting on Guadalcanal," *FMFRP 12-110* (Washington, D.C.: U.S.A. War Office, 1942), p. 35.
62. Memo for the record by H.J. Poole.
63. Nolan, *Operation Buffalo,* p. 68.
64. Memo for the record by H.J. Poole.
65. Ibid.
66. Charles Soard (experienced infantry scout and frequent visitor to Vietnam), in telephone conversation with author, 28 March 2001.
67. Khang, *The Tet Mau Than 1968 Event in South Vietnam*, p. 16.
68. Elford, *Devil's Guard,* pp. 306-310.
69. Maj. Gary Tefler, Lt.Col. Lane Rogers, and V. Keith Fleming, Jr., *U.S. Marines in Vietnam: Fighting the North Vietnamese, 1967* (Washington, D.C.: Hist. & Museums Div., HQMC, 1984), p. 179.
70. Col. Hiromichi Yahara, *The Battle for Okinawa,* trans. Roger Pineau and Masatoshi Uehara (New York: John Wiley & Sons, 1995), pp. 25, 26. (This work will henceforth be cited as Yahara, *The Battle for Okinawa.)*
71. Khang, *The Tet Mau Than 1968 Event in South Vietnam*, p. 13.
72. Ibid., p. 47.
73. Edwin P. Hoyt, *The Marine Raiders* (New York: Pocket Books, 1989), p. 113.
74. Pelli, "Insurgency, Counterinsurgency, and the Marines in Vietnam," p. 5.
75. Goulden, *Korea,* p. 349.
76. Lynn Montross and Capt. Nicholas A. Canzona. *The Chosin Reservoir Campaign,* U.S. Marine Operations in Korea 1950-1953, vol. III (Washington, D.C.: Hist. Branch, HQMC, 1957), p. 174.
77. Gen. Westmoreland, as quoted in *The 30-Year War: 1945-1975,* vol. II (Hanoi: The Gioi Publishers, 2001), p. 189.
78. *The Bear Went over the Mountain,* p. 18.
79. "Handbook On U.S.S.R. Military Forces," *TM 30-340,* p. V-27.
80. Ibid.
81. Khang, *The Tet Mau Than 1968 Event in South Vietnam,* pp. 49, 50.
82. Ibid., p. 76.
83. Bermudez, *North Korean Special Forces,* p. 248.
84. Bill D. Ross, *Iwo Jima: Legacy of Valor* (New York: Vintage, 1986), p. 262.
85. O'Bday conversation.

86. Bermudez, *North Korean Special Forces,* p. 252.
87. *Soviet Combat Regulations of November 1942,* p. 92.
88. Bermudez, *North Korean Special Forces,* p. 231.
89. Lupfer, "The Dynamics of Doctrine," *Leavenworth Papers No. 4,* p. 15.
90. Illustration in *Chinese Communist Reference Manual for Field Fortifications*, trans. Intell. Sect., Gen. Staff, Far East Cmd., 1 May 1951, from "Enemy Field Fortifications in Korea," by U.S. Army Corps of Engineers, *Engineer Intelligence Notes,* No. 15 (Washington, D.C.: Army Map Service, January 1952), as contained in McMichael, "The Chinese Communist Forces in Korea," *Leavenworth Research Survey No. 6,* p. 72.
91. *The German Squad in Combat*, p. 30.
92. Fox, *Marine Rifleman,* p. 221.
93. Ibid., p. 69.
94. *The Bear Went over the Mountain,* p. 34.
95. Suvorov, *Inside the Soviet Army,* p. 275.
96. Poole, *The Last Hundred Yards,* p. 227.
97. Arthur Zich and the editors of Time-Life Books, *The Rising Sun: World War II* (Alexandria, VA: Time-Life Books, 1977), p. 122.
98. "Handbook On U.S.S.R. Military Forces," *TM 30-340,* p. V-121.
99. "Handbook on German Military Forces," *TM-E 30-451,* p. 234.
100. Eric Hammel, *Fire in the Streets: The Battle for Hue, Tet 1968* (Pacifica, CA: Pacifica Press, 1991), pp. 304, 305.

Chapter 11: *One-on-One Tactical Decision Making*

1. Gen. Vo Nguyen Giap, *Peoples War—Peoples Army* (New York: Frederick A. Praeger, 1961), pp. 106-108, from *The Battle of Ap Bac, Vietnam,* by David M. Toczek (Westport, CT: Greenwood Press, 2001), pp. 50, 51.
2. Sun Tzu, *The Art of War,* trans. and with intro. by Samuel B. Griffith, p. 82.
3. Peterson, *Mind of the Ninja,* p. 111.
4. Hayes, *Legacy of the Night Warrior,* p. 153.
5. Sun Zi (Sun Tzu), as quoted in *The Thirty-Six Strategies of Ancient China*, by Stefan H. Verstappen (San Francisco: China Books & Periodicals, 1999), p. 185.
6. *The Strategic Advantage: Sun Zi and Western Approaches to War,* ed. Cao Shan (Beijing: New World Press, 1997), p. 6.
7. Peterson, *Mind of the Ninja,* p. 121.
8. Gen. Vo Nguyen Giap, "Once Again We Will Win," in *The Military Art of People's War,* ed. Russel Stetler (New York: Monthly Review Press, 1970), pp. 264, 265.

9. Phillip B. Davidson, *Vietnam at War—The History: 1946-1975* (New York: Oxford Univ. Press, 1988), p. 64.

10. *Mao Tse-tung,* ed. Fremantle, p. 69.

11. Terrence Maitland and Peter McInerney, *Vietnam Experience: A Contagion of War* (Boston: Boston Publishing, 1968), p. 97.

12. Davidson, *Vietnam at War,* p. 64.

13. Dutch Marine exchange officer, in conversation with author, 16 January 2002.

14. Gudmundsson, *Stormtroop Tactics,* pp. 146, 147.

15. Ibid., p. 94.

16. Fox, *Marine Rifleman,* p. 268.

17. Col. Merritt A. Edson, as quoted in "Fighting on Guadalcanal," *FMFRP 12-110* (Washington, D.C.: U.S.A. War Office, 1942), pp. 14-19.

18. Excerpt from *German Field Service Regulations of 1908,* in Gudmundsson, *Stormtroop Tactics,* p. 1.

19. Gudmundsson, *Stormtroop Tactics,* p. 94.

20. "Handbook on German Military Forces," *TM-E 30-451,* p. 209.

21. *The German Squad in Combat*, p. 15.

22. Ibid.

23. Coox, *Nomonhan,* p. 1000.

24. Drea, "Nomonhan," *Leavenworth Papers No. 2,* p. 88.

25. Boeicho boeikenshujo senshishitsu, ed., *Senshi sosho Kantogun (1) Tai So senbi Nomonhan jiken* [The Kwantung Army, *Preparations for the War against the USSR and the Nomonhan Incident,* Official War Hist. Series, vol. 1] (Tokyo: Asagumo shimbunsha, 1969), p. 36, from Drea, "Nomonhan," *Leavenworth Papers No. 2,* p. 19.

26. Drea, "Nomonhan," *Leavenworth Papers No. 2,* p. 19.

27. *Soviet Combat Regulations of November 1942,* p. 12.

28. Ibid., p. 13.

29. Ibid.

30. Ibid., p. 9.

31. *Sun Bin's Art of War,* p. 190.

32. "Some Interesting Facts about Korea: The Forgotten War," *DAV Magazine,* May/June 2000, p. 28.

33. Philip Warner, *Japanese Army of World War II,* vol. 20, Men-at-Arms Series (London: Osprey Publications Ltd., 1972), pp. 23-25.

34. Le Duan, *Ve chien tranh nhan dan Viet Nam* (Hanoi: Nat. Political Publishing House, 1993), pp. 295, 297, from Khang, *The Tet Mau Than 1968 Event in South Vietnam,* p. 32.

35. Gen. Vo Nguyen Giap, in interview from *Dien Bien Phu: The Epic Battle the U.S. Forgot,* by Howard R. Simpson (Washington, D.C.: Brassey's, 1994), p. xxii.

Chapter 12: *When Told to Hold*

1. Erich Ludendorff, *Ludendorff's Own Story,* vol. 1 (New York, 1919), p. 316, from Lupfer, "The Dynamics of Doctrine," *Leavenworth Papers No. 4,* p. 11.
2. Graeme C. Wyne, *If Germany Attacks* (n.p., n.d.; reprint Westport, CT, 1976), p. 295, from Lupfer, "The Dynamics of Doctrine," *Leavenworth Papers No. 4,* p. 27.
3. Lupfer, "The Dynamics of Doctrine," *Leavenworth Papers No. 4,* p. 15.
4. Ibid., p. 15.
5. Ibid.
6. Gudmundsson, *Stormtroop Tactics,* p. 94.
7. Lupfer, "The Dynamics of Doctrine," *Leavenworth Papers No. 4,* p. 13.
8. Ibid., p. 12.
9. Gudmundsson, *Stormtroop Tactics,* p. 94.
10. *The German Squad in Combat*, pp. 26, 27.
11. "Handbook on German Military Forces," *TM-E 30-451,* p. 231.
12. Ibid.
13. "Handbook on Japanese Military Forces," *TM-E 30-480,* pp. 145-150.
14. "Handbook On U.S.S.R. Military Forces," *TM 30-340,* p. V-47.
15. *Soviet Combat Regulations of November 1942,* p. 75.
16. Ibid., p. 23.
17. Ibid., p. 52.
18. Baxter, *Soviet Airland Battle Tactics,* p. 138.
19. *The Bear Went over the Mountain,* p. 7.
20. Ibid., p. 102.
21. "Handbook on German Military Forces," *TM-E 30-451,* p. 239.
22. *Soviet Combat Regulations of November 1942,* p. 56.
23. "Handbook On U.S.S.R. Military Forces," *TM 30-340,* p. V-145.
24. Illustration from "Closing In: Marines in the Seizure of Iwo Jima," by Col. Joseph H Alexander, *Marines in World War II Commemorative Series* (Washington, D.C.: Hist. & Museums Div., HQMC, 1994), p. 8, from Poole, *Phantom Soldier,* p. 60.
25. Yahara, *The Battle for Okinawa,* p. 228.
26. Poole, *Phantom Soldier,* p. 202.
27. McMichael, "The Chinese Communist Forces in Korea," *Leavenworth Research Survey No. 6,* p. 71.
28. Ibid.
29. Ibid.
30. *Soviet Combat Regulations of November 1942,* p. 23.
31. Ibid.

32. *The Bear Went over the Mountain,* p. 38.
33. *The German Squad in Combat*, p. 28.
34. *The Bear Went over the Mountain,* p. 117.
35. Ibid., p. 38.
36. *Night Movements,* p. 95.
37. *Soviet Combat Regulations of November 1942,* p. 24.
38. Lupfer, "The Dynamics of Doctrine," *Leavenworth Papers No. 4,* p. 57.
39. Ibid.
40. "Handbook on German Military Forces," *TM-E 30-451,* p. 239.
41. Ibid., p. 242.
42. *Soviet Combat Regulations of November 1942,* p. 55.
43. Ibid., p. 24.
44. Warner, *Japanese Army of World War II,* pp. 23-25.
45. Yahara, *The Battle for Okinawa,* p. 81.
46. "North Korea Handbook," *PC-2600-6421-94* (Washington, D.C.: U.S. Dept. of Defense, 1994), p. 3-91.
47. Ibid., p. 3-101.
48. "Handbook on the Chinese Communist Army," *DA Pamphlet 30-51,* p. 26.
49. David M. Glantz, *The Soviet Conduct of Tactical Maneuver: Spearhead of the Offensive* (London: Frank Cass, 1991), pp. 72, 73.
50. *The Bear Went over the Mountain,* p. 127.
51. Baxter, *Soviet Airland Battle Tactics,* pp. 132-136.
52. Ibid., p. 127.
53. Ibid., p. 125.
54. Ibid., pp. 125, 126.
55. Ibid., p. 136.
56. Ibid., p. 129.
57. Ibid., p. 130.
58. *The Bear Went over the Mountain,* p. 135.
59. Ibid., p. 22.
60. Ibid., p. 197.
61. *Night Movements,* p. 103.
62. "Handbook On U.S.S.R. Military Forces," *TM 30-340,* p. V-27.
63. Poole, *Phantom Soldier,* p. 196.
64. *Soviet Combat Regulations of November 1942,* p. 56.
65. Col. Z. Zotov, "Defense of a Town," *Soviet Military Review,* no. 6, 1982, p. 48, from *Soviet Airland Battle Tactics,* by William P. Baxter (Novato, CA: Presidio Press, 1986), p. 142.
66. Baxter, *Soviet Airland Battle Tactics,* p. 142.
67. McMichael, "The Chinese Communist Forces in Korea," *Leavenworth Research Survey No. 6,* p. 70.
68. Baxter, *Soviet Airland Battle Tactics,* p. 134.

69. Michael Lee Lanning and Dan Cragg, *Inside the VC and the NVA: The Real Story of North Vietnam's Armed Forces* (New York: Ivy Books, 1992), pp. 206-208.

70. Ibid.

71. *Mao's Generals Remember Korea,* pp. 153-155.

72. *The Bear Went over the Mountain,* p. 102.

73. Maitland and McInerney, *Vietnam Experience,* p. 100.

74. Khang, *The Tet Mau Than 1968 Event in South Vietnam,* p. 14.

75. Col. Huong Van Ba, as quoted in *Portrait of the Enemy,* by David Chanoff and Doan Van Toai (New York: Random House, 1986), p. 152.

76. Trinh Duc, as quoted in *Portrait of the Enemy,* by David Chanoff and Doan Van Toai (New York: Random House, 1986), p. 101.

77. Grant Evans and Kelvin Rowley, *Red Brotherhood at War: Indochina since the Fall of Saigon* (London: Verso, 1984), p. 161.

78. Nayan Chanda, *Brother Enemy: The War after the War* (New York: Collier Books, 1986), p. 361. (This work will henceforth be cited as Chanda, *Brother Enemy.)*

79. Ibid., p. 350.

80. Can and Thuc, *The War 1858 - 1975 in Vietnam,* fig. 754; Andrew Mollo and Digby Smith, *World Army Uniforms Since 1939,* part 2 (Poole, England: Blandford Press, 1981), pp. 18, 19.

81. Chanda, *Brother Enemy,* p. 360.

82. Lin Man Kin, *Sino-Vietnamese War* (Hong Kong: Kingsway International Publications, 1981), p. 60, from Chanda, *Brother Enemy,* p. 361.

83. Foreign Broadcast Information Service—Asia Pacific, 22 March 1979, L-1, from Chanda, *Brother Enemy,* p. 361.

84. Chanda, *Brother Enemy,* pp. 356, 357.

85. Pike, *PAVN,* p. 268.

86. Dr. R. Perry Bosshart (longtime resident of SE Asia and recent visitor to the Binh Moc tunnels), in telephone conversation with author, 19 April 2003.

87. Guthrie e-mails.

88. Soard telephone conversation.

89. *The German Squad in Combat,* p. 43.

90. *Soviet Combat Regulations of November 1942,* p. 112.

91. Ibid., p. 13.

92. "Handbook on Japanese Forces," *TM-E 30-480,* p. 117.

93. Ross, *Iwo Jima,* p. 256.

94. Ibid., p. 241.

95. U.S. Armed Forces Far East, Mil. Hist. Section, "Record of Operations against Soviet Army on Eastern Front (August 1945)," Japanese Monograph No. 154 (Tokyo, 1954), p. 259, from Glantz, "August Storm," *Leavenworth Papers No. 8,* p. 55.

96. Glantz, "August Storm," *Leavenworth Papers No. 8,* p. 87.
97. U.S. Armed Forces Far East, Mil. Hist. Section, "Record of Operations against Soviet Army on Eastern Front (August 1945)," Japanese Monograph No. 154 (Tokyo, 1954), p. 269, from Glantz, "August Storm," *Leavenworth Papers No. 8,* p. 101.
98. *Mao's Generals Remember Korea,* pp. 70, 71.
99. Memo for the record by H.J. Poole.
100. "Handbook on Japanese Military Forces," *TM-E 30-480,* p. 117.
101. George Mikes, *The Hungarian Revolution* (London: Andre Deutsch Ltd., 1957), p. 97.
102. Bermudez, Jr., *North Korean Special Forces,* p. 231.
103. Gudmundsson, *Stormtroop Tactics,* p. 94.

Chapter 13: *At the Listening Post*

1. Irving Werstein, *Guadalcanal* (New York: Thomas Y. Crowell Co., 1963), p. 150.
2. "Pearl Harbor to Guadalcanal," *FMFRP 12-34-I,* by Lt.Col. Frank O. Hough, Maj. Verle E. Ludwig, and Henry I. Shaw, Jr., vol. I, *History of the U.S. Marine Corps Operations in World War II Series* (Washington, D.C.: Hist. Branch, HQMC, n.d.; reprint Quantico, VA: Marine Corps Combat Develop. Cmd., 1989), p. 334.
3. *The German Squad in Combat*, p. 42.
4. *Night Movements*, p. 66.
5. Ibid., p. 67.
6. Ibid., pp. 67, 68.
7. Ibid., p. 62.
8. Ibid., p. 69.
9. Drea, "Nomonhan," *Leavenworth Papers No. 2,* p. 62.
10. *Night Movements*, pp. 70, 71.
11. Ibid., pp. 72, 73.
12. Ibid., p. 73.
13. Ibid., p. 83.
14. Kim, *Secrets of the Ninja,* p. 83.
15. Still frame from *A Tribute to WWII Combat Cameramen of Japan* (Tokyo: Nippon TV, 1995), 85 min., videocassette, in Poole, *Phantom Soldier,* p. 80.
16. *Soviet Combat Regulations of November 1942,* p. 16.
17. Ibid., pp. 16, 17.
18. Ibid., p. 30
19. *The Bear Went over the Mountain,* p. 129.
20. Kim, *The Invisible Ninja,* p. 15.
21. Hayes, *The Mystic Arts of the Ninja,* pp. 132-138.

22. Kim, *Secrets of the Ninja,* pp. 10-30; Hayes, *The Mystic Arts of the Ninja,* p. 135.
23. "Hypnosis," in *The Complete Manual of Fitness and Well-Being* (Pleasantville, NY: The Reader's Digest Assoc., 1984), p. 330.
24. Kim, *The Invisible Ninja,* p. 56.

Chapter 14: *With Contact Patrolling*

1. Scarborough, "Al-Qaeda Adapts to Pursuit Tactics," p. 10.
2. *The Bear Went over the Mountain,* p. 133.
3. Edwin P. Hoyt, *The Marine Raiders* (New York: Pocket Books, 1989), p. 129.
4. *The German Squad in Combat*, pp. 15, 22.
5. Ibid., p. 15.
6. Ibid.
7. Ibid., p. 9.
8. *Soviet Combat Regulations of November 1942,* p. 24.
9. Ibid., p. 26.
10. Memo for the record by H.J. Poole.
11. O'Brien, *Conscripts and Regulars,* p. 239.
12. Memo for the record by H.J. Poole.
13. Ibid.
14. *The German Squad in Combat*, p. 29.
15. Ibid., p. 20.
16. *Soviet Combat Regulations of November 1942,* pp. 22, 29.
17. M.Sgt. Lester J. Ford, Jr., USMC (Ret.) (Vietnam-era platoon sergeant), in conversations with author, 7 April 2002, 27 November 2002, and 17 January 2003.
18. 2d Lt. Karl Metcalf, as quoted in *Conscripts and Regulars: With the Seventh Battalion in Vietnam,* by Michael O'Brien (St. Leonards, Australia: Allen & Unwin, 1995), p. 199.
19. Taras and Zaruz, *Podgotovka Razvedcheka,* p. 533.
20. *Soviet Combat Regulations of November 1942,* p. 26.
21. "Handbook on Japanese Military Forces," *TM-E 30-480,* p. 86.
22. Maj. John L. Zimmerman, *The Guadalcanal Campaign* (Washington, D.C.: Hist. Div., HQMC, 1949), p. 145.
23. "Handbook on German Military Forces," *TM-E 30-451,* p. 214.
24. *The German Squad in Combat*, pp. 4, 19, 20.
25. Ibid., pp. 15, 22.
26. *The Bear Went over the Mountain,* p. 132.
27. Ibid., pp. 45, 46.
28. Nolan, *Operation Buffalo,* p. 77.
29. *Soviet Combat Regulations of November 1942,* p. 18.
30. Ibid., p. 35.

31. *The German Squad in Combat*, p. 5.
32. Ibid., pp. 20, 21.
33. *Soviet Combat Regulations of November 1942,* pp. 19, 20.
34. Nolan, *Operation Buffalo,* p. 77.
35. "Handbook on German Military Forces," *TM-E 30-451,* p. 228.
36. "Handbook On U.S.S.R. Military Forces," *TM 30-340,* pp. V-2, V-23.
37. "Handbook on Japanese Military Forces," *TM-E 30-480,* p. 86.

Chapter 15: *On Point*

1. James Dotson and Kevin Dockery, *Point Man* (New York: William Morrow and Co., 1993), p. 128.
2. *The Bear Went over the Mountain,* p. 197.
3. Dotson and Dockery, *Point Man,* pp. 200, 201.
4. James Corbett, *Jim Corbett's India,* ed. R.E. Hawkins (London: Oxford Univ. Press, 1978), pp. 112, 113.
5. Ibid., p. 25.
6. Ibid.
7. Corbett, *Jungle Lore,* p. 66.
8. Corbett, *Jim Corbett's India,* p. 111.
9. Ibid., pp. 140, 141.
10. Dotson and Dockery, *Point Man,* p. 189.
11. Ibid., p. 168.
12. Corbett, *Jim Corbett's India,* p. 51.
13. Ibid., p. 181.
14. *The German Squad in Combat*, pp. 35-37.
15. Plt.Sgt. C.C. Arnt, as quoted in "Fighting on Guadalcanal," *FMFRP 12-110* (Washington, D.C.: U.S.A. War Office, 1942), p. 25.
16. *Soviet Combat Regulations of November 1942,* pp. 25-28.
17. Ibid., p. 26.
18. *Night Movements,* pp. 48-51.
19. Scott-Donelan, *Tactical Tracking Operations,* pp. 24, 25.
20. *Night Movements,* p. 52.
21. Scott-Donelan, *Tactical Tracking Operations,* p. 24.
22. Ibid., p. 20.
23. Dotson and Dockery, *Point Man,* p. 273.

Chapter 16: *About Tracking an Intruder*

1. Elford, *Devil's Guard,* p. 177.
2. Scott-Donelan, *Tactical Tracking Operations,* p. 3.
3. Ibid., p. 8.

4. Ibid., p. 7.
5. *Patrolling and Tracking,* p. 69.
6. Scott-Donelan, *Tactical Tracking Operations,* p. 66.
7. Ibid., p. 15.
8. *Patrolling and Tracking,* p. 70.
9. Scott-Donelan, *Tactical Tracking Operations,* p. 15.
10. Ibid., pp. 139, 140.
11. Roland Robbins, *Mantracking* (Montrose, CA: Publishers of *Search and Rescue Magazine*, 1977), pp. 9-20.
12. Scott-Donelan, *Tactical Tracking Operations,* pp. 33, 34.
13. Macfarlan, *Exploring the Outdoors with Indian Secrets,* p. 90.
14. Unidentified member of Onslow County (NC) Search and Rescue Team, in conversation with author, August 1995.
15. Scott-Donelan, *Tactical Tracking Operations,* pp. 29-33.
16. Ibid., pp. 30, 31.
17. Ibid., p. 37.
18. Unidentified member of Onslow County (NC) Search and Rescue Team, in conversation with author, August 1995.
19. *U.S. Special Forces Reconnaissance Manual,* p. 50.
20. Ibid.
21. Corbett, *Jungle Lore,* p. 150.
22. *U.S. Special Forces Reconnaissance Manual,* p. 50.
23. Scott-Donelan, *Tactical Tracking Operations,* p. 38.
24. Corbett, *Jungle Lore,* p. 155.
25. *Patrolling and Tracking,* p. 66.
26. *U.S. Special Forces Reconnaissance Manual,* p. 49.
27. Ibid.
28. Scott-Donelan, *Tactical Tracking Operations,* p. 56.
29. Corbett, *Jungle Lore,* pp. 154, 155.
30. Scott-Donelan, *Tactical Tracking Operations,* p. 25.
31. Ibid.
32. Ford conversations.
33. Scott-Donelan, *Tactical Tracking Operations,* p. 26.
34. Ibid., p. 55.
35. Ibid., p. 56.
36. Ibid., pp. 46-53.
37. *Patrolling and Tracking,* p. 68.
38. Scott-Donelan, *Tactical Tracking Operations,* p. 35.
39. *Combat Tracking* (Paluda, Malaysia: British Army Combat Training Center, n.d.), p. 17.
40. Ibid.
41. Khoi, *The Ho Chi Minh Trail,* p. 46.
42. Scott-Donelan, *Tactical Tracking Operations,* pp. 85-94.
43. Ibid., p. 3.

44. *Combat Tracking,* p. 23.
45. Macfarlan, *Exploring the Outdoors with Indian Secrets,* p. 90.
46. Scott-Donelan, *Tactical Tracking Operations,* p. 3.
47. Elford, *Devil's Guard,* p. 255.
48. Scott-Donelan, *Tactical Tracking Operations,* p. 3.
49. Ibid., p. 123.
50. Ibid., p. 125.

Chapter 17: *While Stalking a Quarry*

1. Macfarlan, *Exploring the Outdoors with Indian Secrets,* p. 81.
2. Ibid., p. 93.
3. Ibid., p. 84.
4. Ibid., p. 91.
5. Ibid., p. 92.
6. James Corbett, *Man-Eaters of Kumaon* (New York: Oxford Univ. Press, 1946), p. 115.
7. Lung, *Knights of Darkness,* p. 68.
8. *Night Movements*, pp. 49, 50.
9. "Pearl Harbor to Guadalcanal," *FMFRP 12-34-I,* p. 334.
10. Hammel, *Chosin,* p. 71.
11. "Light Infantry Platoon/Squad," *FM 7-70,* p. D-13.
12. Ibid., pp. D-13, D-14.
13. Unidentified Marine SNCO, in conversation with author, December 1989.
14. *Night Movements*, p. 25.
15. Ibid., p. 24.
16. Ibid., p. 31.
17. Nolan, *Operation Buffalo,* p. 77.
18. Memo for the record by H.J. Poole.
19. Lung, *Knights of Darkness,* p. 94.
20. Corbett, *Jim Corbett's India,* p. 111.
21. Corbett, *Jungle Lore,* p. 66.

Chapter 18: *To Reconnoiter an Enemy Position*

1. Lt.Col. Whitman S. Bartley, *Iwo Jima: Amphibious Epic* (Washington, D.C.: Hist. Branch, HQMC, 1954), p. 161.
2. *Night Movements*, p. 73.
3. Drea, "Nomonhan," *Leavenworth Papers No. 2,* p. 62.
4. *Night Movements*, p. 83.
5. Ibid., pp. 83, 84.

6. Ibid., p. 131.
7. Scott-Donelan, *Tactical Tracking Operations,* p. 24.
8. McMichael, "The Chinese Communist Forces in Korea," *Leavenworth Research Survey No. 6,* p. 58.
9. Hammel, *Chosin,* p. 87.
10. Lanning and Cragg, *Inside the VC and the NVA,* p. 211.
11. Ibid., pp. 211, 212.
12. Ibid., p. 54.
13. Keith William Nolan, *Sappers in the Wire: The Life and Death of Firebase Mary Ann* (College Station, TX: Texas A&M Univ. Press, 1995), p. 123.
14. *Soviet Combat Regulations of November 1942,* p. 25.
15. Ibid., p. 28.
16. Ibid., p. 29.

Chapter 19: *In the Rural Assault*

1. Baxter, *Soviet Airland Battle Tactics,* p. 91.
2. Ibid.
3. Glantz, *The Soviet Conduct of Tactical Maneuver,* p. 54.
4. Ibid., p. 17.
5. Ibid., p. 19.
6. Glantz, "August Storm," *Leavenworth Papers No. 8,* p. 170.
7. Glantz, *The Soviet Conduct of Tactical Maneuver,* p. 25.
8. Ibid., p. 23.
9. Baxter, *Soviet Airland Battle Tactics,* p. 107.
10. Ibid., p. 113.
11. Ibid.
12. Ibid., p. 119.
13. Glantz, *The Soviet Conduct of Tactical Maneuver,* p. 27.
14. Ibid., p. 28.
15. Ibid., p. 37.
16. Ibid., p. 30.
17. Ibid., p. 52.
18. Ibid., p. 193.
19. Baxter, *Soviet Airland Battle Tactics,* pp. 99-104.
20. Ibid., p. 105.
21. *The Bear Went over the Mountain,* p. 133.
22. *Night Movements,* p. 116.
23. "Handbook On U.S.S.R. Military Forces," *TM 30-340,* p. V-19.
24. Glantz, *The Soviet Conduct of Tactical Maneuver,* p. 24.
25. Lupfer, "The Dynamics of Doctrine," *Leavenworth Papers No. 4,* p. 37.
26. *The German Squad in Combat,* p. 15.

27. "Handbook on German Military Forces," *TM-E 30-451,* p. 214.
28. Ibid., p. 226.
29. Gudmundsson, *Stormtroop Tactics*, pp. 146, 147.
30. Lupfer, "The Dynamics of Doctrine," *Leavenworth Papers No. 4,* p. 53; Gudmundsson, *Stormtroop Tactics,* pp. 147-149.
31. Glantz, "August Storm," *Leavenworth Papers No. 8,* p. 5.
32. Ibid., p. 18.
33. Ibid., p. 16.
34. Ibid., p. 29.
35. Nguyen Van Mo, as quoted in *Portrait of the Enemy,* by David Chanoff and Doan Van Toai (New York: Random House, 1986), p. 162.
36. "North Korea Handbook," *PC-2600-6421-94,* p. 3-121.
37. "Historical Study—Night Combat," *DA Pamphlet 20-236,* p. 22.
38. Canby, "Classic Light Infantry and New Technology," p. 51.
39. Lupfer, "The Dynamics of Doctrine," *Leavenworth Papers No. 4,* p. 42.
40. *British Official History 1918,* vol 1. (n.p., n.d.), p. 166, from Lupfer, "The Dynamics of Doctrine," *Leavenworth Papers No. 4,* p. 50.
41. Lupfer, "The Dynamics of Doctrine," *Leavenworth Papers No. 4,* p. 44.
42. Ibid., p. 43.
43. Gudmundsson, *Stormtroop Tactics,* p. 143.
44. Ibid., pp. 49, 191.
45. Burke Davis, *Marine* (New York: Bantam, 1964), p. 140.
46. "Fighting on Guadalcanal," *FMFRP 12-110* (Washington, D.C.: U.S.A. War Office, 1942), p. 2.
47. Ibid., p. 22.
48. McMichael, "The Chinese Communist Forces in Korea," *Leavenworth Research Survey No. 6,* p. 63.
49. Poole, *Phantom Soldier,* pp. 85, 86.
50. *Night Movements*, p. 122.
51. Hammel, *Chosin,* p. 71.
52. Campbell conversation.
53. Dung, *Our Great Spring Victory,* p. 113.
54. Nolan, *Operation Buffalo,* p. 298.
55. Thao, *The Victorious Tay Nguyen Campaign,* pp. 93, 94.
56. Ibid., p. 53.
57. Ibid.
58. Ibid., p. 88.
59. *The Bear Went over the Mountain,* p. 124.
60. Ibid., p. 117.
61. Yahara, *The Battle for Okinawa,* pp. 36, 213.
62. Ibid., p. 59.
63. Glantz, "August Storm," *Leavenworth Papers No. 8,* p. 69.
64. Ibid., pp. 153-155.

65. Ibid., p. 54.
66. Ibid., p. 132.
67. Ibid., p. 126.
68. Ibid., p. 176.
69. Alexander L. George, *The Chinese Communist Army in Action: The Korean War and Its Aftermath* (New York: Columbia Univ. Press, 1967), p. 3, from McMichael, "The Chinese Communist Forces in Korea," *Leavenworth Research Survey No. 6,* p. 59.
70. Lynn Montross and Nicholas A. Canzona, "U.S. Marine Operations in Korea, 1950-1953," vol. 3 of *The Chosin Reservoir Campaign* (Washington, D.C.: Hist. Branch, HQMC, 1957; reprint St. Clair Shores, MI: Scholarly Press, 1976), p. 92, from McMichael, "The Chinese Communist Forces in Korea," *Leavenworth Research Survey No. 6,* p. 60.
71. "Handbook on the Chinese Communist Army," *DA Pamphlet 30-51,* p. 29.
72. Dunnigan and Nofi, *Dirty Little Secrets of the Vietnam War,* p. 266.
73. Ibid., pp. 279, 280.
74. Lt.Col. Jack Westerman, as quoted in *U.S. Marines in Vietnam: An Expanding War—1966,* by Jack Shulimson (Washington, D.C.: Hist. & Museums Div., HQMC, 1982), p. 186.
75. Nolan, *Sappers in the Wire,* p. 124.
76. Ibid., p. 125.
77. Ibid., front-cover jacket.
78. Suvorov, *Inside the Soviet Army,* p. 275; Lung, *Knights of Darkness,* p. 28.
79. Lung, *Knights of Darkness,* p. 30.
80. Hammel, *Chosin,* p. 55.
81. Ibid., pp. 89, 90.
82. Ibid., p. 68.
83. Nolan, *Sappers in the Wire,* p. 125.
84. Ibid., pp. 124, 125.
85. "Handbook on German Military Forces," *TM-E 30-451,* p. 239.
86. "Handbook on Japanese Military Forces," *TM-E 30-480,* p. 86.
87. Elford, *Devil's Guard,* p. 101.
88. Sasso, "Soviet Night Operations in World War II," *Leavenworth Papers No. 6,* pp. 14, 15; Coox, *Nomonhan,* p. 999; "Handbook On U.S.S.R. Military Forces," *TM 30-340,* p. V-2; *The Bear Went over the Mountain,* p. 18.
89. Glantz, "August Storm," *Leavenworth Papers No. 8,* pp. 60-70.
90. Thao, *The Victorious Tay Nguyen Campaign,* p. 132.
91. Sasso, "Soviet Night Operations in World War II," *Leavenworth Papers No. 6,* p. 43.

92. Ibid., pp. 35, 36.
93. "Handbook on Japanese Military Forces," *TM-E 30-480,* p. 97.
94. Ibid., p. 86.
95. McMichael, "The Chinese Communist Forces in Korea," *Leavenworth Research Survey No. 6,* p. 62.
96. Ibid., p. 63.
97. Poole, *Phantom Soldier,* pp. 85, 86.
98. Elford, *Devil's Guard,* p. 200.
99. *Sun Bin's Art of War,* p. 190.
100. *Mao Tse-tung,* ed. Fremantle, p. 138.
101. David M. Toczek, *The Battle of Ap Bac, Vietnam* (Westport, CT: Greenwood Press, 2001), pp. 50, 51.
102. "Historical Study—Night Combat," *DA Pamphlet 20-236,* p. 22.
103. Gen. Atomiya, as quoted in Yahara, *The Battle for Okinawa,* p. 12.
104. Yahara, *The Battle for Okinawa,* p. 61.
105. Illustration from "Closing In: Marines in the Seizure of Iwo Jima," by Col. Joseph H Alexander, *Marines in World War II Commemorative Series* (Washington, D.C.: Hist. & Museums Div., HQMC, 1994), p. 8, from Poole, *Phantom Soldier,* p. 60.
106. Yahara, *The Battle for Okinawa,* p. 35.
107. Ibid., p. 35.
108. Poole, *Phantom Soldier,* p. 79.
109. Yahara, *The Battle for Okinawa,* p. 15.
110. Ibid., p. 53.
111. Still frame from *A Tribute to WWII Combat Cameramen of Japan* (Tokyo: Nippon TV, 1995), 85 min., videocassette, in Poole, *Phantom Soldier,* p. 79.
112. Poole, *Phantom Soldier,* pp. 81, 82.
113. Ibid., p. 82.
114. Yahara, *The Battle for Okinawa,* pp. 25, 26.
115. Ibid., p. 8.
116. Ibid., p. 15.
117. Ibid., p. 57.
118. Unidentified Japanese POW, as quoted in Yahara, *The Battle for Okinawa,* p. 219.
119. Yahara, *The Battle for Okinawa,* pp. 56, 215.
120. "History of the Sixth Marine Division" (n.p., n.d.), from Yahara, *The Battle for Okinawa,* p. 61.
121. "Handbook on Japanese Military Forces," *TM-E 30-480,* p. 117.
122. Yahara, *The Battle for Okinawa,* p. 60.
123. *Mao's Generals Remember Korea,* p. 154.
124. Ibid.
125. Ibid., p. 168.

126. Ibid., p. 155.
127. Giap, *Dien Bien Phu*, p. 113.
128. Ibid., p. 99.
129. Ibid., p. 121.
130. Ibid., p. 129.
131. Ibid., p. 121.
132. *The Battle of Dien Bien Phu,* videocassette #4010.
133. Ibid.
134. Giap, *Dien Bien Phu*, p. 109.
135. Ibid., p. 116.
136. Ibid.
137. Ibid., p. 109.
138. Lung, *Knights of Darkness,* p. 27.
139. Giap, *Dien Bien Phu*, p. 133.
140. Lung, *Knights of Darkness,* p. 112.
141. Ibid.
142. Ibid., p. 27.
143. Memo for the record by H.J. Poole.
144. Soard telephone conversation.
145. O'Bday conversation; Maj.Gen. Leslie M. Palm USMC (Ret.), in conversation with author, 17 May 2002.
146. *The 30-Year War,* p. 177.
147. O'Bday conversation.
148. Poole, *Phantom Soldier,* p. 83.
149. *The Battle of Dien Bien Phu,* videocassette #4010.
150. Bermudez, Jr., *North Korean Special Forces,* p. 248.
151. Ibid., p. 251.
152. McMichael, "The Chinese Communist Forces in Korea," *Leavenworth Research Survey No. 6,* p. 55.
153. Ibid., p. 56.
154. *The Battle of Dien Bien Phu,* videocassette #4010.

Chapter 20: *For Attacking Cities*

1. Greg Jaffe, "Street Smarts: As Threats Evolve, Marines Should Learn Skill of Combat in Cities," *Wall Street Journal,* 22 August 2002, p. A1.
2. John J. Lumpkin, Associated Press, "Military Visits Urban Tactics against Baghdad," *Jacksonville (NC) Daily News,* 23 September 2002, pp. 1A, 4A.
3. Paul Churchill Hutton III, "Weapons of Restraint: Emphasis on Limiting Non-Combatant Casualties Is Good Reason for Developing New Non-Lethal Weapons," *Armed Forces Journal International,* May 2000, p. 52.

4. Anatol Lieven, "The World Turned Upside Down: Military Lessons of the Chechen War," *Armed Forces Journal International,* August 1998, p. 40.
5. Marshal Georgi K. Zhukov, *Reminiscences and Reflections* (n.p., 1974), as quoted in *Warriors' Words—A Quotation Book,* by Peter G. Tsouras (London: Cassel Arms & Armour, 1992), p. 282.
6. Lieven, "The World Turned Upside Down," p. 40.
7. "Handbook On U.S.S.R. Military Forces," *TM 30-340,* pp. V-121, V-122.
8. *Soviet Combat Regulations of November 1942,* p. 51.
9. Nora Levin, *The Holocaust* (New York: Thomas Y. Crowell Co., 1968), pp. 343-352.
10. "Nepal and Bhutan Country Studies," *DA Pamphlet 550-35,* Area Handbook Series (Washington, D.C.: Hdqts. Dept. of the Army, 1993), pp. 1-30.
11. Dung, *Our Great Spring Victory,* p. 51.
12. Lt.Col. Robert W. Lamont, "'Urban Warrior'—A View from North Vietnam," *Marine Corps Gazette,* April 1999, p. 33.
13. Thai, *How South Vietnam Was Liberated,* pp. 164, 165.
14. Thao, *The Victorious Tay Nguyen Campaign,* pp. 52, 53.
15. Ibid., pp. 52-54.
16. Khang, *The Tet Mau Than 1968 Event in South Vietnam,* p. 66.
17. Thai, *How South Vietnam Was Liberated,* p. 119.
18. Ibid., p. 235.
19. Ibid., p. 232.
20. Ibid., pp. 252, 253.
21. Ibid., p. 254.
22. Ibid., pp. 237, 238.
23. Hutton, "Weapons of Restraint," p. 53.
24. Jaffe, "Street Smarts," p. 1.
25. "U.S. Hopes to Take Baghdad Quickly," from Knight Ridder, *Jacksonville (NC) Daily News,* 26 March 2003, p. 4A.
26. ABC's *Nightly News,* 9 April 2003.
27. Seth Stern, "How the U.S. Plans to Take Control of Baghdad," *Christian Science Monitor,* 7 April 2003, p. 9.
28. Lieven, "The World Turned Upside Down," p. 43.

Chapter 21: *During an Urban Defense*

1. John Shaw and the editors of Time-Life Books, *Red Army Resurgent* (Chicago: Time-Life Books, 1979), p. 158.
2. Ibid., p. 167.
3. Ibid., p. 161.
4. Ibid.

5. Ibid., pp. 163-165.

6. Ibid.

7. Robert R. Smith, "Triumph in the Philippines," *The War in the Pacific,* U.S. Army in World War II Series (Washington, D.C.: U.S. Army Center of Mil. Hist., 1963 [reprint 1993]), pp. 241-242; and Richard Connaughton, John Pimlott, and Duncan Anderson, "The Battle for Manila" (Novato, CA: Presidio Press, 1995), p. 142; from "The Battle of Manila," by Thomas M. Huber, in CSI Home Publications Research MHIST [database on line] (Ft. Leavenworth, KS: Combat Studies Inst., U.S. Army Cmd. & Gen. Staff College, n.d. [updated 30 September 2002; cited 1 January 2003]), available from the CSI website at www-cgsc.army.mil/csi/research/mout/mouthuber. asp, p. 4.

8. Robert R. Smith, "Triumph in the Philippines," *The War in the Pacific,* U.S. Army in World War II Series (Washington, D.C.: U.S. Army Center of Mil. Hist., 1963 [reprint 1993]), pp. 244-245, from "The Battle of Manila," by Thomas M. Huber, in CSI Home Publications Research MHIST [database on line] (Ft. Leavenworth, KS: Combat Studies Inst., U.S. Army Cmd. & Gen. Staff College, n.d. [updated 30 September 2002; cited 1 January 2003]), available from the CSI website at www-cgsc.army.mil/csi/research/mout/mouthuber. asp, p. 4. (The initial work will henceforth be cited as Smith, "Triumph in the Philippines.")

9. Thomas M. Huber, "The Battle of Manila," in CSI Home Publications Research MHIST [database on line] (Ft. Leavenworth, KS: Combat Studies Inst., U.S. Army Cmd. & Gen. Staff College, n.d. [updated 30 September 2002; cited 1 January 2003]), available from the CSI website at www-cgsc.army.mil/csi/research/mout/mouthuber. asp, p. 15. (This work will henceforth be cited as Huber, "The Battle of Manila.")

10. Ibid.

11. Smith, "Triumph in the Philippines," p. 308; and Richard Connaughton, John Pimlott, and Duncan Anderson, "The Battle for Manila" (Novato, CA: Presidio Press, 1995), pp. 189-191; and U.S. Army, "37th Div. Report after Action" (Okinawa: Hdqts., 37th Infantry Div., 1945), p. 51; from Huber, "The Battle of Manila," p. 4.

12. Huber, "The Battle of Manila," p. 8.

13. Ibid.

14. Smith, "Triumph in the Philippines," pp. 287-290, from Huber, "The Battle of Manila," p. 10.

15. Huber, "The Battle of Manila," p. 8.

16. Smith, "Triumph in the Philippines," pp. 280-283; and U.S. Army, "37th Div. Report after Action" (Okinawa: Hdqts., 37th Infantry Div., 1945), pp. 276-280; from Huber, "The Battle of Manila," p. 9.

17. Smith, "Triumph in the Philippines," pp. 287-290, from Huber, "The Battle of Manila," p. 10.

18. Smith, "Triumph in the Philippines," p. 308; and Richard Connaughton, John Pimlott, and Duncan Anderson, "The Battle for Manila" (Novato, CA: Presidio Press, 1995), pp. 189-191; and U.S. Army, "37th Div. Report after Action" (Okinawa: Hdqts., 37th Infantry Div., 1945), p. 51; from Huber, "The Battle of Manila," p. 4.

19. Smith, "Triumph in the Philippines," pp. 280-283; and Stanley L. Frankel, "The 37th Infantry Div. in World War II (Washington, D.C.: Infantry Journal Press, 1948), pp. 276-280; from Huber, "The Battle of Manila," p. 9.

20. Ibid.

21. Huber, "The Battle of Manila," p. 12.

22. U.S. Army, "37th Div. Report after Action" (Okinawa: Hdqts., 37th Infantry Div., 1945), p. 51, from Huber, "The Battle of Manila," p. 8. (The initial work will henceforth be cited as "37th Div. Report after Action.")

23. Ibid.

24. Stanley L. Frankel, "The 37th Infantry Div. in World War II (Washington, D.C.: Infantry Journal Press, 1948), pp. 273, 275; and "37th Div. Report after Action," p. 49; from Huber, "The Battle of Manila," p. 8.

25. Huber, "The Battle of Manila," p. 12.

26. Smith, "Triumph in the Philippines," p. 253, from Huber, "The Battle of Manila," p. 6.

27. Huber, "The Battle of Manila," p. 7.

28. Smith, "Triumph in the Philippines," p. 254; and "37th Div. Report after Action," p. 43; from Huber, "The Battle of Manila," p. 7.

29. Huber, "The Battle of Manila," pp. 2, 3.

30. Smith, "Triumph in the Philippines," pp. 259-260; and Richard Connaughton, John Pimlott, and Duncan Anderson, "The Battle for Manila" (Novato, CA: Presidio Press, 1995), p. 109; and "37th Div. Report after Action," p. 45; and Stanley L. Frankel, "The 37th Infantry Division in World War II" (Washington, D.C.: Infantry Journal Press, 1948), p. 272; from Huber, "The Battle of Manila," p. 7.

31. Huber, "The Battle of Manila," pp. 11, 12.

32. Ross, *Iwo Jima,* p. 333.

33. Still frame from *A Tribute to WWII Combat Cameramen of Japan* (Tokyo: Nippon TV, 1995), 85 min., videocassette, in Poole, *Phantom Soldier,* p. 79.

34. Huber, "The Battle of Manila," p. 12.

35. Read, Anthony and David Fischer, *The Fall of Berlin* (New York: Da Capo Press, 1995), p. 385.

36. Ibid.

37. Editor's footnote, in *From Moscow to Berlin: Marshal Zhukov's Greatest Battles,* by Marshal Georgi K. Zhukov, ed. and intro. by Harrison E. Salisbury, trans. Theodore Shabad, War and Warrior Series (Costa Mesa, CA: The Noontide Press, 1991), pp. 288, 289.

38. Read and Fischer, *The Fall of Berlin,* pp. 387, 388.

39. *The Doomed City of Berlin,* Cities at War Series (Simon & Schuster Video, 1986), 65 min., videocassette #62159-9.

40. Ibid.

41. Ibid.

42. Read and Fischer, *The Fall of Berlin,* p. 385.

43. Marshal Georgi K. Zhukov, *From Moscow to Berlin: Marshal Zhukov's Greatest Battles,* ed. and intro. by Harrison E. Salisbury, trans. Theodore Shabad, War and Warrior Series (Costa Mesa, CA: The Noontide Press, 1991), pp. 279, 286.

44. Read and Fischer, *The Fall of Berlin,* p. 411.

45. Ibid., p. 391.

46. Ibid., p. 387.

47. Ibid., pp. 387, 388.

48. Poole, *The Last Hundred Yards,* p. 272.

49. Read and Fischer, *The Fall of Berlin,* p. 386.

50. *The Doomed City of Berlin,* videocassette #62159-9.

51. Zhukov, *From Moscow to Berlin,* p. 284.

52. *The Doomed City of Berlin,* videocassette #62159-9.

53. Read and Fischer, *The Fall of Berlin,* p. 387.

54. Ibid., p. 396.

55. Ibid., p. 387.

56. Ibid., p. 407.

57. Ibid., p. 393.

58. Ibid., p. 406.

59. Ibid., pp. 393-395.

60. Ibid., p. 415.

61. Ibid., p. 389.

62. Ibid., p. 422.

63. Ibid., pp. 423, 424.

64. Ibid., pp. 449, 450.

65. *The Doomed City of Berlin,* videocassette #62159-9.

66. Zhukov, *From Moscow to Berlin,* p. 288.

67. *The Doomed City of Berlin,* videocassette #62159-9.

68. *The Korean War: History and Tactics,* ed. David Rees (New York: Crescent Books, 1984), p. 45.

69. Ibid.

70. Max Hastings, *The Korean War* (New York: Simon and Schuster, 1987), p. 112. (This work will henceforth be cited as Hastings, *The Korean War.*)

71. Warner, *Japanese Army of World War II,* pp. 23-25.
72. *The Korean War: History and Tactics,* p. 45.
73. S.Sgt. Lee Bergee, as quoted in *The Korean War: Pusan to Chosin, An Oral History,* by Donald Knox (San Diego: Harcourt Brace Jovanovich, 1985), p. 289.
74. PFC Fred Davidson, as quoted in *The Korean War: Pusan to Chosin, An Oral History,* by Donald Knox (San Diego: Harcourt Brace Jovanovich, 1985), pp. 290, 291.
75. PFC Win Scott, as quoted in *The Korean War: Pusan to Chosin, An Oral History,* by Donald Knox (San Diego: Harcourt Brace Jovanovich, 1985), p. 291.
76. PFC Jack Wright, quoted in *The Korean War: Pusan to Chosin, An Oral History,* by Donald Knox (San Diego: Harcourt Brace Jovanovich, 1985), pp. 292, 293.
77. Ross, *Iwo Jima,* p. 149.
78. PFC Win Scott, as quoted in *The Korean War: Pusan to Chosin, An Oral History,* by Donald Knox (San Diego: Harcourt Brace Jovanovich, 1985), p. 293.
79. R.W. Thompson, *Daily Telegraph,* as quoted in Hastings, *The Korean War,* p. 112.
80. Hammel, *Fire in the Streets,* p. 296.
81. George W. Smith, *The Siege at Hue* (New York: Ballantine Publishing, 1999), p. 203.
82. Nicholas Warr, *Phase Line Green: The Battle for Hue, 1968* (Annapolis, MD: Naval Inst. Press, 1997), pp. 155-160.
83. Poole, *Phantom Soldier,* p. 178.
84. Ibid., p. 194.
85. Cooling, "Russia's 1994-96 Campaign for Chechnya," p. 60.
86. Lieven, "The World Turned Upside Down," p. 42.
87. Gen. Mikhail Surkov, as quoted in "The World Turned Upside Down: Military Lessons of the Chechen War," by Anatol Lieven, *Armed Forces Journal International,* August 1998, pp. 40, 41.
88. Cooling, "Russia's 1994-96 Campaign for Chechnya," p. 62.
89. Lieven, "The World Turned Upside Down," p. 43.
90. Cooling, "Russia's 1994-96 Campaign for Chechnya," p. 62.
91. Ibid., p. 61.
92. Ibid.
93. Thomas and Grau, "Russian Lessons Learned from the Battles for Grozny," p. 45.
94. Ibid., p. 46.
95. Ibid., pp. 45, 46.
96. Hutton, "Weapons of Restraint," p. 55.
97. Cooling, "Russia's 1994-96 Campaign for Chechnya," p. 62.
98. Ibid., pp. 63-66.

99. Thomas and Grau, "Russian Lessons Learned from the Battles for Grozny," p. 46.
100. Cooling, "Russia's 1994-96 Campaign for Chechnya," p. 68.
101. Thomas and Grau, "Russian Lessons Learned from the Battles for Grozny," p. 47.
102. Ibid., 48.

Chapter 22: *The Rising Value of the "Little Picture"*

1. Sun Tzu, *The Art of War,* trans. and with intro. by Samuel B. Griffith, p. 84.
2. *The Bear Went over the Mountain,* p. 202.
3. Ibid., p. 139.
4. William S. Lind (author of *The Maneuver Warfare Handbook),* in telephone conversation with author, February 2001.
5. Lt.Col. Ian Thomas (2d Royal Gurkha Rifles Commander), in e-mail from iannathomas@hotmail.com to author, 19 January 2003.
6. *The Bear Went over the Mountain,* p. 75.
7. Shamil Basayev (Chechen guerrilla leader), as quoted in "Russia's 1994-96 Campaign for Chechnya: A Failure in Shaping the Battlespace," by Maj. Norman L. Cooling, *Marine Corps Gazette,* October 2001, p. 58.
8. Fred Weir, "Putin Battles Political Fallout of Chechnya Fight," *Christian Science Monitor,* 16 May 2003.

Chapter 23: *How the Tiger Is Born*

1. Peterson, *Mind of the Ninja,* p. 105.
2. Hayes, *Legacy of the Night Warrior,* p. 152.
3. Gudmundsson, *Stormtroop Tactics,* pp. 147-149.
4. Ibid., p. 143.
5. Ibid., p. 87.
6. Lupfer, "The Dynamics of Doctrine," *Leavenworth Papers No. 4,* p. 56.
7. Lupfer, "The Dynamics of Doctrine," *Leavenworth Papers No. 4,* p. 43.
8. *The German Squad in Combat,* pp. 8-13.
9. *Night Movements,* pp. 37-133.
10. Ibid., p. 29.
11. Ibid., p. 39.
12. Zich et al, *The Rising Sun,* p. 35.
13. Ibid., p. 30.
14. *Night Movements,* p. 33.

15. Ibid., p. 51.
16. "How the Japanese Army Fights," by Lt.Col. Paul W. Thompson, Lt.Col. Harold Doud, Lt.Col. John Scofield, and the editorial staff of "The Infantry Journal," *FMFRP 12-22* (New York: Penguin Books, 1942; reprint Washington, D.C.: HQMC, 1989), p. 51.
17. *Mao's Generals Remember Korea,* pp. 70, 71.
18. Harriet Fast Scott and William F. Scott, *The Armed Forces of the U.S.S.R.,* p. 316.
19. Baxter, *Soviet Airland Battle Tactics,* p. 104.
20. Glantz, "August Storm," *Leavenworth Papers No. 8,* pp. 201, 202.
21. Ibid., pp. 130-132.
22. *The Bear Went over the Mountain,* p. 202.
23. Bermudez, *North Korean Special Forces,* p. 222.
24. Ibid.
25. Dunnigan and Nofi, *Dirty Little Secrets of the Vietnam War,* p. 279.
26. Ibid., pp. 276-279.
27. Lanning and Cragg, *Inside the VC and the NVA,* p. 56.
28. Ibid.
29. Ibid., p. 55.
30. Ibid.
31. William S. Lind, *The Maneuver Warfare Handbook* (Boulder, CO: Westview Press, 1985), p. 25.
32. "Marine Rifle Company/Platoon," *FMFM 6-4* (Washington, D.C.: HQMC, 1978), p. 9.
33. Fox, *Marine Rifleman,* pp. 237, 238.

Chapter 24: *Field Proficiency Has No Substitute*

1. ABC's *Nightly News,* 6 March 2003.
2. Lt.Col. Jeffrey W. Bolander, "The Dragon's New Claws: PLA Publication Sheds New light on Chinese Military Tactics," *Camp Lejeune (NC) Globe,* 8 March 2001, p. 9C.
3. Brad Knickerbocker, "In Era of High-Tech Warfare, 'Friendly Fire' Risk Grows," *Christian Science Monitor,* 14 January 2003, p. 3.
4. Poole, *Phantom Soldier,* p. 144.
5. Pike, *PAVN,* p. 268.
6. *The Bear Went over the Mountain,* p. 3.
7. Todd Pitman, Associated Press, "Coalition Force Explodes Weapons Cache in Caves," *Jacksonville (NC) Daily News,* 11 May 2002, pp. 1-6.

8. Matt Kelley, Associated Press, "Army Gives MIT $50 Million Grant to Develop Futuristic Battle Uniform," Jacksonville (NC) Daily News, 14 March 2002, p. 10A.

9. Lisa Hoffman, Scripps Howard, "Among Troops' Problems in Afghanistan—Their Own Stuff," *Jacksonville (NC) Daily News,* 6 October 2002, p. 9A.

10. *The Bear Went over the Mountain,* p. 176.

11. Lupfer, "The Dynamics of Doctrine," *Leavenworth Papers No. 4,* p. 56.

12. Baxter, *Soviet Airland Battle Tactics,* p. 92.

13. Ibid., p. 93.

14. Lupfer, "The Dynamics of Doctrine," *Leavenworth Papers No. 4,* p. 56.

15. Maitland and McInerney, *Vietnam Experience,* p. 97.

16. Wm. Darryl Henderson, *Cohesion: The Human Element in Combat* (Washington, D.C.: Nat. Defense Univ. Press, 1985), p. 65.

17. Ibid., p. 68.

18. Sasso, "Soviet Night Operations in World War II," *Leavenworth Papers No. 6,* pp. 35, 36.

19. *The Bear Went over the Mountain,* p. 197.

20. Jaffe, "Street Smarts," *Wall Street Journal,* 22 August 2002, p. 1.

21. Evan Thomas and Martha Brant, "The Education of Tommy Franks," *Newsweek,* 19 May 2003, p. 29.

22. ABC's *Nightly News,* 24 March 2003.

23. Brad Knickerbocker, "Guerrilla Tactics vs. U.S. War Plan," *Christian Science Monitor,* 25 March 2003, p. 1.

24. ABC's *Nightly News,* 27 March 2003.

25. ABC's *Morning News,* 28 March 2003.

26. ABC's *Nightly News,* 30 March 2003.

27. National Public Radio's *Morning Edition,* 3 April 2003.

28. Evan Thomas and John Barry, "A Plan under Attack," *Newsweek,* 7 April 2003, p. 29.

29. Burt Herman, Associated Press, "Risky Job for Marines," *Jacksonville (NC) Daily News,* 13 April 2003, p. 2A.

30. Dafna Linzer, Associated Press, "Expecting the Unexpected in Baghdad," *Jacksonville (NC) Daily News,* 18 May 2003.

31. CBS's *60 Minutes,* 4 May 2003.

32. Asprey, *War in the Shadows,* pp. 37, 38; *Random House Encyclopedia,* electronic ed., s.v. "Hun."

33. *The Concise Encyclopedia of Ancient Civilizations,* ed. Janet Serlin Garber (New York: Franklin Watts, 1978), pp. 30, 31.

34. *Random House Encyclopedia,* electronic ed., s.v. "Turkmen."

35. *Webster's New Twentieth Century Dictionary,* 2d ed., s.v. "Turkoman."

36. Cameron W. Barr, "Kirkuk, a Mirror of Iraq's Schisms," *Christian Science Monitor,* 4 March 2003, p. 6.

37. Asprey, *War in the Shadows,* p. 49.

38. Ibid., p. 61.

39. Ibid., p. 51.

40. Scott Peterson, "Iraq: Saladin to Saddam," *Christian Science Monitor,* 4 March 2003, p. 12; "Kublai Khan and the Mongol Empire," World of Wonder Series, by Newspaper in Education, *Jacksonville (NC) Daily News,* 24 March 2003, p. 8B.

41. *The Travels of Marco Polo,* revised from Marsden's trans., ed. and intro. by Manuel Komroff (New York: Liveright Publishing Corp., 1926), p. 95.

42. Peterson, "Iraq," p. 12.

43. Samuel B. Griffith, in translator's intro. to *The Art of War,* by Sun Tzu (New York: Oxford Univ. Press, 1963), p. xi.

44. Lung, *Knights of Darkness,* pp. 14, 15.

45. Ibid.

46. Hatsumi, *The Essence of Ninjutsu,* pp. 12, 55; Hayes, *Ninjutsu: The Art of the Invisible Warrior*, p. 151.

47. Hayes, *Ninjutsu: The Art of the Invisible Warrior*, p. 151.

48. Hatsumi, *The Essence of Ninjutsu,* pp. 33, 39.

49. Hatsumi, *Ninjutsu: History and Tradition,* p. 7; Hatsumi, *The Essence of Ninjutsu,* pp. 101, 102; Kim, *The Invisible Ninja,* p. 13.

50. Hayes, *Legacy of the Night Warrior,* p. 26.

51. Hayes, *Ninjutsu: The Art of the Invisible Warrior*, p. 153.

52. Dr. David H. Reinke (expert on parapsychology and Eastern religions), in telephone conversation with author, June 2001.

53. Hayes, *The Mystic Arts of the Ninja,* p. 102.

54. Hatsumi, *The Essence of Ninjutsu,* p. 39.

55. Ibid., p. 40.

56. "Hypnosis," p. 330.

57. *The Travels of Marco Polo,* revised from Marsden's trans., ed. Manuel Komroff (New York: Modern Library, 1953), pp. 44, 45.

58. Editor's footnote, *The Travels of Marco Polo,* revised from Marsden's trans., ed. and intro. by Manuel Komroff (New York: Liveright Publishing Corp., 1926), p. 45.

59. Kim, *The Invisible Ninja,* pp. 1, 56.

60. "Hypnosis," p. 330.

61. Ibid.

62. *Night Movements,* p. 69.

63. Reinke e-mails.

64. Hayes, *The Mystic Arts of the Ninja,* p. 135.

65. Kim, *Secrets of the Ninja,* p. 10.

66. Hayes, *The Mystic Arts of the Ninja,* p. 1.

67. Hatsumi, *Ninjutsu: History and Tradition,* p. 3.
68. Ibid., author's preface page.
69. Hayes, *Legacy of the Night Warrior,* p. 20; Hatsumi, *Ninjutsu: History and Tradition,* p. 181.
70. Hayes, *Ninjutsu: The Art of the Invisible Warrior,* p. 153.
71. Ibid.
72. Hayes, *The Mystic Arts of the Ninja,* p. 1.
73. Hatsumi, *Ninjutsu: History and Tradition,* p. 5.
74. Ibid., p. 14.
75. Hatsumi, *The Essence of Ninjutsu,* p. 39.
76. Hatsumi, *Ninjutsu: History and Tradition,* p. 13; Hayes, *Ninjutsu: The Art of the Invisible Warrior*, p. 154.
77. Hayes, *The Mystic Arts of the Ninja,* p. 136.
78. Hatsumi, *Ninjutsu: History and Tradition,* p. 12.
79. Hayes, *The Mystic Arts of the Ninja,* pp. 132-138.
80. Ibid., p. 134.
81. Kim, *Secrets of the Ninja,* pp. 5-31.
82. Hatsumi, *Ninjutsu: History and Tradition,* author's preface page.
83. Hayes, *Ninjutsu: The Art of the Invisible Warrior,* p. 154.
84. Hayes, *The Mystic Arts of the Ninja,* p. 137.
85. Hayes, *Ninjutsu: The Art of the Invisible Warrior,* p. 157.
86. Ibid.
87. Ibid.
88. Ibid.
89. Hayes, *The Mystic Arts of the Ninja,* p. 139.
90. Kim, *The Invisible Ninja,* pp. 1, 56; Kim, *Secrets of the Ninja,* pp. 10-29; Hayes, *The Mystic Arts of the Ninja,* pp. 133-135.
91. Kim, *The Invisible Ninja,* p. 56.
92. Hayes, *The Mystic Arts of the Ninja,* p. 135; Kim, *Secrets of the Ninja,* pp. 10-30.
93. Hatsumi, *The Essence of Ninjutsu,* p. 39.
94. Hayes, *The Mystic Arts of the Ninja,* p. 133; Hatsumi, *Ninjutsu: History and Tradition,* pp. 231, 232.
95. Hatsumi, *The Essence of Ninjutsu,* p. 40.
96. Kim, *Secrets of the Ninja,* p. 83.
97. Kim, *The Invisible Ninja,* p. 13.
98. Hayes, *The Mystic Arts of the Ninja,* p. 102.
99. Kim, *Secrets of the Ninja,* p. 95.
100. Ibid., p. 149.
101. Ibid.
102. Ibid.
103. Ibid., p. 150.
104. Hayes, *Legacy of the Night Warrior,* p. 154.

105. Kevin Peraino and Evan Thomas, "Father, Where Art Thou," *Newsweek,* 27 January 2003, p. 56.

106. Todd Petit (Lee Malvo's court-appointed guardian), interview by Clare Shipman, ABC's *Good Morning America,* 30 April 2003.

107. William S. Lind, "Fourth Generation Warfare's First Blow: A Quick Look," *Marine Corps Gazette,* November 2001, p. 72; William S. Lind, Maj. John F. Schmitt, and Col. Gary I. Wilson, "Fourth Generation Warfare: Another Look," *Marine Corps Gazette,* December 1994, reprint November 2001, pp. 69-71; William S. Lind, Col. Keith Nightengale, Capt. John F. Schmitt, Col. Joseph W. Sutton, and Lt.Col. Gary I. Wilson, "The Changing Face of War: Into the Fourth Generation," *Marine Corps Gazette,* October 1989, reprint November 2001, pp. 65-68.

108. Bolander, "The Dragon's New Claws," p. 9C.

109. Memo for the record by H.J. Poole.

110. Vijai K. Nair, "America's War on Terrorism Impinges on China's Grand Strategy" (N.p., n.d.), as forwarded by e-mail to author from Remcol@aol.com, 5 January 2002.

111. Tom Raum, Associated Press, "War with Iraq Has Wild Cards," *Jacksonville (NC) Daily News,* 19 March 2003, p. 8A.

112. Brent Israelsen, "Depot Stumped Over Intruder: Army Doesn't Know How He Got Away," *The Salt Lake Tribune,* 7 September 2002, p. B1.

113. Foster Zeh (American nuclear power plant security guard), interview by Dianne Sawyer, ABC's *Good Morning America,* 9 December 2002.

114. Bermudez, *North Korean Special Forces,* p. 234.

115. Robert Marquand, "GIs in South Korea Unfazed by Crisis," *Christian Science Monitor,* 11 March 2003, pp. 6, 7.

116. Lung, *Knights of Darkness,* p. 128.

117. Ibid., p. 126.

118. Ibid., p. 117.

119. Jack R. White, *The Invisible World of Infrared* (New York: Dodd, Mead & Co., 1984), p. 36.

120. Allan Maurer, *Lasers: Light Wave of the Future* (New York: Arco, 1982), p. 50.

121. Ibid., p. 17.

122. Scarborough, "Al-Qaeda Adapts to Pursuit Tactics," p. 10.

123. Lung, *Knights of Darkness,* p. 125.

124. Ibid.

125. Maj.Gen. Robert H. Scales, Jr., "From Korea to Kosovo," *Armed Forces Journal International,* December 1999, p. 41.

126. Lung, *Knights of Darkness,* p. 125.

127. Unidentified Marine infantry Lt., in conversation with author, August 2000.

128. White, *The Invisible World of Infrared,* p. 41.
129. Lung, *Knights of Darkness,* p. 125.
130. Ibid., p. 127.
131. Mark Wheeler, "Shadow Wolves," *Smithsonian,* January 2003, p. 42.
132. "Article Details Everyday Life in Army," *Chungang Ilbo,* 15 April 1993, p. 1, as cited in FBIS-EAS-93-092, 14 May 1993, pp. 32, 33; and U.S. Defense Intell. Agency, "North Korean Army Training Is Steadily Progressing," *Defense Intelligence Digest,* June 1968, pp. 29-31; and U.S. Army, "North Korean People's Army Operations," *FC 100-2-99* (Washington, D.C.: Hdqts. Dept. of the Army, 5 December 1986), chapt. 16 ["Education and Training System"]; from Bermudez, *North Korean Special Force,* p. 219.
133. Maj. Jon T. Hoffman USMCR, "The Legacy and the Lessons of Tarawa," *Marine Corps Gazette,* November 1993, p. 64.

Appendix A: Casualty Comparisons

1. *Mao's Generals Remember Korea,* p. 6.
2. Millett and Maslowski, *For the Common Defense,* p. 504.
3. Hastings, *The Korean War,* p. 329.
4. *Encyclopedia of the Korean War: A Political, Social, and Military History,* ed. Spencer C. Tucker, vol. I (Santa Barbara, CA: ABC-CLIO, Inc., 2000), p. 101 (this work will henceforth be cited as *Encyclopedia of the Korean War.);* Xu Yan, "Chinese Forces and Their Casualties in the Korean War," trans. Xiaobing Li, *Chinese Historians,* vol. 2, fall 1993, pp. 51-54, from *Mao's Generals Remember Korea,* p. 6; editors' substantive endnote 12, *Mao's Generals Remember Korea,* p. 254;
5. Hastings, *The Korean War,* p. 329; Millett and Maslowski, *For the Common Defense,* p. 504; editors' substantive endnote 12, *Mao's Generals Remember Korea,* p. 254.
6. Hastings, *The Korean War,* p. 329.
7. *Encyclopedia of the Korean War,* p. 101; editors' substantive endnote 12, *Mao's Generals Remember Korea,* p. 254.
8. Editors' substantive endnote 12, *Mao's Generals Remember Korea,* p. 254.
9. Xu Yan, "Chinese Forces and Their Casualties in the Korean War," trans. Xiaobing Li, *Chinese Historians,* issue 2, Fall 1993, pp. 51-54, from *Mao's Generals Remember Korea,* p. 6.
10. *Encyclopedia of the Korean War,* p. 98.
11. Ibid., p. 100.
12. Editors' substantive endnote 40, *Mao's Generals Remember Korea,* p. 249.

13. *The Korean War: An Encyclopedia,* ed. Stanley Sandler (New York: Garland Publishing, 1995), p. 266; editors' substantive endnote 40, *Mao's Generals Remember Korea,* p. 249; *Encyclopedia of the Korean War,* p. 101.

14. "U.S. Forces Casualty Table," *Encyclopedia of the Korean War,* p. 100; Hastings, *The Korean War,* p. 329; editors' substantive endnote 12, *Mao's Generals Remember Korea,* p. 254.

15. Hastings, *The Korean War,* p. 329.

16. Editors' substantive endnote 40, *Mao's Generals Remember Korea,* p. 249.

17. *The Korean War: An Encyclopedia,* ed. Sandler, p. 266.

18. *Encyclopedia of the Korean War,* p. 101.

19. Ibid.

20. Ibid.

21. Editor's substantive endnote 6, *Mao's Generals Remember Korea,* p. 246.

22. Jonathan Kandell, "Korea: A House Divided," *Smithsonian,* July 2003, p. 42.

23. Editors' substantive endnote 40, *Mao's Generals Remember Korea,* p. 249.

24. *Encyclopedia of the Korean War,* p. 101; Kandell, "Korea: A House Divided," p. 42.

25. Hastings, *The Korean War,* p. 329; Millett and Maslowski, *For the Common Defense,* p. 504.

26. *Encyclopedia of the Korean War,* p. 101.

27. Ibid.

28. Ibid.

29. Khang, *The Tet Mau Than 1968 Event in South Vietnam,* pp. 119, 120.

30. Ibid., p. 89.

31. *Cuoc Tong cong kich-Tong knoi nghia cua Viet cong Mau Than 1968* (Saigon: ARVN Hdqts., August 1968), from Khang, *The Tet Mau Than 1968 Event in South Vietnam,* p. 89.

32. Ibid.

33. Millett and Maslowski, *For the Common Defense,* p. 560.

34. Khang, *The Tet Mau Than 1968 Event in South Vietnam,* p. 89.

35. Ibid.

36. Ibid.

37. Millett and Maslowski, *For the Common Defense,* p. 560.

38. Khang, *The Tet Mau Than 1968 Event in South Vietnam,* pp. 119-120.

39. *Cuoc khang chien chong My cuu muoc 1954-1975—nhung su kien quan su* (Hanoi: Vietnamese Mil. Hist. Inst., 1988), from Khang, *The Tet Mau Than 1968 Event in South Vietnam,* p. 89.

40. Ibid.
41. Ibid.

Appendix B: *Enemy Entry-Level Training*

1. Wilhelm Balck, *Development of Tactics—World War,* trans. Harry Bell (Ft. Leavenworth, 1922), p. 13; and Crown Prince Wilhelm of Germany, *My War Experiences* (London, 1923), p. 295; and Eric Ludendorff, *Ludendorff's Own Story,* vol. 2 (New York, 1919), p. 209; from Lupfer, "The Dynamics of Doctrine," *Leavenworth Papers No. 4,* p. 46.
2. Lupfer, "The Dynamics of Doctrine," *Leavenworth Papers No. 4,* p. 46.
3. Ibid., p. 56.
4. Ibid., p. 43.
5. "Handbook on German Military Forces," *TM-E 30-451,* p. 69.
6. Ibid., p. 71.
7. Ibid., p. 8.
8. Japanese Research Div., Mil. Hist. Sect., Hdqts. U.S. Army Forces Far East, "Japanese Night Combat," part 1, *Principles of Night Combat,* 1955, charts 1-a-d, 2-1-f, and 3-a-e, from Drea, "Nomonhan," *Leavenworth Papers No. 2,* p. 20.
9. "Handbook on Japanese Military Forces," *TM-E 30-480,* p. 5.
10. Ibid., pp. 5, 6.
11. Ibid., p. 8.
12. Suvorov, *Inside the Soviet Army,* pp. 272, 273.
13. Ibid., pp. 274, 275.
14. Ibid., pp. 281, 282.
15. McMichael, "The Chinese Communist Forces in Korea," *Leavenworth Research Survey No. 6,* p. 60.
16. "Article Details Everyday Life in Army," *Chungang Ilbo,* 15 April 1993, p. 1, as cited in FBIS-EAS-93-092, 14 May 1993, pp. 32, 33; and U.S. Defense Intell. Agency, "North Korean Army Training Is Steadily Progressing," *Defense Intelligence Digest,* June 1968, pp. 29-31; and U.S. Army, "North Korean People's Army Operations," *FC 100-2-99* (Washington, D.C.: Hdqts. Dept. of the Army, 5 December 1986), chapt. 16 ["Education and Training System"]; from Bermudez, *North Korean Special Force,* p. 219.
17. Bermudez, *North Korean Special Forces,* p. 219.
18. Maitland and McInerney, *Vietnam Experience,* p. 94.
19. Lanning and Cragg, *Inside the VC and the NVA,* p. 46.
20. Dunnigan and Nofi, *Dirty Little Secrets of the Vietnam War,* pp. 273, 274.

21. Lanning and Cragg, *Inside the VC and the NVA,* p. 43.
22. Ibid., p. 54.

Appendix C: Advised U.S. Battledrills

1. Lung, *Knights of Darkness,* p. 56.
2. Kim, *Secrets of the Ninja,* pp. 34, 75.
3. Hayes, *Ninjutsu: The Art of the Invisible Warrior*, p. 157.

Glossary

ABC	American Broadcasting Company (one of the U.S. television networks)
AC-130	Gunship (fixed-wing aircraft that can provide extremely accurate fire from a rapid-firing Vulcan cannon, 105mm howitzer, and other weapons)
AK-47	Assault rifle (Communist Bloc assault rifle)
ARVN	Army of the Republic of Vietnam (U.S. ally during the Vietnam War)
AP	Associated Press (American news service)
AT	Antitank (armor killing weapon)
BMP	(Russian acronym for armored personnel carrier)
BOF	Base of Fire (support element that provides covering fire for the maneuver element in an attack)
CBS	Columbia Broadcasting System (one of the U.S. television networks)
CCF	Chinese Communist Forces (the Chinese Communist Army)
CO	Commanding Officer (commander of company or larger)
DA	Dept. of the Army (headquarters for the U.S. Army)
DMZ	De-Militarized Zone (the buffer zone between nations)
DoD	Dept. of Defense (administrative headquarters for the various branches of the U.S. military)

EEG	Electroencephalogram (a device that measures brain wave activity)
ESP	Extra-Sensory Perception (that beyond normal senses)
FAE	Fuel Air Explosive (fire bomb that explodes above ground and sucks up all the oxygen beneath it)
FBI	Federal Bureau of Investigation (U.S. law enforcement agency)
FCL	Final Coordination Line (last covered position before a maneuver force assaults a daytime objective)
FEBA	Forward Edge of the Battle Area (front of the defensive zone)
FM	Field Manual (U.S. Army publication)
FMFRP	Fleet Marine Force Reference Publication (U.S. Marine Corps lessons-learned document)
GC	Grid Coordinates (way to establish one's longitudinal and latitudinal location on a map)
GI	Government Issue (an American enlisted soldier)
GPS	Global Positioning System (device that resects satellite signals to determine one's location)
HIND	Heavily armored Soviet attack helicopter
HQMC	Headquarters Marine Corps (highest echelon of the U.S. Marine Corps)
IJA	Imperial Japanese Army (Japan's army prior to the end of WWII)
IR	Infrared (electromagnetic energy of wave length between visible light and microwaves)
KPA	Korean People's Army (army of North Korea)
M16A2	Rifle (small arm carried by U.S. infantryman)

MG	Machinegun (fully automatic small arm)
MIT	Massachusetts Institute of Technology (renowned university)
MLR	Main Line of Resistance (front of defensive zone)
MOUT	Military Operations in Urban Terrain (city fighting)
MP	Military Police (an army's self-policing agency)
MPLA	Malayan People's Liberation Army (predominantly Chinese insurgents in Malaysia)
NBC	Nuclear/Biological/Chemical (a type of warfare)
NCO	Noncommissioned Officer (enlisted man of pay grade E-4 or above)
NKA	North Korean Army (ground combat force of North Korea)
NKPA	North Korean Peoples' Army (army of North Korea)
NPR	National Public Radio (nonprofit American radio network)
NVA	North Vietnamese Army (America's adversary during the Vietnam War)
NVG	Night Vision Goggles (eyewear to enhance ambient light at night)
OP	Observation Post (normally called a sentry post in the daytime and a listening post at night)
PAVN	People's Army of Vietnam (North Vietnamese Army)
PFC	Private First Class (U.S. pay grade E-2)
PLA	People's Liberation Army (ground forces of Communist China)
POW	Prisoner of War (captive combatant of another nation)

PRC — People's Republic of China (Communist China)

ROK(A) — Republic of Korea Army (U.S. ally during Korean War)

RPG — Rocket Propelled Grenade (Communist Bloc hand-held anti-armor or antibunker weapon)

S-2 — Intelligence officer (member of the U.S. infantry battalion staff)

SEAL — Sea, Air, and Land (U.S. Navy commando)

SMAW — Shoulder-launched Multipurpose Assault Weapon (bunker buster organic to infantry company)

SNCO — Staff Noncommissioned Officer (enlisted man of grade E-6 and above)

TAOR — Tactical Area of Responsibility (a unit's defensive sector)

TM — Technical Manual (publication of the U.S. War Dept.)

TNT — Trinitrotoluene (high explosive)

UH-60 — Helicopter ("Blackhawk" designation)

UN — United Nations (world organization)

USMC — United States Marine Corps (America's amphibious landing force)

USSR — Union of Soviet Socialist Republics (the old Soviet Union)

UV — Ultraviolet (electromagnetic energy of wave length between X-rays and visible light)

VC — Viet Cong (Communist insurgents in South Vietnam)

WP — White Phosphorus (mortar and artillery round that sprays burning phosphorus over a wide area)

WWI — World War One (the global conflict from 1914 to 1918)

WWII — World War Two (the global conflict from 1939 to 1945)

Bibliography

U.S. Government Manuals and Chronicles

Alexander, Col. Joseph H. *Closing In: Marines in the Seizure of Iwo Jima.* Marines in World War II Commemorative Series. Washington, D.C.: Hist. & Museums Div., HQMC, 1994.

Bartley, Lt.Col. Whitman S. *Iwo Jima: Amphibious Epic.* Washington, D.C.: Hist. Branch, HQMC, 1954.

Bauman, Dr. Robert F. "Russian-Soviet Unconventional Wars in the Caucasus, Central Asia, and Afghanistan." *Leavenworth Papers No. 20.* Ft. Leavenworth, KS: Combat Studies Inst., U.S. Army Cmd. & Gen. Staff College, 1993.

Cash, John A., John Albright, and Allan W. Sandstrum. *Seven Firefights in Vietnam.* Washington, D.C.: Center of Mil. Hist., U.S. Army, 1985.

Chinese Communist Reference Manual for Field Fortifications. Translated by Intell. Sect. Gen. Staff, Far East Cmd., 1 May 1951. From "Enemy Field Fortifications in Korea," by U.S. Army Corps of Engineers. No. 15, *Engineer Intelligence Notes.* Washington, D.C.: Army Map Service, January 1952. As contained in "A Historical Perspective on Light Infantry," *Leavenworth Research Survey No. 6,* by Maj. Scott R. McMichael. Ft. Leavenworth, KS: Combat Studies Inst., U.S. Army Cmd. & Gen. Staff College, 1987.

Drea, Edward J. "Nomonhan: Japanese—Soviet Tactical Combat, 1939." *Leavenworth Papers No. 2.* Ft. Leavenworth, KS: Combat Studies Inst., U.S. Army Cmd. & Gen. Staff College, 1981.

"Fighting on Guadalcanal." *FMFRP 12-110.* Washington, D.C.: U.S.A. War Office, 1942.

Glantz, David M. "August Storm: Soviet Tactical and Operational Combat in Manchuria, 1945." *Leavenworth Papers No. 8.* Ft. Leavenworth, KS: Combat Studies Inst., U.S. Army Cmd. & Gen. Staff College, 1983.

"Handbook on German Military Forces." *TM-E 30-451.* Washington, D.C.: U.S. War Dept., 1945. Reprint, Baton Rouge, LA: LSU Press, 1990.

"Handbook on Japanese Military Forces." *TM-E 30-480.* Washington, D.C.: U.S. War Dept., 1944. Reprint, Baton Rouge, LA: LSU Press, 1991.

"Handbook on the Chinese Communist Army." *DA Pamphlet 30-51.* Washington, D.C.: Hdqts. Dept. of the Army, 1960.

"Handbook On U.S.S.R. Military Forces," *TM 30-340.* Washington, D.C.: U.S. War Dept., 1945. Republished as *Soviet Tactical Doctrine in WWII,* with foreword by Shawn Caza. West Chester, OH: G.F. Nafziger, 1997.

Henderson, Wm. Darryl. *Cohesion: The Human Element in Combat.* Washington, D.C.: Nat. Defense Univ. Press, 1985.

"Historical Study—Night Combat." *DA Pamphlet 20-236.* Washington, D.C.: Hdqts. Dept. of the Army, 1953.

"Historical Study—Russian Combat Methods in World War II." *DA Pamphlet 20-230.* Washington, D.C.: Hdqts. Dept. of the Army, 1950.

"How the Japanese Army Fights," by Lt.Col. Paul W. Thompson, Lt.Col. Harold Doud, Lt.Col. John Scofield, and the editorial staff of "The Infantry Journal." *FMFRP 12-22.* New York: Penguin Books, 1942. Reprint Washington, D.C.: HQMC, 1989.

Hubler, Thomas M. "Japan's Battle for Okinawa, April - June 1945." *Leavenworth Papers No. 18.* Ft. Leavenworth, KS: Combat Studies Inst., U.S. Army Cmd. & Gen. Staff College, 1990.

Huber, Thomas M. "The Battle of Manila." In CSI Home Publications Research MHIST [database on line]. Ft. Leavenworth, KS: Combat Studies Inst., U.S. Army Cmd. & Gen. Staff College, n.d. [updated 30 September 2002; cited 1 January 2003]. Available from the CSI website at www-cgsc.army.mil/csi/research/mout/mouthuber. asp.

"Light Infantry Platoon/Squad." *FM 7-70.* Washington, D.C.: Hdqts. Dept. of the Army, 1986.

Lupfer, Timothy L. "The Dynamics of Doctrine: The Changes in German Tactical Doctrine during the First World War." *Leavenworth Papers No. 4.* Ft. Leavenworth, KS: Combat Studies Inst., U.S. Army Cmd. & Gen. Staff College, 1981.

"Marine Rifle Company/Platoon." *FMFM 6-4.* Washington, D.C.: HQMC, 1978.

McMichael, Maj. Scott R. "The Chinese Communist Forces in Korea." In "A Historical Perspective on Light Infantry." *Leavenworth Research Survey No. 6.* Ft. Leavenworth, KS: Combat Studies Inst., U.S. Army Cmd. & Gen. Staff College, 1987.

Montross, Lynn and Capt. Nicholas A. Canzona. *The Chosin Reservoir Campaign.* U.S. Marine Operations in Korea 1950-1953, vol. III. Washington, D.C.: Hist. Branch, HQMC, 1957.

"Nepal and Bhutan Country Studies." *DA Pamphlet 550-35.* Area Handbook Series. Washington, D.C.: Hdqts. Dept. of the Army, 1993.

"North Korea Handbook." *PC-2600-6421-94.* Washington, D.C.: Department of Defense, 1994.

"North Korean People's Army Operations." *FC 100-2-99.* Washington, D.C.: Hdqts. Dept. of the Army, 1986.

NVA-VC Small Unit Tactics & Techniques Study. Edited by Thomas Pike. Part I. N.p.: U.S.A.R.V., n.d.

"Pearl Harbor to Guadalcanal." *FMFRP 12-34-I.* By Lt.Col. Frank O. Hough, Maj. Verle E. Ludwig, and Henry I. Shaw, Jr. Vol I. of *History of the U.S. Marine Corps Operations in World War II Series.* Washington, D.C.: Hist. Branch, HQMC. Reprint Quantico, VA: Marine Corps Combat Develop. Cmd., 1989.

Sasso, Maj. Claude R. "Soviet Night Operations in World War II." *Leavenworth Papers No. 6.* Leavenworth, KS: Combat Studies Inst., U.S. Army Cmd. & Gen. Staff College, 1982.

Tefler, Maj. Gary, Lt.Col. Lane Rogers, and V. Keith Fleming, Jr. *U.S. Marines in Vietnam: Fighting the North Vietnamese, 1967.* Washington, D.C.: Hist. & Museums Div., HQMC, 1984.

The Bear Went over the Mountain: Soviet Combat Tactics in Afghanistan. Translated and edited by Lester W. Grau, Foreign Mil. Studies Office, U.S. Dept. of Defense, Fort Leavenworth, KS. Washington, D.C.: Nat. Defense Univ. Press, 1996. Originally published under its Russian title. Soviet Union: Frunze Mil. Academy, n.d.

Updegraph, Charles L., Jr. *U.S. Marine Corps Special Units in World War II.* Washington, D.C.: Hist. & Museums Div., HQMC, 1972.

U.S. Marines in Vietnam: An Expanding War—1966. By Jack Shulimson. Washington, D.C.: Hist. & Museums Div., HQMC, 1982.

U.S. Special Forces Reconnaissance Manual. N.p., n.d. Reprint, Sims, AR: Lancer Militaria, 1982.

Zimmerman, Maj. John L. *The Guadalcanal Campaign.* Washington, D.C.: Hist. Div., HQMC, 1949.

Civilian Books, Magazine Articles, and Video/Film Presentations

Asprey, Robert B. *War in the Shadows.* Garden City, NY: Doubleday & Co., 1975.

Barr, Cameron W. "Kirkuk, a Mirror of Iraq's Schisms." *Christian Science Monitor,* 4 March 2003.

Barry, John and Evan Thomas. "The Kosovo Cover-Up." *Newsweek,* 15 May 2000.

The Battle of Dien Bien Phu. Visions of War Series. Vol. 10. New Star Video, 1988. 50 min. Videocassette #4010.

Baxter, William P. *Soviet Airland Battle Tactics.* Novato, CA: Presidio Press, 1986.

Benson, Merritt E. "Seventeen Days at Iwo Jima." *Marine Corps Gazette,* February 2000.

Bermudez, Joseph S., Jr. *North Korean Special Forces.* Annapolis, MD: Naval Inst. Press, 1998.

Bernard, Joe *(ninjutsu* student). In e-mails from JBernard29 to author, 27 November 2002 and 17 January 2003.

Bolander, Lt.Col Jeffrey W. "The Dragon's New Claws." *Camp Lejeune (NC) Globe,* 8 March 2001.

Bosshart, Dr. R. Perry (longtime resident of SE Asia and recent visitor to the Binh Moc tunnels). In a telephone conversation with the author, 19 April 2003.

Bowden, Mark. "Blackhawk Down." *Philadelphia Inquirer On Line,* www3.phillynews.com/packages/somalia/ncv16/rang16.asp, 16 November 1997.

Burns, Robert. Associated Press. "Retired General: Exercise Rigged." *Jacksonville (NC) Daily News,* 17 August 2002.

Campbell, Maj. Donald H. USMC (Ret.) (platoon sergeant during Korean hill battles of 1952-53), in conversation with author, 26 March 2001.

Can, Nguyen Khac and Pham Viet Thuc. *The War 1858 - 1975 in Vietnam.* Hanoi: Nha Xuat Ban Van Hoa Dan Toc, n.d.

Canby, Steven L. "Classic Light Infantry and New Technology." Produced under contract MDA 903-81-C-0207 for the Defense Advanced Research Project Agency (Potomac, MD: C&L Associates, 1983).

Chanda, Nayan. *Brother Enemy: The War after the War.* New York: Collier Books, 1986.

Channon, John and Robert Hudson. *The Penguin Historical Atlas of Russia.* London: Penguin Books, 1996.

Chanoff, David and Doan Van Toai. *Portrait of the Enemy.* New York: Random House, 1986.

Chapman, Spencer. *The Jungle Is Neutral.* London: Chatto & Windus, 1952.

Collins, Lt.Gen. Arthur S., Jr., U.S. Army (Ret.). *Common Sense Training—A Working Philosophy for Leaders.* Novato, CA: Presidio Press.

Combat Tracking. Paluda, Malaysia: British Army Combat Training Center, n.d.

The Concise Encyclopedia of Ancient Civilizations. Edited by Janet Serlin Garber. New York: Franklin Watts, 1978.

Cooling, Maj. Norman L. "Russia's 1994-96 Campaign for Chechnya: A Failure in Shaping the Battlespace." *Marine Corps Gazette,* October 2001.

Corbett, James. *Jim Corbett's India,* edited by R.E. Hawkins. London: Oxford Univ. Press, 1978.

Corbett, James. *Jungle Lore.* London: Oxford Univ. Press, 1953.

Corbett, James. *Man-Eaters of Kumaon.* London: Oxford Univ. Press, 1946.

Coox, Alvin D. *Nomonhan: Japan against Russia, 1939.* Stanford, CA: Stanford Univ. Press, 1985.

Craig, William. *The Enemy at the Gates: The Battle for Stalingrad.* New York: Readers Digest Press, 1973.

Cumings, Bruce. In "The Battle of the Minds" segment of *Korea — the Unknown War Series.* London: Thames TV in assoc. with WGBH Boston, 1990. NC Public TV.

Dahmus, Joseph. *A Popular History of the Middle Ages.* New York: Barnes & Noble Books, 1968.

Davidson, Phillip B. *Vietnam at War—The History: 1946-1975.* New York: Oxford Univ. Press, 1988.

Davis, Burke. *Marine.* New York: Bantam Books, 1964.

The Doomed City of Berlin. Cities at War Series. Simon & Schuster Video, 1986. 65 min. Videocassette #62159-9.

Dotson, James and Kevin Dockery. *Point Man.* New York: William Morrow and Co., 1993.

Dung, Gen. Van Tien. *Our Great Spring Victory: An Account of the Liberation of South Vietnam.* Translated by John Spragens, Jr. Hanoi: The Gioi Publishers, 2001.

Dunnigan, James F. and Albert A. Nofi. *Dirty Little Secrets of the Vietnam War.* New York: Thomas Dunne Books, 1999.

Dupuy, Trevor N., David L. Bongard, and Richard C. Anderson Jr. *Hitler's Last Gamble.* New York: Harper Perennial, 1994.

Elford, George Robert. *Devil's Guard.* New York: Dell, 1985.

The Encyclopedia of the Korean War: A Political, Social, and Military History. Edited by Spencer C. Tucker. Santa Barbara, CA: ABC-CLIO Publishing, 2000.

Encyclopedia of the Vietnam War. Edited by Stanley I. Kutler. New York: Macmillan Reference, 1996.

English, John A. *On Infantry.* New York: Praeger, 1981.

Evans, Grant and Kelvin Rowley. *Red Brotherhood at War: Indochina since the Fall of Saigon.* London: Verso, 1984.

Fehrenbach, T.R. *This Kind of War: The Classic Korean War History.* Dulles, VA: Brassey's, 1963.

Ford, M.Sgt. Lester J., Jr., USMC (Ret.) (Vietnam-era platoon sergeant), in conversations with author, 7 April 2002, 27 November 2002, and 17 January 2003.

Fournier, Ron. Associated Press. "Bush: Russian Firms Aiding Iraqi War Effort." *Jacksonville (NC) Daily News,* 25 March 2003.

Fox, Col. Wesley. L. *Marine Rifleman: Forty-Three Years in the Corps.* Washington, D.C.: Brassey's, 2002.

The German Squad in Combat. Translated and edited by the U.S. Mil. Intell. Service. N.p., 1943. From a German manual of similar title. N.p.: German Army, 1940-42. Republished as *German Squad Tactics in WW,* edited by Matthew Gajkowski. West Chester, OH: Nafziger, 1995.

Giap, Gen. Vo Nguyen. *Dien Bien Phu.* 6th edition, supplemented. Hanoi: The Gioi Publishers, 1999.

Giap, Gen. Vo Nguyen. "Once Again We Will Win." In *The Military Art of People's War,* edited by Russel Stetler. New York: Monthly Review Press, 1970.

Giap, Gen. Vo Nguyen. *Peoples War—Peoples Army.* New York: Frederick A. Praeger, 1961.

Glantz, David M. *The Soviet Conduct of Tactical Maneuver: Spearhead of the Offensive.* London: Frank Cass, 1991.

Goulden, Joseph C. *Korea: The Untold Story of the War.* New York: Times Books, 1982.

Grant, Reg G. "Fighting the VC Way." Chapt. 29 in *NAM: The Vietnam Experience 1965-75.* London: Orbis Publishing, 1987.

Grau, Lester W. "Technology and the Second Chechen Campaign: Not All New and Not That Much." Chapt. in *The Second Chechen War,* edited by Mrs. A.C. Aldis. Surrey, England: Royal Military Academy Sandhurst, Conflict Studies Research Centre, June 2000. From www.da.mod.uk/CSRC/Home/Caucasus/P31.

Griffith, Samuel B. Introduction to *The Art of War,* by Sun Tzu. New York: Oxford Univ. Press, 1963.

Gudmundsson, Bruce I. *Stormtroop Tactics: Innovation in the German Army 1914-1918.* Westport, CT: Praeger Pubs., 1989.

Guthrie, CWO-4 Charles "Tag" USMC (Ret.) (Vietnam-era infantryman and intell. analyst). In various e-mail and telephone conversations with author, August 2002 - March 2003.

Hammel, Eric M. *Chosin: Heroic Ordeal of the Korean War.* Novato, CA: Presidio Press, 1981.

Hammel, Eric M. *Fire in the Streets: The Battle for Hue, Tet 1968.* Pacifica, CA: Pacifica Mil. Hist., 1991.

Hastings, Max. *The Korean War.* New York: Simon & Schuster, 1987.

Hatsumi, Dr. Masaaki. *Ninjutsu: History and Tradition.* Burbank, CA: Unique Publications, 1981.

Hatsumi, Dr. Masaaki. *The Essence of Ninjutsu.* Chicago: Contemporary Books, 1988.

Hatsumi, Dr. Masaaki. *The Grandmaster's Book of Ninja Training.* Translated by Chris W. P. Reynolds. Chicago: Contemporary Books, 1988.

Hayes, Stephen K. *Legacy of the Night Warrior.* Santa Clarita, CA: Ohara Publications, 1985.

Hayes, Stephen K. *Ninjutsu: The Art of the Invisible Warrior.* Chicago: Contemporary Books, 1984.

Hayes, Stephen K. *The Mystic Arts of the Ninja: Hypnotism, Invisibility, and Weaponry.* Chicago: Contemporary Books, 1985.

Hayes, Stephen K. *The Ninja and Their Secret Fighting Art.* Rutland, VT: Charles E. Tuttle Co., 1981.

Hayes, Stephen K. *Warrior Path of Togakure.* Santa Clarita, CA: Ohara Publications, 1983.

Herman, Bert. Associated Press. "Risky Job for Marines." *Jacksonville (NC) Daily News,* 13 April 2003.

His Holiness John Paul II. *Crossing the Threshold of Hope.* New York: Alfred A. Knopf, 1995.

Hoffman, Lisa. Scripps Howard. "Among Troops' Problems in Afghanistan—Their Own Stuff." *Jacksonville (NC) Daily News,* 6 October 2002.

Hoyt, Edwin P. *The Marine Raiders.* New York: Pocket Books, 1989.

Hutton, Paul Churchill III. "Weapons of Restraint: Emphasis on Limiting Non-Combatant Casualties Is Good Reason for Developing New Non-Lethal Weapons." *Armed Forces Journal International,* May 2000.

"Hypnosis." In *The Complete Manual of Fitness and Well-Being.* Pleasantville, NY: The Reader's Digest Assoc., 1984.

Jaffe, Greg. "Street Smarts: As Threats Evolve, Marines Should Learn Skill of Combat in Cities." *Wall Street Journal,* 22 August 2002.

Journey into China. Edited by Kenneth C. Danforth. Washington, D.C.: Nat. Geographic Society, n.d.

Kandell, Jonathan. "Korea: A House Divided." *Smithsonian,* July 2003.

Kelley, Matt. Associated Press. "Army Gives MIT $50 Million Grant to Develop Futuristic Battle Uniform." *Jacksonville (NC) Daily News,* 14 March 2002.

Khang, Ho. *The Tet Mau Than 1968 Event in South Vietnam.* Hanoi: The Gioi Publishers, 2001.

Khoi, Hoang. *The Ho Chi Minh Trail.* Hanoi: The Gioi Publishers, 2001.

Kim, Ashida. *Secrets of the Ninja.* New York: Citadel Press, 1981.

Kim, Ashida. *The Invisible Ninja: Ancient Secrets of Surprise.* New York: Citadel Press, 1983.

Knickerbocker, Brad. "Guerrilla Tactics vs. U.S. War Plan." *Christian Science Monitor,* 25 March 2003.

Knickerbocker, Brad. "In Era of High-Tech Warfare, 'Friendly Fire' Risk Grows." *Christian Science Monitor,* 14 January 2003.

Knickmeyer, Ellen and Ravi Nes. Associated Press. "A Cargo Plane Lands at Airport outside Capital," *Jacksonville (NC) Daily News,* 7 April 2003.

Knox, Donald. *The Korean War: Pusan to Chosin, An Oral History.* San Diego: Harcourt Brace Jovanovich, 1985.

The Korean War: An Encyclopedia. Edited by Stanley Sandler. New York: Garland Publishing, 1995.

The Korean War: History and Tactics. Edited by David Rees. New York: Crescent Books, 1984.

"Kublai Khan and the Mongol Empire," from Newspaper in Education. World of Wonder Series. *Jacksonville (NC) Daily News,* 24 March 2003.

Kuryacheba, A.A. *Voina v Koree: 1950-1953 (War in Korea: 1950-1953).* Voenno-Estorecheskaya Biblioteka. Saint Petersburg, Russia: Polygon Publishing, 2003.

Lamont, Lt.Col. Robert W. "'Urban Warrior'—A View from North Vietnam." *Marine Corps Gazette,* April 1999.

Lanning, Michael Lee and Dan Cragg. *Inside the VC and the NVA: The Real Story of North Vietnam's Armed Forces.* New York: Ivy Books, 1992.

Levin, Nora. *The Holocaust.* New York: Thomas Y. Crowell, 1968.

Liang, Qiao and Wang Xiangsui. *Unrestricted Warfare: Thoughts on War and Strategy in a Global Era.* Beijing: People's Liberation Army Arts Publishers, 1999.

Lieven, Anatol. "The World Turned Upside Down: Military Lessons of the Chechen War." *Armed Forces Journal International,* August 1998.

Lind, William S. "Fourth Generation Warfare's First Blow: A Quick Look." *Marine Corps Gazette,* November 2001.

Lind, William S. (author of *The Maneuver Warfare Handbook).* In telephone conversation with author, February 2001.

Lind, William S. *The Maneuver Warfare Handbook.* Boulder, CO: Westview Press, 1985.

Lind, William S., Maj. John F. Schmitt, and Col. Gary I. Wilson. "Fourth Generation Warfare: Another Look." *Marine Corps Gazette,* December 1994, reprint November 2001.

Lind, William S., Col. Keith Nightengale, Capt. John F. Schmitt, Col. Joseph W. Sutton, and Lt.Col. Gary I. Wilson. "The Changing Face of War: Into the Fourth Generation." *Marine Corps Gazette,* October 1989, reprint November 2001.

Linzer, Dafna. Associated Press. "Expecting the Unexpected in Baghdad." *Jacksonville (NC) Daily News,* 18 May 2003.

Lumpkin, John J. Associated Press. "Military Visits Urban Tactics against Baghdad." *Jacksonville (NC) Daily News,* 23 September 2002.

Lung, Dr. Haha. *Knights of Darkness: Secrets of the World's Deadliest Night Fighters.* Boulder, CO: Paladin Press, 1998.

Macfarlan, Allan A. *Exploring the Outdoors with Indian Secrets.* Harrisburg, PA: Stackpole Books, 1971.

Maitland, Terrence and Peter McInerney. *Vietnam Experience: A Contagion of War.* Boston: Boston Publishing, 1968.

Mao Tse-tung: An Anthology of His Writings. Edited by Anne Fremantle. New York: Mentor, 1962.

Mao's Generals Remember Korea. Translated and edited by Xiaobing Li, Allan R. Millett, and Bin Yu. Lawrence, KS: Univ. Press of Kansas, 2001.

Marquand, Robert. "GIs in South Korea Unfazed by Crisis." *Christian Science Monitor,* 11 March 2003.

Maurer, Allan. *Lasers: Light Wave of the Future.* New York: Arco, 1982.

Mikes, George. *The Hungarian Revolution.* London: Andre Deutsch Ltd., 1957.

Millett, Allan R. and Peter Maslowski. *For the Common Defense: A Military History of the United States of America.* New York: The Free Press, 1984.

Mollo, Andrew and Digby Smith. *World Army Uniforms Since 1939.* Poole, England: Blandford Press, 1981.

Nair, Vijai K. "America's War on Terrorism Impinges on China's Grand Strategy." N.p., n.d. As forwarded by e-mail to author from Remcol@aol.com, 5 January 2002.

Night Movements. Translated by C. Burnett. Port Townsend, WA: Loompanics Unlimited, n.d. Originally published as a Japanese training manual. Tokyo: Imperial Japanese Army, 1913.

Nolan, Keith William. *Operation Buffalo: USMC Fight for the DMZ.* New York: Dell Publishing, 1991.

Nolan, Keith William. *Sappers in the Wire: The Life and Death of Firebase Mary Ann.* College Station, TX: Texas A&M Univ. Press, 1995.

O'Brien, Michael. *Conscripts and Regulars: With the Seventh Battalion in Vietnam.* St. Leonards, Australia: Allen & Unwin, 1995.

Palm, Maj.Gen. Leslie M. USMC (Ret.). In conversation with author, 17 May 2002.

Patrolling and Tracking. Boulder, CO: Paladin Press, n.d.

Pelli, Maj. Frank D. "Insurgency, Counterinsurgency, and the Marines In Vietnam." Extracted on 10 November 2002 from http://www.ehistory.com/vietnam/essays/insurgency/index.cfm.

Peraino, Kevin and Evan Thomas. "Father, Where Art Thou." *Newsweek,* 27 January 2003.

Peterson, Kirkland C. *Mind of the Ninja: Exploring the Inner Power.* Chicago: Contemporary Books, 1986.

Peterson, Patrick. Knight Ridder. "Marines Shoot at Anything that Moves." *Jacksonville (NC) Daily News,* 26 March 2003.

Peterson, Scott. "Iraq: Saladin to Saddam." *Christian Science Monitor,* 4 March 2003.

Petit, Todd (Lee Malvo's court-appointed guardian). Interview by Clare Shipman, ABC's *Good Morning America,* 30 April 2003.

Pike, Douglas. *PAVN: People's Army of Vietnam.* Novato, CA: Presidio Press, 1986.

Pitman, Todd. Associated Press. "Coalition Force Explodes Weapons Cache in Caves," *Jacksonville (NC) Daily News,* 11 May 2002.

Raum, Tom. Associated Press. "War with Iraq Has Wild Cards." *Jacksonville (NC) Daily News,* 19 March 2003.

Read, Anthony and David Fischer. *The Fall of Berlin.* New York: Da Capo Press, 1995.

Reinke, Dr. David H. (expert on parapsychology and Eastern religions). In numerous telephone calls and e-mails from DRDAVIDHREINKE to author, June 2001 - August 2003.

Resnick, Abraham. *Russia: A History to 1917.* Chicago: Children's Press, 1983.

Robbins, Roland. *Man Tracking: Introduction to the Step-by-Step Method.* Montrose, CA: Search and Rescue Magazine, 1977.

Ross, Bill D. *Iwo Jima: Legacy of Valor.* New York: Vintage, 1986.

Scales, Maj.Gen. Robert H., Jr. "From Korea to Kosovo." *Armed Forces Journal International,* December 1999.

Scarborough, Rowan. "Al Qaeda Adapts to Pursuit Tactics." *Washington Times,* 15 January 2003.

Scharfen, Col. John C. and Michael J. Deane. "To Fight Russians in Cities, Know Their Tactics." *Marine Corps Gazette,* January 1977.

Scott, Harriet Fast and William F. Scott. *The Armed Forces of the USSR.* Boulder, CO: Westview Press, 1979.

Scott-Donelan, David. *Tactical Tracking Operations: The Essential Guide for Military and Police Trackers.* Boulder, CO: Paladin Press, 1998.

Severance, Col. Dave E. "Night Patrol on Iwo Jima," *Marine Corps Gazette,* March 2003.

Shaw, John and the editors of Time-Life Books. *Red Army Resurgent.* Chicago: Time-Life Books, 1979.

Shipunov, Arkady and Gennady Filimonov. "Field Artillery to be Replaced with Shmel Infantry Flamethrower." *Military Parade (Moscow),* Issue 29, Sept.-Oct. 1998, *www.milparade.com / security / 29.*

Simpson, Howard R. *Dien Bien Phu: The Epic Battle the U.S. Forgot.* Washington, D.C.: Brassey's, 1994.

Smalley, Suzanne. "Counting Civilians: Iraq's War Dead," *Newsweek,* 7 April 2003.

Smith, George W. *The Siege at Hue.* New York: Ballantine Publishing, 1999.

Soard, Charles (experienced infantry scout and frequent visitor to Vietnam). In telephone conversation with author, 28 March 2001.

"Some Interesting Facts about Korea: The Forgotten War." *DAV Magazine,* May/June 2000.

Soviet Combat Regulations of 1942. Moscow: Stalin, 1942. Republished as *Soviet Infantry Tactics in World War II: Red Army Infantry Tactics from Squad to Rifle Company from the Combat Regulations.* With translation, introduction, and notes by Charles C. Sharp. West Chester, OH: George Nafziger, 1998.

Stern, Seth. "How the U.S. Plans to Take Control of Baghdad." *Christian Science Monitor,* 7 April 2003.

The Strategic Advantage: Sun Zi and Western Approaches to War. Edited by Cao Shan. Beijing: New World Press, 1997.

Sun Bin's Art of War: World's Greatest Military Treatise. Translated by Sui Yun. Singapore: Chung Printing, 1999.

Sun Tzu. *The Art of War.* Translated and with introduction by Samuel B. Griffith, foreword by B.H. Liddell Hart. New York: Oxford Univ. Press, 1963.

Sun Tzu's Art of War: The Modern Chinese Interpretation, by Gen. Tao Hanzhang. Translated by Yuan Shibing. New York: Sterling Publishing, 1990.

Suvorov, Viktor. *Inside the Soviet Army.* New York: Berkley Books, 1983.

Taras, A.E. and F.D. Zaruz. *Podgotovka Razvegchika: Sistema Spetsnaza GRU (Training of Agents: Special Forces of the GRU).* Minsk, Belarus: AST Publishing, 2002.

Thai, Gen. Hoang Van. *How South Vietnam Was Liberated.* Hanoi: The Gioi Publishers, 1996.

Thao, Maj.Gen. Hoang Min. *The Victorious Tay Nguyen Campaign.* Hanoi: Foreign Languages Publishing House, 1979.

The 30-Year War: 1945-1975. Vol. II. Hanoi: The Gioi Publishers, 2001.

Thomas, Evan and John Barry. "A Plan under Attack." *Newsweek,* 7 April 2003.

Thomas, Evan and John Barry. "Saddam's War." *Newsweek,* 17 March 2003.

Thomas, Lt.Col. Ian (2d Royal Gurkha Rifles Commander). In e-mail from iannathomas@hotmail.com to author, 19 January 2003.

Thomas, Lt.Col. Timothy L. and Lester W. Grau. "Russian Lessons Learned from the Battles for Grozny." *Marine Corps Gazette,* April 2000.

Toczek, David. M. *The Battle of Ap Bac, Vietnam.* Westport, CT: Greenwood Press, 2001.

The Travels of Marco Polo. Revised from Marsden's translation. Edited and with introduction by Manuel Komroff. New York: Liveright Publishing Corp., 1926.

The Travels of Marco Polo. Revised from Marsden's translation. Edited by Manuel Komroff. New York: Modern Library, 1953.

A Tribute to WWII Combat Cameramen of Japan. Tokyo: Nippon TV, 1995. 85 min. Videocassette.

"Urban Battle in Nasiriyah Hurts Troops from Lejeune." From Associated Press. *Jacksonville (NC) Daily News,* 27 March 2003.

"U.S. Hopes to Take Baghdad Quickly." From Knight Ridder. *Jacksonville (NC) Daily News,* 26 March 2003.

Verstappen, Stefan H. *The Thirty-Six Strategies of Ancient China.* San Francisco: China Books & Periodicals, 1999.

Warner, Philip. *Japanese Army of World War II.* Vol. 20. Men-at-Arms Series. London: Osprey Publications Ltd., 1972.

Warr, Nicholas. *Phase Line Green: The Battle for Hue, 1968* (Annapolis, MD: Naval Inst. Press, 1997.

Weir, Fred. "Putin Battles Political Fallout of Chechnya Fight." *Christian Science Monitor,* 16 May 2003.

Werstein, Irving. *Guadalcanal.* New York: Thomas Y. Crowell Co., 1963.

Wheeler, Mark. "Shadow Wolves." *Smithsonian,* January 2003.
White, Jack R. *The Invisible World of Infrared.* New York: Dodd, Mead & Co., 1984.
"The World Turned Upside Down: Military Lessons of the Chechen War." *Armed Forces Journal International,* August 1998.
Yahara, Col. Hiromichi Yahara. *The Battle for Okinawa.* Translated by Roger Pineau and Masatoshi Uehara. New York: John Wiley & Sons, 1995.
Zeh, Foster (nuclear power plant security guard). Interview by Dianne Sawyer. ABC's *Good Morning America,* 9 December 2002.
Zhukov, Marshal Georgi K. *From Moscow to Berlin: Marshal Zhukov's Greatest Battles.* Edited and introduction by Harrison E. Salisbury. Translated by Theodore Shabad. War and Warrior Series. Costa Mesa, CA: The Noontide Press, 1991.
Zhukov, Marshal Georgi K. *Reminiscences and Reflections.* N.p., 1974. As quoted in *Warriors' Words—A Quotation Book,* by Peter G. Tsouras. London: Cassel Arms & Armour, 1992.
Zich, Arthur and the editors of Time-Life Books. *The Rising Sun: World War II.* Alexandria, VA: Time-Life Books, 1977.

About the Author

After almost 28 years as a commissioned and noncommissioned infantry officer, John Poole retired from the United States Marine Corps in April 1993. On active duty, he studied small-unit tactics for nine years: (1) six months at the Basic School in Quantico (1966), (2) seven months as a platoon commander in Vietnam (1966-67), (3) three months as a rifle company commander at Camp Pendleton (1967), (4) five months as a regimental headquarters company commander in Vietnam (1968), (5) eight months as a rifle company commander in Vietnam (1968-69), (6) five and a half years as an instructor with the Advanced Infantry Training Company (AITC) at Camp Lejeune (1986-92), and (7) one year as the SNCOIC of the 3rd Marine Division Combat Squad Leaders Course (CSLC) on Okinawa (1992-93).

While at AITC, he developed, taught, and refined courses on maneuver warfare, land navigation, fire support coordination, call for fire, adjust fire, close air support, M203 grenade launcher, movement to contact, daylight attack, night attack, infiltration, defense, offensive Military Operations in Urban Terrain (MOUT), defensive MOUT, Nuclear/Biological/Chemical (NBC) defense, and leadership. While at CSLC, he further refined the same periods of instruction and developed others on patrolling.

He has completed all of the correspondence school requirements for the Marine Corps Command and Staff College, Naval War College (1000-hour curriculum), and Marine Corps Warfighting Skills Program. He is a graduate of the Camp Lejeune Instructional Management Course, the 2nd Marine Division Skill Leaders in Advanced Marksmanship (SLAM) Course, and the East-Coast School of Infantry Platoon Sergeants' Course.

Since retirement, he has researched the small-unit tactics of other nations and written three previous books. Published by Posterity Press in 1997 was *The Last Hundred Yards: The NCO's Contribution to Warfare*—a squad combat study based on the consensus opinions of 1200 NCOs and the casualty statistics of hundreds of field trials at AITC and CSLC. Then from Posterity Press in late 1999 came *One More Bridge to Cross: Lowering the Cost of War*—a treatise on enemy assault tactics and how to counter them. Finally in August of 2001, Posterity Press produced *Phantom Soldier: The Enemy's Answer to U.S. Firepower*—an in-depth look at the highly deceptive Asian style of war.

As of September 2003, John Poole had conducted multiday training sessions for 32 Marine battalions (24 of them infantry), one Naval Special Warfare Group, and nine Marine schools on how to acquire common-sense warfare capabilities at the small-unit level. He has been stationed twice each in South Vietnam and Okinawa. He has visited Japan, Taiwan, the Philippines, Indonesia, South Korea, Mainland China, Hong Kong, Macao, North Vietnam, Myanmar (Burma), Thailand, Cambodia, Malaysia, Singapore, Tibet, Nepal, Bangladesh, India, Russia, East Germany, West Germany, and Israel.

Name Index

A

Aidid, Mohammed Farrah 110
An Hoa (base) 123, 262
Anaconda (operation) 323
Arnt, Plt.Sgt. C.C. 198
Atomiya, Gen. 256
Attila the Hun xxiii, xxiv, xxv, xxvi, 331

B

Ba, Col. Huong Van 167
Baghdad (battle) 110, 277, 278, 328, 329, 331
Basayev, Shamil 308
Basra (battle) 331
Bergee, S.Sgt. Lee 291
Berlin (battle) 106, 269, 286, 287, 288, 289, 290, 329
Bin Laden, Osama 322
Binh Moc (tunnel complex) 173
Budapest (battle) 175
Buffalo (operation) 97, 122, 191, 192, 247
Bulge (battle) 106, 140, 238, 251
Buon Me Thuot (battle) 102, 272

C

Cai Be Post (battle) 233
Cam Lo (base) 79, 249
Cam Ranh Bay (base) 250
Camrai (battle) 245, 311
Carlson, Brig.Gen. Evans 189
Chosin Reservoir (battle) 13, 96, 105, 107, 108, 125, 247, 251
Chu Lai (base) 122
Chuikov, Lt.Gen. Vasily 281, 289
Collins, Lt.Gen. Arthur S. xxii
Con Thien Base (battle) 94, 173
Corbett, James 61, 62, 66, 68, 195, 196, 197, 211, 213, 222, 228
Cu Chi (tunnel complex) 167, 168, 173, 275, 323

D

Da Nang (base) 250
Dac To (battle) 113
Davidson, PFC Fred 291, 301
Deckhouse VI (operation) 122
Dien Bien Phu Base (battle) 53, 78, 113, 114, 258, 260, 261, 266
Dong Ha (base) 79, 262
Duc, Trinh 61, 168

E

Eban-Emael Fort (battle) 51
Edson, Col. Merritt A. 39, 58, 136

F

Ford, M.Sgt. Lester J. "the Rock" 188, 214
Fox, Col. Wesley L. 70, 128, 135, 320
Fujibayashi, Yasuyoshi 26

G

Gangle, Col. Randolph 278
Gardez (city) 323
Garrison, Maj.Gen. William F. 110
Gia Binh (village) 83
Giap, Gen. Vo Nguyen 125, 133, 139, 254
Gordon, C.D. xxiii
Griffith, Brig.Gen. Samuel B. 114, 333
Grozny (battle) 6, 98, 268, 278, 293, 294, 296, 308
Guadalcanal (battle) 38, 39, 58, 95, 108, 109, 122, 136, 187, 198, 247

H

Hagenbeck, Maj.Gen. Franklin L. 342
Hailar (battle) 248
Hasan-ibn-Sabah 334
Hashishin (sect) 334
Hatsumi, Dr. Masaaki 27, 28, 57, 63, 65, 66
Henderson Airfield (battle) 95, 122, 223, 246
Hoa Binh Airfield (battle) 247
Hong, Hoang Tat 87
Hue City (battle) 14, 94, 113, 126, 130, 156, 157, 267, 270, 272, 292, 293, 295
Hussein, Saddam xxi, 307, 329, 331
Hutier, Gen. Oskar von 245, 349
Hutou Fort (battle) 315
Hwarang (sect) 26, 33

I

Ia Drang (battle) 97
Ionin, Col. G. 155
Iwabuchi, Rear Admiral Sanji 286
Iwo Jima (battle) 53, 55, 70, 98, 127, 139, 140, 150, 174, 255, 256, 286, 292, 323

J

Jahalabad (city) 167

K

Kabul (city) 331
Kandahar (city) 186
Kathmandu (battle) 270
Kerch Peninsula (battle) 244
Kesselring, Field Marshal Albert 71
Khan, Batu xxv, xxvi
Khan, Genghis xxv, xxvi
Khan, Kublai xxv
Khe Sahn Base (battle) 113, 114, 120, 127, 244, 250, 262
Khost (city) 186
Killer (operation) 292
Kontum (base) 272
Krebitz, Sgt. 124
Kreh, PFC 83
Kuribayashi, Gen. Tadamichi 255, 286
Kut (city) 331

L

Lam Son 719 (operation) 250
Lang Son City (battle) 171
Lang Vei (battle) 114
Lind, William S. 319
Lomov, Gen. N.A. 11
Long Son (village) 79, 85
Ludendorff, Gen. Erich 16
Lung, Dr. Haha 341

M

Mai Hac De Supply Depot (battle) 247, 272
Malvo, Lee Boyd 338
Manila (battle) 267, 284, 285, 290, 291
Marble Mountain (helicopter base) 250
Marshal (maybe S.L.A. Marshall) 151
Mary Ann Firebase (battle) 235, 250, 251
Maui Peak (operation) 53
McMichael, Maj. Scott R. 77
Meade River (operation) 97
Meckel, Maj. Jakob xxvi
Merzylak, Col. L. 155
Metcalf, 2d Lt. Karl 188
Millennium Challenge (operation) 93
Minh, Ho Chi 1
Mitanchiang (battle) 175
Mo, Nguyen Van 90, 95, 244
Mogadishu (battle) 110
Moshuh Nanren (sect) 26, 34, 333

N

Nasiriyah (battle) 69
"Nightsider" (sect) 26
Nivelle, Gen. Robert 325
Nomonhan (battle) 41, 86, 136, 179, 232
Novgorod (city) xxvi

O

O'Bday, M.Gy.Sgt. Robert R. 262
Okinawa (battle) 53, 98, 116, 124, 139, 150, 153, 255, 256, 257, 323
Oman, C.W.C. 333

P

Page, Col. Mitchell 177
Parker, Sgt. 83
Peleliu (battle) 98, 323
Pelli, Maj. Frank D. 102
Perekop Isthmus (battle) 253
Perry, PFC 83
Phat Diem (battle) 271
Phu Bai (base) 83
Phuoc An (battle) 253
Pike, John 322
Pleiku (base) 250, 272
Polo, Marco 333, 334, 335
Pope John Paul II xxix
Puller, Lt.Gen. Lewis B. "Chesty" 95, 122, 177, 186, 219, 223

R

Reinke, Dr. David H. 76, 336
Reznichenko, Maj.Gen. V.G. 241
Ridgway, Gen. Matthew B. 151
Rinpoche, Chogyal 76
Ripper (operation) 292
Roach, 1stLt. William L. 83
Rogers, Maj. Robert 217, 219
Rohr, Capt. Willy 246

Rommel, Field Marshal Erwin 45, 99, 297

S

Saigon (battle) 86, 117, 118, 123, 124, 167, 270, 275, 276, 323, 325
Sakhalin Island (battle) 253
Scott, PFC Win 291, 292, 301
Scott-Donelan, David 205, 207
Seoul (battle) 270, 290, 291, 292, 301
Shea, Brig.Gen. Robert M. 338
Shuri (battle) 150, 153, 256, 257, 258
Simon, Max 71
Skorning, Maj. 289
Smith, Gen. O.P. 290
Southworth, Alaine 339
Stalingrad (battle) 11, 138, 269, 279, 280, 286, 293
Sun Tzu xxiii, xxv, xxvi, 22, 26, 58, 94, 101, 111, 132, 141, 302, 333
Sung, Kim Il 265, 290
Surkov, Gen. Mikhail 269, 294
Sykes, Maj. 335

T

Tae-yun, Lt. Yu 10
Takamatsu, Toshitsugu 28
Tamerlane xxvi
Tan An (village) 83
Tan Son Nhut Airfield (battle) 275
Tay Nguyen (battle) 116
Thai, Gen. Hoang Van 86, 119, 276
Tham Than Khe (battle) 124
Thomas, Lt.Col. Ian 307
Thompson, R.W. 292
Tse-tung, Mao 9, 15, 21, 102, 103, 115, 117, 124, 125, 134, 138, 139, 161, 258, 290, 292, 314, 323

U

Umm Qasr (battle) 331
Unsan (battle) 44

V

Vandegrift, Maj.Gen. Alexander A. 38
Viborg (battle) 253

W

Warsaw Ghetto (battle) 270, 279
Warsaw City (battle) 286
Wright, PFC Jack 292, 301

X

Xuey, Ghia 203

Y

Yahara, Col. Hiromichi 150, 153, 256, 257, 258
Yudam-ni (battle) 251
Yule, Col. Henry 335

Z

Zeh, Foster 340
Zhukov, Marshal Georgi K. xxvi, 136, 269, 286, 288